ERDC/CERL TR-10-11
March 2010

Military Training Lands Historic Context
Small Arms Ranges

Dan Archibald, Adam Smith, Sunny Adams, and Manroop Chawla

Construction Engineering Research Laboratory (CERL)
U.S. Army Engineer Research and Development Center
2902 Newmark Dr.
Champaign, IL 61822-1076

Final Report

Approved for public release; distribution is unlimited.

Prepared for Legacy Resource Management Program
Cultural Resources Management
1225 South Clark Street, Suite 1500
Arlington, VA 22202

Abstract: This work provides an historic context for military training lands, written to satisfy a part of Section 110 of the National Historic Preservation Act (NHPA) of 1966 as amended. Cultural resources personnel at the installation level and their contractors will use this historic context to determine whether military training resources are eligible for the National Register of Historic Places (NRHP), and whether an adverse effect will take place. This overall project covered five types of military training: small arms ranges, large arms ranges, training villages and sites, bivouac areas, and large-scale operation areas. This document provides an historic context of small arms ranges on military training lands for the U.S. Army, U.S. Navy, U.S. Army Air Corps/U.S. Air Force, and the U.S. Marines, with a focus on the landscape outside the developed core of military installations. This work determined that that military training lands are significant enough in our nation's history to be surveyed for eligibility to the NRHP. However, training lands must be viewed as a whole; individual buildings on a training range are rarely eligible for the NRHP; buildings in their larger context (and the integrity of that larger context) are important.

DISCLAIMER: The contents of this report are not to be used for advertising, publication, or promotional purposes. Citation of trade names does not constitute an official endorsement or approval of the use of such commercial products. All product names and trademarks cited are the property of their respective owners. The findings of this report are not to be construed as an official Department of the Army position unless so designated by other authorized documents.

DESTROY THIS REPORT WHEN NO LONGER NEEDED. DO NOT RETURN IT TO THE ORIGINATOR.

Table of Contents

List of Figures vi

Preface ix

Unit Conversion Factors xxxvi

1 Introduction 11
    Background 11
    Objectives 2
    Approach 2
        Literature review 2
        Archival research 13
        Site visits 13
        Analysis 13
    Scope 13
    Mode of technology transfer 4

2 Small Arms Ranges 15
    General information 6
        Training procedures 6
        Weapons 7
        Safety fans and distances 8
        Multiple range layouts 11
        Firing lines 13
        Targets 21
        Embankments/trenches/etc. 43
        Buildings 62
    Fixed firing points and fixed targets 86
        1,000-in. fixed target type range 86
        Known distance ranges 89
        Machine gun squares 91
        Pistol ranges 95
        Preliminary rifle instruction circle 107
        Rifle ranges 113
    Fixed firing points and moving targets 141
        Anti-aircraft towed target range 141
        Machine gun ranges 145
        1,000-in. machine gun rolling target range 155
        1000-in. miniature anti-tank range 160
        Miniature anti-aircraft range 162
        Moving target range 164
        Night firing range 165
        Skeet range 170

Submarine target range	174
Moving firing points and fixed targets	177
Class B or combat range	177
Dismounted submachine gun practice course	183
Mounted pistol range	189
Moving vehicle machine gun range	190
Moving vehicle submachine gun range	195
Transition range	199
Moving firing points & moving targets	216
Moving base range	216
Moving target submachine gun range	218
Hand and rifle grenade ranges	225
Hand grenade ranges	225
Rifle grenade ranges	246
Trainfire ranges (layouts)	263
Firing lines	277
Targets	282
Embankments/trenches/etc.	287
Buildings	290

3   Evaluating Properties Under the Military Training Lands Historic Context ... 299

Criteria for evaluation	299
Criterion consideration G	299
Army cold war guidelines and contexts	300
Air Force cold war guidelines and context	301
Significance	302
Aspects of integrity	304
Character defining features	305
Context example photographs	305
Trainfire ranges	306
Transition range	308
Firing lines	309
Cleared firing positions	309
Sandbag supports	309
Foxholes	310
Covered firing lines	311
Firing pits	311
Firing trenches	312
Machine gun firing platforms	312
Simulated window, door, and rooftop firing positions	313
Targets	21
Stationary panel targets	21
Stationary silhouette targets	314
Raised panel targets	315
Pop-up silhouette targets	316
Moving target tracks	319

Embankments and Trenches...321
    Embankments *321*
    Trenches *323*
Walls 324
Target butts 31
Buildings 327
    Observation towers *327*
    Storage buildings *335*
    Latrines *340*
    Bleachers *343*
    Mess halls *344*
    Weapons cleaning point *345*
    Firing towers *346*

4 Conclusions 347

References 348

Report Documentation Page 349

List of Figures

Figure		Page
1	Color key for plan drawings	6
2	Small arms range fans, circa 1942 (AR 750-10, Range regulations for firing ammunition for training and target practice, 14 February 1942, pp 7-9)	9
3	Small arms range fans, circa 1955 (AR 385-63, Regulations for firing ammunition for training, target practice, and combat, 8 December 1955, pp 12, 13)	9
4	Small arms range fan, circa 1983 (AR 385-63, Regulations for firing ammunition for training, target practice, and combat, 15 October 1983, Chapter 6)	10
5	Typical division range layout, circa 1950 (SR 210-20-20, Installations training areas and facilities for ground troops, 23 May 1950, p 4)	11
6	Typical division range layout, circa 1964 (AR 210-21, Training areas and facilities for ground troops, 18 December 1964, p.10)	12
7	Firing ranges, general layout plan, Murch Range Fort Bragg, NC, 1944 (Record drawing FBR 150, "General layout, firing ranges," 15 March 1944)	13
8	Rifle range firing line at Finger Bay, Adak, AK, 23 June 1945 (NARA College Park, RG 80-G, box 1285, photo 342163)	14
9	Platoon 381 on the pistol range at Camp Matthews, MCRD San Diego, CA, 29 November 1951 (NARA College Park, RG 127-GC, box 5, photo A77625)	15
10	Firing position on pistol range during training at Amphibious Training Base Fort Pierce, FL, 23 November 1943 (NARA College Park, RG 80-G, box 861, photo 264381)	15
11	Recruit using aiming device at MCRD Parris Island, SC, 1970 (NARA College Park, RG 127-GG-919, box 32, photo A601880)	16
12	Small arms fire course, rifle support, Fort Bragg, NC, 1952 (Standard drawing 28-13-37 sheet 2 of 2, "Small arms fire course, details," 20 June 1952)	16
13	Typical rifle range section showing firing point, Fort Bragg, NC, 1952 (Standard drawing No. 28-13-09 drawing 1 of 7, "Range, rifle, known distance, plans and details," 5 January 1952)	17
14	ATC facilities, night firing range "N", firing line, Fort Bragg, NC, 1966 (Standard drawing 28-13-117 sheet 9, "Construction of ranges, phase 1, U.S. ATC Facilities, Fort Bragg, NC, range "N", night firing range," 28 June 1966)	17
15	ATC facilities, range details, foxhole detail, Fort Bragg, NC, 1966 (Standard drawing 28-13-117 sheet 14, Construction of ranges, phase 1, U.S. ATC Facilities, Fort Bragg, NC, range details, 28 June 1966)	17
16	Trench firing line at Camp Wheeler, GA, 1918 (New York Public Library, digital No. 117146)	18
17	Throwing live fragmentation hand grenades at the dummy targets at Fort Jackson, SC, 1 November 1943 (NARA College Park, RG 111-SC WWII, box 681, photo SC324452)	18
18	In their first contact with guns (that work by air power) that keep on firing, students pepper plane silhouettes (that are mounted on a track developed for shooting galleries) at Tyndall Army Air Field (Panama City), FL, 29 December 1942 (NARA College Park, RG 342-FH, box 2202, photo 4A-17270)	19

Figure		Page
19	Students fire swivel mount shot guns at Laredo Army Air Field, TX, 16 July 1943 (NARA College Park, RG 342-FH, box 2202, photo 4A-17294)	19
20	Marines operating machine gun at MCAS Cherry Point, NC, 21 August 1943 (NARA College Park, RG 80-G, box 1360, photo 358872)	20
21	A Drill Instructor teaches recruits the proper method of firing from inside a building at MCRD San Diego (Camp Pendleton), CA, December 1972 (NARA College Park, RG 127-GG-936, photo 230641)	20
22	Panel target, circa 1942 (Standard drawing No. 1600-135, Moving vehicle ranges, 23 October 1942)	21
23	CMTC students on the pistol range at Camp Vail, NJ, 1923 (NARA College Park, RG 111-SC WWI, box 700, photo 94912)	22
24	CMTC students on machine gun range at Camp Del Monte, CA, 13 August 1925 (NARA College Park, RG 111-SC WWI, box 700, photo 94896)	22
25	Target Line at Sioux Falls Army Air Field, SD, 1942 (NARA College Park, RG 342-FH, box 2202, photo 4A-17256)	23
26	The circular slow-fire targets are being ripped full of holes, Marines observe each shot and mark its value with colored discs, at the bottom of each frame is a rapid fire target, these will soon be hoisted into place at MCB San Diego, CA, February 1944 (NARA College Park, RG 127-GC, box 5, photo 35366)	23
27	Here is the rapid-fire target complete with markers, the first one shows the results of aiming too low, those firing on targets 2 and 3 registered better patterns at MCB San Diego, CA, February 1944 (NARA College Park, RG 127-GC, box 5, photo 35364)	24
28	Butts at rifle range at Camp Matthews, MCRD San Diego, CA, 6 February 1946 (NARA College Park, RG 127-GC, box 5, photo 401016)	24
29	Recruits watch their targets during firing practice at MCRD Parris Island, SC, 6 December 1951 (NARA College Park, RG 127-GC, box 34, photo A60616)	25
30	Recruits watching their targets for hits at MCRD Parris Island, SC, 6 December 1951 (NARA College Park, RG 127-GC, box 34, photo A60628)	26
31	Platoon 23 works in the butts while Platoons 532 and 533 fire on their record day at Camp Matthews, MCRD San Diego, CA, 14 February 1952 (NARA College Park, RG 127-GC, box 5, photo A219158)	26
32	"Baker" targets cover the landscape near "Echo" range as they are laid out to dry at Camp Matthews, MCRD San Diego, CA, 4 May 1959 (NARA College Park, RG 127-GG-2019, box 40, photo A229012)	27
33	Production-line teamwork is employed by a detail of Marines to build some 2000 targets at Camp Matthews, MCRD San Diego, CA, 4 May 1959 (NARA College Park, RG 127-GG-2019, box 40, photo A229014)	27
34	Target handlers work in the butts marking and discing targets at Camp Matthews, MCRD San Diego, CA, 4 May 1959 (NARA College Park, RG 127-GG-2019, box 40, photo A229013)	28
35	Butt of X Course target in the air at MCB Camp Lejeune, NC, 1966 (NARA College Park, RG 127-GG-2019, box 40, photo A342975)	29
36	Sustained fire stationary paper target, circa 1941 (TM 9-855, Targets, target material, and rifle range construction, 19 June 1941, pp 98)	29
37	Perhaps the most important day in the recruit's boot training is qualification day, he will not only fire the rifle for an official score to be entered in his record book, he will also assist in operating some part of the firing line like phoning to the butts or keeping score at Camp Matthews, MCRD San Diego, CA, 6 June 1955 (NARA College Park, RG 127-GC, box 5, photo A227310)	30

Figure		Page
38	Silhouette targets for prone (F), kneeling (E), and standing (M) positions, circa 1943 (FM 23-41, Basic field manual, Submachinegun caliber .45, M3, 30 October 1943, p 60)	30
39	Moving target, circa 1940 to 1943 (FM 23-5, Basic field manual, U.S. rifle, caliber .30, M1, 30 July 1943, pp 194, 195, FM 23-7. Basic field manual, U.S. carbine, caliber .30, M1, 20 May 1942, pp 95, 96, FM 23-60, Basic field manual browning machine gun, caliber .50, HB, M2 ground, 25 September 1940, p 134)	31
40	Dismounted submachine gun practice course, sled target, Fort Bragg, NC, 1951 (Standard drawing 28-13-13, sheet 1 of 2, Range, submachine gun, dismounted practice course, plans and details, 7 December 1951)	32
41	Moving target, circa 1942-1949 (FM 23-30, Hand and rifle grenades, rocket, AT, HE, 2.36-in., 14 February 1944, p 54, FM 23-30, Hand and rifle grenades, rocket, AT, HE, 2.36-in., 14 February 1944, p 89, FM 23-30, Hand and rifle grenades, 14 April 1949, p 92)	32
42	Motorized movable target plan with drum detail, circa 1943 (FM 23-41, Basic field manual, Submachine gun caliber .45, M3, 30 October 1943, pp 68-70, TM 9-855, Targets, target material, and rifle range construction, changes No. 1, 5 August 1942, p 5)	33
43	Moving target assembly, Fort Bragg, NC, June 1949 (Standard drawing 28-09-01 sheet 7 of 8, Training aids, moving target assembly, 6 June 1949)	33
44	Moving target assembly, Fort Bragg, NC, June 1949 (Standard drawing 28-09-01 sheet 7 of 8, Training aids, moving target assembly, 6 June 1949)	34
45	Moving target assembly, Fort Bragg, NC, June 1949 (Standard drawing 28-09-01 sheet 7 of 8, Training aids, moving target assembly, 6 June 1949)	34
46	Silhouette targets, moving target assembly, Fort Bragg, NC, June 1949 (Standard drawing 28-09-01 sheet 8 of 8, Training aids, moving target assembly, 6 June 1949)	35
47	Looking down on target track on submarine target range at Camp Wissahickon, NJ, undated (circa 1918) (NARA College Park, RG 24-TC, box 1, Folder K)	35
48	Motor driven buda target car on the aviation free gunnery range at NAS Kaneohe Bay, HI, March 1944 (NARA College Park, RG 80-G, box 1705, photo 419200)	36
49	Overhead rifle target, circa 1943 (FM 23-5, Basic field manual, U.S. rifle, .30 caliber, M1, 30 July 1943, p 204)	37
50	Aerial rifle target layout, circa 1942 (TM 9-855, Targets, target material, and rifle range construction, changes No. 1, 5 August 1942, p 3)	37
51	Miniature anti-aircraft range, target pulley system plan and elevation, circa 1942-1951 (Standard drawing No. 1600-130/131, Training aids, A. A. Range miniature, 14 October 1942, TM 9-855, Targets, target material, and training course layouts, 1 November 1951, pp 28, 29)	38
52	Miniature anti-aircraft range, target pulley system plan and elevation, circa 1942-1951 (Standard drawing No. 1600-130/131, Training aids, A. A. range miniature, 14 October 1942, TM 9-855, Targets, target material, and training course layouts, 1 November 1951, pp 28, 29)	38
53	High tower and turret range NAGS shot gun range at NAS Jacksonville, FL, 27 February 1944 (NARA College Park, RG 80-G, box 740, photo 241829)	39
54	Dismounted submachine gun practice course, surprise target diagram, Fort Bragg, NC, 1951 (Standard drawing 28-13-13 sheet 2 of 2, Range, submachine gun, dismounted practice course, details, 7 December 1951)	39
55	Small arms range, field target, isometric view of target installation, Fort Bragg, NC, 1951 (Standard drawing 28-13-18 sheet 1 of 1, Range, field target, small arms, plan and details, 21 November 1951)	40

Figure		Page
56	Small arms range, field target, details of control rack, Fort Bragg, NC, 1951 (Standard drawing 28-13-18 sheet 1 of 1, Range, field target, small arms, plan and details, 21 November 1951)	40
57	Electrical targets at Camp Lejeune, NC, 2 May 1958 (NARA College Park, RG 127-GC, box 22, photo 340848)	41
58	OQ-14/TDD-3/TDD-4 Radioplane, circa 1944 (TR 140-5, http://www.designation-systems.info/dursm/app1/oq-14.html)	42
59	Radio controlled airplane target OQ-2A taking off from catapult at the AA Range at Fort Jackson, SC, 1 November 1943 (NARA College Park, RG 111-SC WWII, box 415, photo SC246692)	42
60	Small arms target practice for cadets at NAS Corpus Christi, TX, 23 July 1941 (NARA College Park, RG 80-G, box 1978, photo 463689)	43
61	A 1000-in. fixed target track type range, section through salvage wall, Fort Bragg, NC, 1952 (Standard drawing 28-13-06 sheet 1 of 2, Range, 1000-in. fixed target, track type, plans and details, 20 June 1952)	44
62	Horses digging target trench, Fort Knox, KY, undated (scans from Knox, training 5, vol. 5)	44
63	Horses digging target trench, Fort Knox, KY, undated (scans from Knox, training 6, vol. 5)	45
64	Target complex built on top of trench, Fort Knox, KY, undated (scans from Knox, training 7, vol. 5)	45
65	Looking up out of target trench (rifle range pits), Fort Knox, KY, undated (scans from Knox, training 8, vol. 5, #358, Kirkpatrick photo)	46
66	Target line at the USMC Winthrop rifle range, Indian Head, MD, 1915 (NARA College Park, RG 127-G-100b, box 23, photo 521528)	46
67	Typical rifle range target butts concrete retaining wall, Fort Bragg, NC, 1952 (Standard drawing No. 28-13-09 drawing 2 of 7, Range, rifle, known distance, details, 5 January 1952)	47
68	Typical rifle range target butts section detail, Fort Bragg, NC, 1952 (Standard drawing No. 28-13-09 drawing 2 of 7, Range, rifle, known distance, details, 5 January 1952)	48
69	Typical rifle range target butts section elevation, Fort Bragg, NC, 1952 (Standard drawing No. 28-13-09 drawing 2 of 7, Range, rifle, known distance, details, 5 January 1952)	48
70	Typical rifle range target butts timber wall (on hillside), Fort Bragg, NC, 1952 (Standard drawing No. 28-13-09 drawing 2 of 7, Range, rifle, known distance, details, 5 January 1952)	48
71	Typical rifle range target butts timber wall (on level ground), Fort Bragg, NC, 1952 (Standard drawing No. 28-13-09 drawing 2 of 7, Range, rifle, known distance, details, 5 January 1952)	49
72	Typical rifle range target butts log crib retaining wall, Fort Bragg, NC, 1952 (Standard drawing No. 28-13-09 drawing 2 of 7, Range, rifle, known distance, details, 5 January 1952)	49
73	Dismounted submachine gun practice course, surprise target diagram, Fort Bragg, NC, 1951 (Standard drawing 28-13-13 sheet 2 of 2, Range, submachine gun, dismounted practice course, details, 7 December 1951)	50
74	ATC facilities, range details, detail of concrete M-31-A1 pop-up target emplacement, Fort Bragg, NC, 1966 (Standard drawing 28-13-117 sheet 14, Construction of ranges, phase 1, U.S. ATC Facilities, Fort Bragg, NC, range details, 28 June 1966)	50

Figure		Page
75	Typical rifle range, M31A1 target device, section through target butt and pop-up target enclosure, Fort Bragg, NC, 1963 (Standard drawing No. 28-13-09 drawing 5 of 7, Range, rifle, known distance, layout, and details for installation of M31A1 target device, 6 November 1963)	51
76	Moving target range, circa 1951 (TM 9-855, Targets, target material, and training course layouts, 1 November 1951, p 35)	51
77	Moving target submachine gun range, plan, Fort Bragg, NC, 1952 (Standard drawing 28-13-14 sheet 1 of 3, Range, moving target, submachine gun, plans and details, 20 June 1952)	52
78	Moving target submachine gun range, section thru firing area, Fort Bragg, NC, 1952 (Standard drawing 28-13-14 sheet 2 of 3, Range, moving target, submachine gun, repair pit and parapet details, 20 June 1952)	52
79	Above ground moving target pit, section and plan, Fort Bragg, NC, 1955 (Standard drawing 28-13-05 sheet 1 of 1, Transition ranges, M1 rifle and MG, above ground pits, construction details, 6 June 1955)	53
80	Typical rifle range section showing firing point, Fort Bragg, NC, 1952 (Standard drawing No. 28-13-09 drawing 1 of 7, Range, rifle, known distance, plans and details, 5 January 1952)	53
81	Superimposed transition range, firing line, Fort Bragg, NC, 1954 (Standard drawing 28-13-107 sheet 1 of 1, Superimposed transition range, table VII, 28 March 1954)	54
82	Marines on firing line at unknown location, undated (NARA College Park, RG 127-GC, box 5, photo 227310)	54
83	Training of marines on the firing line at NAS Jacksonville, 20 May 1942 (NARA College Park, RG 80-G, box 283, photo 64650)	54
84	Machine gun crews in action on range at MCAS Cherry Point, NC, 21 August 1943 (NARA College Park, RG 80-G, box 1360, photo 358873)	55
85	Castner Range #2, a .30 caliber known distance range at Fort Bliss, TX, 31 August 1953 (NARA College Park, RG 111-SC WWII, box 279, photo SC460727)	55
86	Typical rifle range diagram of multiple ranges, Fort Bragg, NC, 1952 (Standard drawing No. 28-13-09 drawing 1 of 7, Range, rifle, known distance, plans and details, 5 January 1952)	56
87	Trench firing line at Camp Wheeler, GA, 1918 (New York Public Library, digital No. 117146)	56
88	Throwing live fragmentation hand grenades at the dummy targets at Fort Jackson, SC, 1 November 1943 (NARA College Park, RG 111-SC WWII, box 681, photo SC324452)	57
89	ATC facilities, night firing range "N", firing line, Fort Bragg, NC, 1966 (Standard drawing 28-13-117 sheet 9, Construction of ranges, phase 1, U.S. ATC facilities, Fort Bragg, NC, range "N", night firing range, 28 June 1966)	57
90	Throwing pits course, circa 1942 (FM 23-30, Grenades, 15 June 1942, p 23)	57
91	Practice grenade course, Foxhole Court, Fort Bragg, NC, 1951 (Standard drawing 28-13-43 sheet 1 of 1, Practice grenade course layout and details, 21 November 1951)	58
92	Shell hole target details, Fort Bragg, NC, 1957 (Standard drawing 28-13-05, Close combat course, plan and details, 8 September 1957)	58
93	Superimposed transition range, plan of targets in foxholes and section, Fort Bragg, NC, 1954 (Standard drawing 28-13-108 sheet 1 of 1, Superimposed transition range, table VIII, 22 March 1954)	59

Figure		Page
94	Live grenade practice course, plan and profile, circa 1944-1949 (FM 23-30, Hand and rifle grenades, rocket, AT, HE, 2.36-in., 14 February 1944, pp 35, 36, FM 23-30, Hand and rifle grenades, 14 April 1949, pp 40, 41)	60
95	Superimposed transition range, section thru control pit, Fort Bragg, NC, 1954 (Standard drawing 28-13-108 sheet 1 of 1, Superimposed transition range, table VIII, 22 March 1954)	61
96	Transition range moving target detail, Fort Bragg, NC, 1955 (Standard drawing 28-13-04 sheet 1 of 1, Range, transition, details, 6 June 1955)	61
97	ATC Facilities, range details, demolition pit, Fort Bragg, NC, 1966 (Standard drawing 28-13-117 sheet 14, Construction of ranges, phase 1, U.S. ATC Facilities, Fort Bragg, NC, range details, 28 June 1966)	62
98	Range at NAS Deland, FL, 23 December 1944 (NARA College Park, RG 80-G, box 1382, photo 363872)	63
99	Aerial of gunnery range at NAS Beaufort, SC, 24 January 1945 (NARA College Park, RG 80-G, box 1376, photo 362569)	63
100	Rifle range observation tower elevation, Fort Bragg, NC, 1952 (Standard drawing No. 28-13-09A drawing 4 of 4, Range, rifle, known distance, observation tower and latrine, 5 January 1952)	64
101	Rifle range observation tower foundation and platform plans, Fort Bragg, NC, 1952 (Standard drawing No. 28-13-09A drawing 4 of 4, Range, rifle, known distance, observation tower and latrine, 5 January 1952)	64
102	Mendick tollgate range control tower side and rear elevations, Fort Knox, KY, 1981 (Standard drawing 28-13-13, Control tower details, mendick-tollgate range, table V, September 1981)	65
103	Mendick tollgate range section through control tower, Fort Knox, KY, 1981 (Standard drawing 28-13-13, Control tower details, Mendick-Tollgate Range, table V, September 1981)	66
104	Mendick-Tollgate Range control tower plans, Fort Knox, KY, 1981 (Standard drawing 28-13-13, Control tower details, Mendick-Tollgate Range, table V, September 1981)	66
105	The combat rifleman environmental range tower #1 including the "C" portion and part of "B" portion as seen from Tower #2 at Range 208 at MCB Camp Pendleton, CA, 24 July 1962 (NARA College Park, RG 127-GG-2058, box 42, photo A353462)	67
106	Controls that operate pop-up targets at Range 208 at MCB Camp Pendleton, CA, 24 July 1962 (NARA College Park, RG 127-GG-2021, box 40, photo A353453)	67
107	The combat rifleman environmental range tower #1 and bleachers at Range 208 at MCB Camp Pendleton, CA, 24 July 1962 (NARA College Park, RG 127-GG-2058, box 42, photo A353466)	68
108	New rifle range control tower that is an essential component of the rifle ranges at Fort Jackson, SC, 14 January 1964 (NARA College Park, RG 111-SC post-1955, box 384, photo SC607734)	69
109	Typical rifle range portable control tower, Fort Bragg, NC, 1952 (Standard drawing No. 28-13-09 drawing 5 of 7, Range, rifle, known distance, details, 5 January 1952)	70
110	Rifle range latrine front elevation and section, Fort Bragg, NC, 1952 (Standard drawing No. 28-13-09A drawing 4 of 4, Range, rifle, known distance, observation tower and latrine, 5 January 1952)	70
111	Rifle range latrine elevation of urinal trough and section of latrine box, Fort Bragg, NC, 1952 (Standard drawing No. 28-13-09A drawing 4 of 4, Range, rifle, known distance, observation tower and latrine, 5 January 1952)	71

Figure		Page
112	Rifle range latrine floor plan, Fort Bragg, NC, 1952 (Standard drawing No. 28-13-09A drawing 4 of 4, "Range, rifle, known distance, observation tower and latrine, 5 January1952")	71
113	Range latrine, roof detail, Fort Bragg, NC, 1983 (DEH 4122 sheet 2 of 5, Range latrine, 21 December 1983)	72
114	Range latrine, front and side elevations, Fort Bragg, NC, 1983 (DEH 4122 sheet 2 of 5, Range latrine, 21 December 1983)	72
115	Typical rifle range target storage and latrine building, elevation A, Fort Bragg, NC, 1952 (Standard drawing No. 28-13-09A drawing 3 of 4, Range, rifle, known distance, target storage and latrine Bldg, 5 January 1952)	72
116	Typical rifle range target and storage building, elevation B and Section G-G, Fort Bragg, NC, 1952 (Standard drawing No. 28-13-09A drawing 3 of 4, Range, rifle, known distance, target storage and latrine Bldg, 5 January 1952)	73
117	Typical rifle range target storage and latrine building, Section EE, Fort Bragg, NC, 1952 (Standard drawing No. 28-13-09A drawing 3 of 4, Range, rifle, known distance, target storage and latrine Bldg, 5 January 1952)	73
118	Typical rifle range target storage and latrine building, section FF, Fort Bragg, NC, 1952 (Standard drawing No. 28-13-09A drawing 3 of 4, Range, rifle, known distance, target storage and latrine Bldg, 5 January 1952)	74
119	Typical rifle range target and storage building floor plan, Fort Bragg, NC, 1952 (Standard drawing No. 28-13-09A drawing 3 of 4, Range, rifle, known distance, target storage and latrine Bldg, 5 January 1952)	74
120	Range target storage and repair shed, south and west elevations, Fort Knox, KY, 1946 (FK-9-2 (Adapted From Standard drawing No. T.O. 700-6013), Range target storage and repair shed, Type S-A-T (modified), plans, elevations, and details, 13 August 1946)	75
121	Range target storage and repair shed floor plan, Fort Knox, KY, 1946 (FK-9-2 (adapted from Standard drawing No. T.O. 700-6013), Range target storage and repair shed, type S-A-T (modified), plans, elevations, and details, 13 August 1946)	75
122	Target shelter, front and side elevations, Fort Bragg, NC, undated (DEH-4154 sheet o5101, Range 51, target shelter, undated)	75
123	Target shelter, section, Fort Bragg, NC, undated (DEH-4154 sheet o5101, Range 51, target shelter, undated)	76
124	Target shelter, floor plan, Fort Bragg, NC, undated (DEH-4154 sheet o5101, Range 51, target shelter, undated)	76
125	Masonry target repair and range house, elevation A, Fort Bragg, NC, 1952 (Standard drawing 28-13-96 sheet 1 of 3, Target repair and range house, masonry, plan, elevations, and section, 11 April 1952)	77
126	Masonry target repair and range house, elevation A & C, and Section A, Fort Bragg, NC, 1952 (Standard drawing 28-13-96 sheet 1 of 3, Target repair and range house, masonry, plan, elevations, and section, 11 April 1952)	77
127	Masonry target repair and range house plan, Fort Bragg, NC, 1952 (Standard drawing 28-13-96 sheet 1 of 3, Target repair and range house, masonry, plan, elevations, and section, 11 April 1952)	77
128	Target repair and range house, elevation, Fort Bragg, NC, 1952 (Standard drawing No. 28-13-29 sheet 1 of 3, Target repair and range house, frame construction, plan, elevations, and section, 11 April 1952)	78
129	Target repair and range house, elevations, Fort Bragg, NC, 1952 (Standard drawing No. 28-13-29 sheet 1 of 3, Target repair and range house, frame construction, plan, elevations, and section, 11 April 1952)	78

Figure		Page
130	Target repair and range house, plan, Fort Bragg, NC, 1952 (Standard drawing No. 28-13-29 sheet 1 of 3, Target repair and range house, frame construction, plan, elevations, and section, 11 April 1952)	78
131	Bleachers, plan, rear elevation, and side elevation, Fort Bragg, NC, 1966 (Standard drawing 28-13-117 sheet 22, Construction of ranges, phase I, U.S. ATC facilities, Fort Bragg, NC, bleachers with shed, 28 June 1966)	79
132	Bleacher end view and details, Fort Bragg, NC, 1966 (Standard drawing 28-13-117 sheet 22, Construction of ranges, phase I, U.S. ATC Facilities, Fort Bragg, NC, Bleachers with shed, 28 June 1966)	79
133	Bleacher shed front and side elevations, Fort Bragg, NC, 1966 (Standard drawing 28-13-117 sheet 22, Construction of ranges, phase I, U.S. ATC Facilities, Fort Bragg, NC, bleachers with shed, 28 June 1966)	80
134	Bleachers layout plan, Fort Bragg, NC, 1966 (Standard drawing 28-13-117 sheet 22, Construction of ranges, phase I, U.S. ATC Facilities, Fort Bragg, NC, bleachers with shed, 28 June 1966)	80
135	ATC facilities, range details, ammo storage details, Fort Bragg, NC, 1966 (Standard drawing 28-13-117 sheet 14, Construction of ranges, phase 1, U.S. ATC facilities, Fort Bragg, NC, range details, 28 June 1966)	81
136	Moving target submachine gun range, winch house plan, Fort Bragg, NC, 1952 (Standard drawing 28-13-14 sheet 3 of 3, Range, moving target, submachine gun, winch house, 20 June 1952)	81
137	Moving target submachine gun range, winch house section, Fort Bragg, NC, 1952 (Standard drawing 28-13-14 sheet 3 of 3, Range, moving target, submachine gun, winch house, 20 June 1952)	82
138	Moving target submachine gun range, winch house rear and front elevations, Fort Bragg, NC, 1952 (Standard drawing 28-13-14 sheet 3 of 3, Range, moving target, submachine gun, winch house, 20 June 1952)	82
139	Moving target submachine gun range, winch house side elevation, Fort Bragg, NC, 1952 (Standard drawing 28-13-14 sheet 3 of 3, Range, moving target, submachine gun, winch house, 20 June 1952)	83
140	Moving target submachine gun range, counterweight tower, Fort Bragg, NC, 1952 (Standard drawing 28-13-14, sheet 1 of 3, Range, moving target, submachine gun, plans and details, 20 June 1952)	83
141	ATC Facilities, range building, elevations, Fort Bragg, NC, 1966 (Standard drawing 28-13-117 sheet 24, Construction of ranges, phase 1, U.S. ATC Facilities, Fort Bragg, NC, range building, 28 June 1966)	84
142	ATC facilities, range building, plan and section, Fort Bragg, NC, 1966 (Standard drawing 28-13-117, sheet 24, Construction of ranges, phase 1, U.S. ATC Facilities, Fort Bragg, NC, range building, 28 June 1966)	84
143	Prefabricated metal range building (range 9), front and rear elevations, Fort Bragg, NC, 1981 (DFE 3629 sheet 3 of 4, Construct metal prefab building [range 9], plan, elevations, and details, 7 May 1981)	85
144	Prefabricated metal range building (range 9), elevations and details, Fort Bragg, NC, 1981 (DFE 3629 sheet 3 of 4, Construct metal prefab building [range 9], plan, elevations, and details, 7 May 1981)	85
145	Prefabricated metal range building (range 9), plan, Fort Bragg, NC, 1981 (DFE 3629 sheet 3 of 4, Construct metal prefab building [range 9], plan, elevations, and details, 7 May 1981)	85
146	A 1000-in. fixed target type range, danger area, 1952 (Standard drawing No 28-13-07, sheet 1 of 1, Range, 1000-in. fixed target type, 26 February 1952)	86

Figure		Page
147	A 1000-in. fixed target type range, plan, 1952 (Standard drawing No 28-13-07 sheet 1 of 1, Range, 1000-in. fixed target type, 26 February 1952)	87
148	1000-in. fixed target type range, section, 1952 (Standard drawing No 28-13-07 sheet 1 of 1, Range, 1000-in. fixed target type, 26 February 1952)	87
149	A 1000-in. fixed target type range, target frame, 1952 (Standard drawing No 28-13-07 sheet 1 of 1, Range, 1000-in. fixed target type, 26 February 1952)	88
150	A 1000-in. fixed target type range, section, 1952 (Standard drawing No 28-13-07 sheet 1 of 1, Range, 1000-in. fixed target type, 26 February 1952)	89
151	Known distance range plan, circa 1906 (War Department Document No. 261, Small arms firing regulations – 1906, 1 January 1906, p 142)	90
152	Castner Range #2, A .30 caliber known distance range at Fort Bliss, TX, 31 August 1953 (NARA College Park, RG 111-SC WWII, box 279, photo SC460727)	90
153	Machine gun squares, layout, Fort Bragg, NC, 1951 (Standard drawing No. 28-13-27 sheet 1 of 1, Machine gun squares, plan and details, 21 November 1951)	92
154	Machine gun squares, target frame, Fort Bragg, NC, 1951 (Standard drawing No. 28-13-27 sheet 1 of 1, Machine gun squares, plan and details, 21 November 1951)	93
155	Machine gun squares, target, Fort Bragg, NC, 1951 (Standard drawing No. 28-13-27 sheet 1 of 1, Machine gun squares, plan and details, 21 November 1951)	93
156	Machine gun squares, instructors platform, Fort Bragg, NC, 1951 (Standard drawing No. 28-13-27 sheet 1 of 1, Machine gun squares, plan and details, 21 November 1951)	94
157	Pistol range danger area safety fan, Fort Bragg, NC, 1951 (Standard drawing No. 28-13-12 drawing 1 of 1, Range, pistol, landscape targets, plans and details, 21 November 1951)	95
158	Pistol range plan, Fort Bragg, NC, 1951 (Standard drawing No. 28-13-12 drawing 1 of 1, Range, pistol, landscape targets, plans and details, 21 November 1951)	96
159	CMTC students on the pistol range at Camp Vail, NJ, 1923 (NARA College Park, RG 111-SC WWI, box 700, photo 94912)	97
160	Perhaps the most important day in the recruit's boot training is qualification day, he will not only fire the rifle for an official score to be entered in his record book, he will also assist in operating some part of the firing line like phoning to the butts or keeping score at Camp Matthews, MCRD San Diego, CA, 6 June 1955 (NARA College Park, RG 127-GC, box 5, photo A227310)	97
161	High angle view of the electric pistol range at MCRD Parris Island, SC, 10 February 1958 (NARA College Park, RG 127-GC, box 34, photo A600627)	98
162	CMTC students during pistol practice at Fort Sam Houston, TX, undated (NARA College Park, RG 111-SC WWI, box 700, photo 94878)	98
163	A .45 caliber pistol firing line at Fort Jackson, SC, 23 April 1943 (NARA College Park, RG 111-SC WWII, box 155, photo SC173966)	99
164	Firing position on pistol range during training at Amphibious Training Base Fort Pierce, FL, 23 November 1943 (NARA College Park, RG 80-G, box 861, photo 264381)	99
165	Pistol firing range at NATC Corpus Christi, TX, November 1942 (NARA College Park, RG 80-G, box 1694, photo 417646)	100
166	Pistol firing line on the 50-yd range at Finger Bay, Adak, AK, 23 June 1945 (NARA College Park, RG 80-G, box 1285, photo 342166)	100
167	Pistol Firing Range at San Antonio Aviation Cadet Center, TX, 9 April 1943 (NARA College Park, RG 342-FH, box 2207, photo 4A-18406)	101

Figure		Page
168	Platoon 23 firing the .22 caliber pistol on L Range at Camp Matthews, MCRD San Diego, CA, 11 February 1952 (NARA College Park, RG 127-GC, box 5, photo A219142)	101
169	Platoon 381 on the pistol range at Camp Matthews, MCRD San Diego, CA, 29 November 1951 (NARA College Park, RG 127-GC, box 5, photo A77625)	102
170	Pistol range plan, Fort Bragg, NC, 1951 (Standard drawing No. 28-13-12 drawing 1 of 1, Range, pistol, landscape targets, plans and details, 21 November 1951)	102
171	Pistol range details for bobbing targets, Fort Bragg, NC, 1951 (Standard drawing No. 28-13-12 drawing 1 of 1, Range, pistol, landscape targets, plans and details, 21 November 1951)	103
172	Pistol range typical target frames, Fort Bragg, NC, 1951 (Standard drawing No. 28-13-12 drawing 1 of 1, Range, pistol, landscape targets, plans and details, 21 November 1951)	103
173	Pistol range landscape target details, Fort Bragg, NC, 1951 (Standard drawing No. 28-13-12 drawing 1 of 1, Range, pistol, landscape targets, plans and details, 21 November 1951)	104
174	Pistol range landscape targets plan, Fort Bragg, NC, 1951 (Standard drawing No. 28-13-12 drawing 1 of 1, Range, pistol, landscape targets, plans and details, 21 November 1951)	104
175	Official scores and markers inspecting the targets on the pistol range at Camp Matthews, MCRD San Diego, CA, March 1949 (NARA College Park, RG 127-GC, box 5, photo A224845)	105
176	Small arms target practice for cadets at NAS Corpus Christi, TX, 23 July 1941 (NARA College Park, RG 80-G, box 1978, photo 463689)	105
177	Range at NAS Deland, FL, 23 December 1944 (NARA College Park, RG 80-G, box 1382, photo 363872)	106
178	Each recruit must be proficient with basic infantry weapons, here familiarity with the .22 caliber pistol at Camp Matthews, MCRD San Diego, CA, 6 June 1955 (NARA College Park, RG 127-GC, box 5, photo A227316)	106
179	Preliminary rifle instruction circle, plan, Fort Bragg, NC, 1951 (Standard drawing 28-13-28 sheet 1 of 1, Preliminary rifle instruction circle, plan and details, 7 December 1951)	107
180	A platoon of recruits "snap in" during their first week at the rifle range at MCRD Parris Island, SC, 1967 (NARA College Park, RG 127-GG-921, box 33, photo A601744)	108
181	Preliminary rifle instruction circle, numbered firing position pegs, Fort Bragg, NC, 1951 (Standard drawing 28-13-28 sheet 1 of 1, Preliminary rifle instruction circle, plan and details, 7 December 1951)	108
182	Preliminary rifle instruction circle, target details, Fort Bragg, NC, 1951 (Standard drawing 28-13-28 sheet 1 of 1, Preliminary rifle instruction circle, plan and details, 7 December 1951)	109
183	Preliminary rifle instruction circle, system for operating sustained fire targets, Fort Bragg, NC, 1951 (Standard drawing 28-13-28 sheet 1 of 1, Preliminary rifle instruction circle, plan and details, 7 December 1951)	110
184	Preliminary rifle instruction circle, instructors platform plan, Fort Bragg, NC, 1951 (Standard drawing 28-13-28 sheet 1 of 1, Preliminary rifle instruction circle, plan and details, 7 December 1951)	111
185	Preliminary rifle instruction circle, bleachers, Fort Bragg, NC, 1951 (Standard drawing 28-13-28 sheet 1 of 1, Preliminary rifle instruction circle, plan and details, 7 December 1951)	112

Figure		Page
186	Preliminary rifle instruction circle, bleachers layout plan, Fort Bragg, NC, 1951 (Standard drawing 28-13-28 sheet 1 of 1, Preliminary rifle instruction circle, plan and details, 7 December 1951)	112
187	Typical rifle range danger area safety fan, Fort Bragg, NC, 1952 (Standard drawing No. 28-13-09 drawing 1 of 7, Range, rifle, known distance, plans and details, 5 January 1952)	113
188	Typical rifle range basic plan and section, Fort Bragg, NC, 1952 (Standard drawing No. 28-13-09 drawing 1 of 7, Range, rifle, known distance, plans and details, 5 January 1952)	114
189	Winners of the Marine Corps rifle match at USMC Winthrop Rifle Range, Indian Head, MD, 1913 (NARA College Park, RG 127-G-100b, box 23, photo 516689)	114
190	View of Winthrop Rifle Range, Indian Head, MD, 1915 (NARA College Park, RG 127-G-100b, box 23, photo 521532)	115
191	Rifle matches at Camp Curtis Guild, Wakefield, MA, August 1927 (NARA College Park, RG 127-G-100c, box 23, photo 524720)	115
192	Rifle team at Camp Curtis Guild, Wakefield, MA, 1929 (NARA College Park, RG 127-G-100c, box 23, photo 527257)	116
193	Rifle range and pistol range at Finger Bay, Adak, AK, 23 June 1945 (NARA College Park, RG 80-G, box 1285, photo 342161)	116
194	Typical "boot" camp scene at MCRD San Diego, CA, June 1946 (NARA College Park, RG 127-GC, box 5, photo 400734)	117
195	Rifle range "A" at Range Area looking toward the butts and New River with personnel firing on the 200-yd line at Camp Lejeune, NC, July 1957 (NARA College Park, RG 127-GC, box 22, photo 340630)	117
196	Rifle range "B" at Range Area looking toward the butts and New River with personnel firing on the 200-yd line at Camp Lejeune, NC, July 1957 (NARA College Park, RG 127-GC, box 22, photo 340620)	118
197	One of the newer techniques in recruit rifle marksmanship training is the 900-in. firing line that is designed to familiarize the recruit with the M-14 rifle and to give him the proper windage and elevation for the 200-yd line at MCRD Parris Island, SC, 1967 (NARA College Park, RG 127-GG-921, box 33, photo A601726)	118
198	Typical rifle range section showing firing point, Fort Bragg, NC, 1952 (Standard drawing No. 28-13-09 drawing 1 of 7, Range, rifle, known distance, plans and details, 5 January 1952)	119
199	Firing line at Camp Meade, MD, 1918 (New York Public Library, digital No. 117098)	119
200	Dr. Samuel Scott served as private-gunnery sergeant at the USMC Winthrop Rifle Range, Indian Head, MD, 1915 (NARA College Park, RG 127-G-100b, box 23, photo 523539)	120
201	Rifle range at Camp Curtis Guild, Wakefield, MA, 1917, http://is.noblenet.org/images/wak/camp_curtis_wakefield_rifle_range.jpg	120
202	Major J.C. Smith, USMC rifle team captain at Camp Curtis Guild, Wakefield, MA, 1929 (NARA College Park, RG 127-G, box 32, photo 527261)	121
203	Firing a Springfield 03 rifle at Camp Perry, Port Clinton, OH, 1939 (NARA College Park, RG 127-G-126N, box 32, photo 521798)	121
204	Rifle range firing line at Finger Bay, Adak, AK, 23 June 1945 (NARA College Park, RG 80-G, box 1285, photo 342163)	122
205	Marines on firing line at unknown location, undated (NARA College Park, RG 127-GC, box 5, photo 227310)	122

Figure		Page
206	Training of Marines on the firing line at NAS Jacksonville, 20 May 1942 (NARA College Park, RG 80-G, box 283, photo 64650)	123
207	Looking down the firing line at Camp Lejeune, NC, 18 May 1949 (NARA College Park, RG 127-GC, box 22, photo 506751)	123
208	Recruits in the sitting position fire from the 200-yd line at MCRD Parris Island, SC, 9 February 1961 (NARA College Park, RG 127-GG-925, photo 601432)	124
209	The control tower NCO tells the shooters how much time they have to finish their string of slow fire at the 300-yd line at MCRD Parris Island, SC,1967 (NARA College Park, RG 127-GG-921, box 33, photo A601829)	124
210	A rifle range coach assists a recruit in marking a sighting change on the 500-yd range at MCRD Parris Island, SC, 1967 (NARA College Park, RG 127-GG-921, box 33, photo A601729)	125
211	Trainees participating in the M-16 rifle qualification on Range #12 at Fort Ord, CA, 26 September 1969 (NARA College Park, RG 111-CRB box 86, photo SC67552)	125
212	Small arms fire course, rifle support, Fort Bragg, NC, 1952 (Standard drawing 28-13-37 sheet 2 of 2, Small arms fire course, details, 20 June 1952)	126
213	Platoon 23 at an exercise on the sighting bars, a mirror is at the end of the bar and the recruits obtain a sight picture by moving the target with his right hand while sighting at the reflection in the mirror at MCRD San Diego, CA, 6 February 1952 (NARA College Park, RG 127-GC, box 5, photo A219137)	126
214	Every person who fires the rifle in the Marine Corps gets practice at the sighting and aiming bar which determines if a man is getting the correct sight picture at MCRD Parris Island, SC, 10 February 1958 (NARA College Park, RG 127-GC, box 34, photo A600628)	127
215	Recruits are given instruction in the use of aiming devices at MCRD Parris Island, SC, 9 February 1961 (NARA College Park, RG 127-GG-925, photo 610427)	127
216	Recruit using aiming device at MCRD Parris Island, SC, 1970 (NARA College Park, RG 127-GG-919, box 32, photo A601880)	128
217	A Marine recruit utilizes the M-16 rifle equipped with laser marksmanship rifle trainer device at MCRD Parris Island, SC, 1 March 1977 (NARA College Park, RG 127-GG-913, box 32, photo A602764)	128
218	Typical rifle range target arrangement, Fort Bragg, NC, 1952 (Standard drawing No. 28-13-09 drawing 1 of 7, Range, rifle, known distance, plans and details, 5 January 1952)	129
219	Typical rifle range target number, Fort Bragg, NC, 1952 (Standard drawing No. 28-13-09 drawing 2 of 7, Range, rifle, known distance, details, 5 January 1952)	129
220	Rifle range with butts and target shed at Finger Bay, Adak, AK, 23 June 1945 (NARA College Park, RG 80-G, box 1285, photo 342162)	130
221	Target line at Sioux Falls Army Air Field, SD, 1942 (NARA College Park, RG 342-FH, box 2202, photo 4A-17256)	130
222	Butts at rifle range at Camp Matthews, MCRD San Diego, CA, 6 February 1946 (NARA College Park, RG 127-GC, box 5, photo 401016)	131
223	Typical rifle range, M31A1 target device, basic plan for fifty targets enlarged, Fort Bragg, NC, 1963 (Standard drawing No. 28-13-09 drawing 5 of 7, Range, rifle, known distance, layout and details for installation of M31A1 Target Device, 6 November 1963)	132
224	Typical rifle range, M31A1 target device, elevation of target arrangement, Fort Bragg, NC, 1963 (Standard drawing No. 28-13-09 drawing 5 of 7, Range, rifle, known distance, layout and details for installation of M31A1 target device, 6 November 1963)	132

Figure		Page
225	Typical rifle range, M31A1 target device, section through target butt and pop-up target enclosure, Fort Bragg, NC, 1963 (Standard drawing No. 28-13-09 drawing 5 of 7, Range, rifle, known distance, layout and details for installation of M31A1 target device, 6 November 1963)	132
226	Typical rifle range, M31A1 target device, plan of pop-up target enclosure, Fort Bragg, NC, 1963 (Standard drawing No. 28-13-09 drawing 5 of 7, Range, rifle, known distance, layout and details for installation of M31A1 target device, 6 November 1963)	133
227	Typical rifle range, M31A1 target device, portable control tower, Fort Bragg, NC, 1963 (Standard drawing No. 28-13-09 drawing 5 of 7, Range, rifle, known distance, layout and details for installation of M31A1 target device, 6 November 1963)	133
228	Typical rifle range target butts timber wall (on hillside), Fort Bragg, NC, 1952 (Standard drawing No. 28-13-09 drawing 2 of 7, Range, rifle, known distance, details, 5 January 1952)	134
229	Typical rifle range target butts timber wall (on level ground), Fort Bragg, NC, 1952 (Standard drawing No. 28-13-09 drawing 2 of 7, Range, rifle, known distance, details, 5 January 1952)	135
230	Typical rifle range target butts log crib retaining wall, Fort Bragg, NC, 1952 (Standard drawing No. 28-13-09 drawing 2 of 7, Range, rifle, known distance, details, 5 January 1952)	135
231	Typical Rifle Range Target Butts Concrete Retaining Wall, Fort Bragg, NC, 1952 (Standard drawing No. 28-13-09 drawing 2 of 7, Range, rifle, known distance, details, 5 January 1952)	136
232	Typical rifle range target butts section detail, Fort Bragg, NC, 1952 (Standard drawing No. 28-13-09 drawing 2 of 7, Range, rifle, known distance, details, 5 January 1952)	136
233	Typical rifle range target butts section elevation, Fort Bragg, NC, 1952 (Standard drawing No. 28-13-09 drawing 2 of 7, Range, rifle, known distance, details, 5 January 1952)	137
234	Target line at the USMC Winthrop Rifle Range, Indian Head, MD, 1915 (NARA College Park, RG 127-G-100b, box 23, photo 521528)	137
235	Ground rifle practice in standing position at Miami Army Air Field, FL, 2 October 1943 (NARA College Park, RG 342-FH, box 2207, photo 4A-18407)	138
236	Horses digging target trench, Fort Knox, KY, undated (scans from Knox, training 5, vol. 5)	138
237	Horses digging target trench, Fort Knox, KY, undated (scans from Knox, training 6, vol. 5)	139
238	Target complex built on top of trench, Fort Knox, KY, undated (scans from Knox, training 7, vol. 5)	139
239	Looking up out of target trench (rifle range pits), Fort Knox, KY, undated (scans from Knox, training 8, vol. 5, #358, Kirkpatrick photo)	140
240	Typical rifle range diagram of multiple ranges, Fort Bragg, NC, 1952 (Standard drawing No. 28-13-09 drawing 1 of 7, Range, rifle, known distance, plans and details, 5 January 1952)	140
241	Anti-aircraft towed target range, circa 1942 (Standard drawing No. 1600-140, Field target range A-A towed target range, 17 November 1942)	141
242	Firing range at Anti-Aircraft Training and Test Center Dam Neck, VA, December 1943 (NARA College Park, RG 80-G, box 8, photo 2316)	142
243	Beach firing range at NAS Fort Lauderdale, FL, 28 February 1944 (NARA College Park, RG 80-G, box 1513, photo 388275)	142

Figure		Page
244	Aerial of machine gun range at MCAS Cherry Point, NC, 9 August 1943 (NARA College Park, RG 80-G, box 1495, photo 384685)	143
245	Machine gun crews in action on range at MCAS Cherry Point, NC, 21 August 1943 (NARA College Park, RG 80-G, box 1360, photo 358873)	144
246	Marines operating machine gun at MCAS Cherry Point, NC, 21 August 1943 (NARA College Park, RG 80-G, box 1360, photo 358872)	144
247	Sailors at gunnery range at NAS Wildwood, NJ, 30 May 1944 (NARA College Park, RG 80-G, box 1487, photo 383360)	145
248	CMTC students on machine gun range at Camp Del Monte, CA, 13 August 1925 (NARA College Park, RG 111-SC WWI, box 700, photo 94896)	146
249	Machine gun course, circa 1952 (Standard drawing No. 28-13-05, Machine gun course, table II, course A, 4 January 1952)	147
250	Circular machine gun ranges at Naval Air Test Center Corpus Christi, VA, 12 November 1942 (NARA College Park, RG 80-G, box 276, photo 63318)	148
251	.30 caliber machine gun Harmonization Range #3 at Tyndall Army Air Field (Panama City), FL, 11 June 1942 (NARA College Park, RG 342-FH, box 2202, photo 4A-17261)	149
252	Machine gun range at Coast Guard Air Station Port Angeles, WA, 8 January 1957 (NARA College Park, RG 80-G, box 39, photo 7163)	149
253	Small arms fire course, machine gun platform, Fort Bragg, NC, 1952 (Standard drawing 28-13-37 sheet 2 of 2, Small arms fire course, details, 20 June 1952)	150
254	Training enlisted personnel on the Machine Gun Range at Amphibious Training Base Fort Pierce, FL, 23 November 1943 (NARA College Park, RG 80-G, box 862, photo 264377)	150
255	Firing a Lewis machine gun from shoulder at Camp Wheeler, GA, 1917 (New York Public Library, digital No. 117105)	151
256	Soldiers operating a Lewis machine gun at unknown location, February 1918 (New York Public Library, digital No. 117103)	151
257	Fixed machine gun range at NAS Wildwood, NJ, 29 April 1944 (NARA College Park, RG 80-G, box 1487, photo 383347)	152
258	Machine gun practice for cadets at NAS Corpus Christi, TX, 23 July 1941 (NARA College Park, RG 80-G, box 1978, photo 463691)	152
259	Trainees fire the 30 caliber machine guns (in background) on swivel at fixed targets 200 yds away first, then trainees fire the 50 caliber machine guns at unknown location, July 1942 (NARA College Park, RG 342-FH, box 2202, photo 4A-17268)	153
260	Machine gun firing range at unknown location, undated (NARA College Park, RG 342-FH, box 2230, photo 4A-24093)	154
261	A group of trainees fire .30 caliber light machine guns on one of the ranges at Fort Knox, KY, 1947 (NARA College Park, RG 111-SC WWII, box 602, photo SC299052)	154
262	A 1000-in. fixed target track type range, danger area, Fort Bragg, NC, 1952 (Standard drawing 28-13-06 sheet 1 of 2, Range, 1000-in. Fixed target, track type, plans and details, 20 June 1952)	156
263	A 1000-in. fixed target track type range, plan, Fort Bragg, NC, 1952 (Standard drawing 28-13-06 sheet 1 of 2, Range, 1000-in. fixed target, track type, plans and details, 20 June 1952)	157
264	A 1000-in. fixed target track type range, section, Fort Bragg, NC, 1952 (Standard drawing 28-13-06 sheet 1 of 2, Range, 1000-in. Fixed target, track type, plans and details, 20 June 1952)	157

Figure		Page
265	1000-in. fixed target track type range, target assembly, Fort Bragg, NC, 1952 (Standard drawing 28-13-06 sheet 1 of 2, Range, 1000-in. fixed target, track type, plans and details, 20 June 1952)	158
266	A 1000-in. fixed target track type range, target track details, Fort Bragg, NC, 1952 (Standard drawing 28-13-06 sheet 1 of 2, Range, 1000-in. fixed target, track type, plans and details, 20 June 1952)	159
267	A 1000-in. fixed target track type range, section through salvage wall, Fort Bragg, NC, 1952 (Standard drawing 28-13-06 sheet 1 of 2, Range, 1000-in. fixed target, track type, plans and details, 20 June 1952)	159
268	A 1000-in. miniature anti-tank range, circa 1951 (Standard drawing No. 1600-120, 1000-in. Miniature anti-tank range, 28 October 1942, TM 9-855, Targets, target material and training course layouts, 1 November 1951, p 32)	160
269	A 1000-in. miniature anti-tank range, circa 1951 (Standard drawing No. 1600-120, 1000-in. miniature anti-tank range, 28 October 1942, TM 9-855, Targets, target material and training course layouts, 1 November 1951, p 32)	161
270	Miniature anti-aircraft range, plan, circa 1942-1951 (Standard drawing No. 1600-130/131, Training aids, A. A. range miniature, 14 October 1942, TM 9-855, Targets, target material and training course layouts, 1 November 1951, pp 28, 29)	162
271	Miniature anti-aircraft range, target pulley system plan and elevation, circa 1942-1951 (Standard drawing No. 1600-130/131, Training aids, A. A. range miniature, 14 October 1942, TM 9-855, Targets, target material and training course layouts, 1 November 1951, pp 28, 29)	163
272	Miniature anti-aircraft range, target pulley system plan and elevation, circa 1942-1951 (Standard drawing No. 1600-130/131, Training aids, A. A. range miniature, 14 October 1942, TM 9-855, Targets, target material and training course layouts, 1 November 1951, pp 28, 29)	163
273	Moving target range, circa 1951 (TM 9-855, Targets, target material and training course layouts, 1 November 1951, p 35)	164
274	ATC facilities, night firing range "N", danger area, Fort Bragg, NC, 1966 (Standard drawing 28-13-117 sheet 8, Construction of ranges, phase 1, U.S. ATC Facilities, Fort Bragg, NC, range "N", night firing range, 28 June 1966)	166
275	ATC facilities, night firing range "N", plan and typical section, Fort Bragg, NC, 1966 (Standard drawing 28-13-117 sheet 8, Construction of ranges, phase 1, U.S. ATC Facilities, Fort Bragg, NC, range "N", night firing range, 28 June 1966)	167
276	Night firing range, circa 1992 (TC 25-8, Training ranges, 25 February 1992, p 6-14)	168
277	ATC facilities, night firing range "N", firing line, Fort Bragg, NC, 1966 (Standard drawing 28-13-117 sheet 9, Construction of ranges, phase 1, U.S. ATC Facilities, Fort Bragg, NC, range "N", night firing range, 28 June 1966)	169
278	Night fire range electrical layout, Fort Bragg, NC, 1966 (Standard drawing 28-13-117 sheet 28, Electrical distribution, 28 June 1966)	169
279	ATC facilities, night firing range "N", range extension plan, Fort Bragg, NC, undated (Standard drawing 28-13-117 sheet 9A, construction of ranges, phase 1, U.S. ATC facilities, Fort Bragg, NC, range "N", night firing range extension (70 points to 110 points), undated)	170
280	Night range light, Fort Bragg, NC, 1966 (Standard drawing 28-13-117 sheet 28, Electrical distribution, 28 June 1966)	170
281	Skeet range, circa 1945 (AAF 85-0-1, Army Air Forces gunnery and bombardment ranges, 15 June 1945, p 2-1-3)	171

Figure		Page
282	On one of a dozen skeet ranges, aerial gunnery students learn to fire rapidly and to lead their target at Fort Myers Army Air Field, FL, December 1942 (NARA College Park, RG 342-FH, box 2207, photo 4A-18414)	172
283	Skeet target practice for cadets at NAS Corpus Christi, TX, 23 July 1941 (NARA College Park, RG 80-G, box 1978, photo 463690)	172
284	Close-up of tower on the 7-mile skeet range at Geiger Field (near Spokane), WA, 3 March 1942 (NARA College Park, RG 342-FH, box 2202, photo 4A-18382)	173
285	Mile skeet range at Geiger Field (near Spokane), WA, 4 March 1942 (NARA College Park, RG 342-FH, box 2207, photo 4A-18423)	174
286	Overall view of submarine target range at Camp Wissahickon, NJ, undated (circa 1918) (NARA College Park, RG 24-TC, box 1, Folder P)	174
287	Looking down on target track on submarine target range at Camp Wissahickon, NJ, undated (circa 1918) (NARA College Park, RG 24-TC, box 1, folder K)	175
288	Sailor operating targets on track at submarine target range at Camp Wissahickon, NJ, undated (circa 1918) (NARA College Park, RG 24-TC, box 1, folder K)	176
289	End of target track on submarine target range at Camp Wissahickon, NJ, undated (circa 1918) (NARA College Park, RG 24-TC, box 1, folder K)	176
290	Class "B" range, circa 1913 (War Department Document No. 422, Small arms firing manual – 1913 (corrected to 15 April 1917), 28 February 1913, p 199)	178
291	Typical combat range, Fort Bragg, NC, 1951 (Standard drawing 28-13-18 sheet 1 of 1, Range, field target, small arms, plan and details, 21 November 1951)	179
292	Combat firing range electrical layout, Fort Bragg, NC, 1966 (Standard drawing 28-13-117 sheet 32, Electrical distribution, 28 June 1966)	180
293	Small arms range, field target, isometric view of target installation, Fort Bragg, NC, 1951 (Standard drawing 28-13-18 sheet 1 of 1, Range, field target, small arms, plan and details, 21 November 1951)	181
294	Small arms range, field target, details of control rack, Fort Bragg, NC, 1951 (Standard drawing 28-13-18 sheet 1 of 1, Range, field target, small arms, plan and details, 21 November 1951)	181
295	Small arms range, field target, details of wire support, Fort Bragg, NC, 1951 (Standard drawing 28-13-18 sheet 1 of 1, Range, field target, small arms, plan and details, 21 November 1951)	182
296	Combat firing range electrical layout, Fort Bragg, NC, 1966 (Standard drawing 28-13-117 sheet 32, Electrical distribution, 28 June 1966)	183
297	ATC facilities, combat firing range "Q" plan, Fort Bragg, NC, 1966 (Standard drawing 28-13-117 sheet 10, Construction of ranges, phase 1, U.S. ATC facilities, Fort Bragg, NC, range "Q", combat firing range, 28 June 1966)	183
298	Dismounted submachine gun practice course, danger area, Fort Bragg, NC, 1951 (Standard drawing 28-13-13, sheet 1 of 2, Range, submachine gun, dismounted practice course, plans and details, 7 December 1951)	184
299	Dismounted submachine gun practice course, firing and target areas, Fort Bragg, NC, 1951 (Standard drawing 28-13-13, sheet 1 of 2, Range, submachine gun, dismounted practice course, plans and details, 7 December 1951)	185
300	Dismounted submachine gun practice course, pivot target, Fort Bragg, NC, 1951 (Standard drawing 28-13-13 sheet 1 of 2, Range, submachine gun, dismounted practice course, plans and details, 7 December 1951)	186
301	Dismounted submachine gun practice course, surprise target diagram, Fort Bragg, NC, 1951 (Standard drawing 28-13-13 sheet 2 of 2, Range, submachine gun, dismounted practice course, details, 7 December 1951)	187

Figure		Page
302	Dismounted submachine gun practice course, surprise 2-target group isometric, Fort Bragg, NC, 1951 (Standard drawing 28-13-13 sheet 2 of 2, Range, submachine gun, dismounted practice course, details, 7 December 1951)	187
303	Dismounted submachine gun practice course, surprise 3-target group isometric, Fort Bragg, NC, 1951 (Standard drawing 28-13-13 sheet 2 of 2, Range, submachine gun, dismounted practice course, details, 7 December 1951)	188
304	Dismounted submachine gun practice course, sled target, Fort Bragg, NC, 1951 (Standard drawing 28-13-13 sheet 1 of 2, Range, submachine gun, dismounted practice course, plans and details, 7 December 1951)	188
305	Mounted pistol course, circa 1932 (Basic field manual, volume III – Basic weapons, Part one rifle company, Chapter 3 – Automatic pistol marksmanship, 5 April 1932, p 40)	189
306	Moving vehicle range, machine gun, circa 1942 (Standard drawing No. 1600-135, Moving vehicle ranges, 23 October 1942)	191
307	Trainees firing the 50 caliber machine gun mounted in the half-track at Fort Knox, KY, August 1942 (NARA College Park, RG 111-SC WWII, box 85, photo SC114295)	192
308	Machine gun firing range at Casper Army Air Field, WY, 5 May 1943 (NARA College Park, RG 342-FH, box 2202, photo 4A-17283)	192
309	Students operate machine guns from an E-9 gunnery truck at Harvard Army Air Field, NE, 22 April 1945 (NARA College Park, RG 342-FH, box 2202, photo 4A-17254)	193
310	Kneeling pasteboard target E, circa 1951, and panel target, circa 1942 (TM 9-855, Targets, target materials, and rifle range construction, 17 November 1951, pp 174, 176, "Framing for panel targets" RO-1 fig 22 pg. 36, Standard drawing No. 1600-135, Moving vehicle ranges, 23 October 1942)	193
311	Moving vehicle range, machine gun, circa 1942 (Standard drawing No. 1600-135, Moving vehicle ranges, 23 October 1942)	194
312	Moving vehicle range, machine gun, obstacles, circa 1942 (Standard drawing No. 1600-135, Moving vehicle ranges, 23 October 1942)	194
313	Submachine gun moving vehicle range, danger area, Fort Bragg, NC, 1951 (Standard drawing 28-13-21 sheet 1 of 1, Range, moving vehicle, submachine gun, plan and details, 21 November 1951)	196
314	Moving Vehicle range, submachine gun, circa 1942 to 1951 (Standard drawing No. 1600-135, Moving vehicle ranges, 23 October 1942)	197
315	Submachine gun moving vehicle range, plan, Fort Bragg, NC, 1951 (Standard drawing 28-13-21 sheet 1 of 1, Range, moving vehicle, submachine gun, plan and details, 21 November 1951)	198
316	Transition range danger area table V, Fort Bragg, NC, 1955 (Standard drawing 28-13-04 sheet 1 of 1, Range, transition, details, 6 June 1955)	200
317	Superimposed transition range, danger areas, Fort Bragg, NC, 1954 (Standard drawing 28-13-106 sheet 1 of 1, Superimposed transition range, table VIII, 22 March 1954, Standard drawing 28-13-107 sheet 1 of 1, Superimposed transition range, table VII, 28 March 1954)	200
318	Typical rifle transition range target layout, circa 1943 to 1952 (Standard drawing No. 1600-200, Transition firing course, 18 August 1943)	201
319	Transition range layouts, table IV and V, Fort Bragg, NC, 1955 (Standard drawing 28-13-04 sheet 1 of 1, Range, transition, details, 6 June 1955)	202
320	Superimposed transition range, Fort Bragg, NC, 1955 (Standard drawing 28-13-04 sheet 1 of 1, Range, transition, details, 6 June 1955)	203

Figure		Page
321	Transition firing range, machine gun range 53, overall site plan, Fort Bragg, NC, 1978 (Standard drawing DFE-3278 sheet 3 of 10, Transition firing range, machine gun-range 53, overall site plan, 19 January 1978)	204
322	Aerial photograph of Heins Rifle Transition Range, Fort Knox, KY, undated (Training 3a, vol. 1, Stock Shot #296, Fort Knox, KY, Heins Rifle Transition Range, Fort Knox, photographer Unk., undated, USAARMC Photo Branch, Fort Knox, KY)	205
323	Superimposed transition range, firing line, Fort Bragg, NC, 1954 (Standard drawing 28-13-107 sheet 1 of 1, Superimposed transition range, VII, 28 March 1954)	
324	Transition firing range, machine gun range 53, Detail site plan, Fort Bragg, NC, 1978 (Standard drawing DFE-3278 sheet 6 of 10, Transition firing range, machine gun-range 53, range details, 19 January 1978)	206
325	Superimposed transition range, firing line, Fort Bragg, NC, 1954 (Standard drawing 28-13-107 sheet 1 of 1, Superimposed transition range, table VII, 28 March 1954)	207
326	Marines at the transition range at MCB Camp Lejeune, NC, 1961 (NARA College Park, RG 127-GG-958, photo 341782)	207
327	A drill instructor teaches recruits the proper method of firing from inside a building at MCRD San Diego (Camp Pendleton), CA, December 1972 (NARA College Park, RG 127-GG-936, photo 230641)	208
328	Superimposed transition range, Fort Bragg, NC, 1955 (Standard drawing 28-13-04 sheet 1 of 1, Range, transition, details, 6 June 1955)	209
329	Transition range detail of M.G. Panel target, Fort Bragg, NC, 1955 (Standard drawing 28-13-04 sheet 1 of 1, Range, transition, details, 6 June 1955)	209
330	Transition range moving target detail, Fort Bragg, NC, 1955 (Standard drawing 28-13-04 sheet 1 of 1, Range, transition, details, 6 June 1955)	210
331	Superimposed transition range, section thru control pit, Fort Bragg, NC, 1954 (Standard drawing 28-13-108 sheet 1 of 1, Superimposed transition range, table VIII, 22 March 1954)	210
332	Above ground moving target pit, section and plan, Fort Bragg, NC, 1955 (Standard drawing 28-13-05 sheet 1 of 1, Transition ranges, M1 rifle and MG, above ground pits, construction details, 6 June 1955)	211
333	Above ground deep pit, sections and plan, Fort Bragg, NC, 1955 (Standard drawing 28-13-05 sheet 1 of 1, Transition ranges, M1 Rifle and MG, above ground pits, construction details, 6 June 1955)	212
334	Shell hole target details, Fort Bragg, NC, 1957 (Standard drawing 28-13-05, close combat course, plan and details, 8 September 1957)	213
335	Superimposed transition range, plan of targets in foxholes, and section, Fort Bragg, NC, 1954 (Standard drawing 28-13-108 sheet 1 of 1, Superimposed transition range, table VIII, 22 March 1954)	213
336	Superimposed transition range, vertical target pit dimensions, Fort Bragg, NC, 1954 (Standard drawing 28-13-108 sheet 1 of 1, Superimposed transition range, table VIII, 22 March 1954)	214
337	Transition firing range, machine gun range 53, detail of concrete M-30 pop-up target emplacement, Fort Bragg, NC, 1978 (Standard drawing DFE-3278 sheet 7 of 10, Transition firing range, machine gun-range 53, range details, 19 January 1978)	214
338	Superimposed transition range, target layout plan and window target details, Fort Bragg, NC, 1954 (Standard drawing 28-13-108 sheet 1 of 1, Superimposed transition range, table VIII, 22 March 1954)	215

Figure		Page
339	Transition firing range, machine gun range 53, centerline profiles for lanes 1 to 3, Fort Bragg, NC, 1978 (Standard drawing DFE-3278 sheet 4 of 10, transition firing range, machine gun-range 53, centerline profiles, lanes 1 through 3, 19 January 1978)	215
340	Transition firing range, machine gun range 53, centerline profiles for lanes 4 to 6, Fort Bragg, NC, 1978 (Standard drawing DFE-3278 sheet 5 of 10, Transition firing range, machine gun-range 53, centerline profiles, lanes 4 through 6, 19 January 1978)	216
341	Moving base range, circa 1945 (AAF Manual 85-0-1, Army Air Forces gunnery and bombardment ranges, 15 June 1945, pp 2-2-1 to 2-2-3)	217
342	Moving target submachine gun range, danger area, Fort Bragg, NC, 1952 (Standard drawing 28-13-14 sheet 1 of 3, Range, moving target, submachine gun, plans and details, 20 June 1952)	218
343	Submachine gun moving target range, circa 1952 (Standard drawing No. 28-13-14, Range, moving target, submachine gun, sheets 1 to 3, 20 June 1952)	219
344	Moving target submachine gun range, plan, Fort Bragg, NC, 1952 (Standard drawing 28-13-14 sheet 1 of 3, Range, moving target, submachine gun, plans and details, 20 June 1952)	219
345	Moving target submachine gun range, target and target rack assembly, Fort Bragg, NC, 1952 (Standard drawing 28-13-14 sheet 1 of 3, Range, moving target, submachine gun, plans and details, 20 June 1952)	220
346	Moving target submachine gun range, section thru firing area, Fort Bragg, NC, 1952 (Standard drawing 28-13-14 sheet 2 of 3, Range, moving target, submachine gun, repair pit and parapet details, 20 June 1952)	221
347	Moving target submachine gun range, center repair pit elevation, Fort Bragg, NC, 1952 (Standard drawing 28-13-14 sheet 2 of 3, Range, moving target, submachine gun, repair pit and parapet details, 20 June 1952)	221
348	Moving Target submachine gun range, sections thru repair pits and wing walls, Fort Bragg, NC, 1952 (Standard drawing 28-13-14 sheet 2 of 3, Range, moving target, submachine gun, repair pit and parapet details, 20 June 1952)	222
349	Moving target submachine gun range, wing walls for end pits, Fort Bragg, NC, 1952 (Standard drawing 28-13-14 sheet 2 of 3, Range, moving target, submachine gun, repair pit and parapet details, 20 June 1952)	223
350	Moving target submachine gun range, winch house plan, Fort Bragg, NC, 1952 (Standard drawing 28-13-14 sheet 3 of 3, Range, moving target, submachine gun, winch house, 20 June 1952)	223
351	Moving target submachine gun range, winch house section, Fort Bragg, NC, 1952 (Standard drawing 28-13-14 sheet 3 of 3, Range, moving target, submachine gun, winch house, 20 June 1952)	224
352	Moving target submachine gun range, winch house rear and front elevations, Fort Bragg, NC, 1952 (Standard drawing 28-13-14 sheet 3 of 3, Range, moving target, submachine gun, winch house, 20 June 1952)	224
353	Moving target submachine gun range, winch house side elevation, Fort Bragg, NC, 1952 (Standard drawing 28-13-14 sheet 3 of 3, Range, moving target, submachine gun, winch house, 20 June 1952)	225
354	Moving target submachine gun range, counterweight tower, Fort Bragg, NC, 1952 (Standard drawing 28-13-14 sheet 1 of 3, Range, moving target, submachine gun, plans and details, 20 June 1952)	225
355	Live grenade course, plan, danger area, Fort Bragg, NC, 1951 (Standard drawing 28-13-44 sheet 1 of 1, Live grenade course, plans and details, 21 November 1951)	227

Figure

		Page
356	Fragmentation and offensive hand grenade range surface danger area, circa 1943-1986 (AR 385-63, Regulations for firing ammunition for training, target practice, and combat, 17 June 1968, Chapter 8, AR 385-63, Safety regulations for firing ammunition for training, target practice, and combat, 28 February 1973, p 8-3, AR 385-63/MCO P3570.1, Policies and Procedures for firing ammunition for training, target practice, and combat, 22 February 1978, p 7-3)	228
357	Fragmentation and offensive hand grenade range modified surface danger area, circa 1983-1986 (FM 23-30, AR 385-63/MCO P3570.1A, Policies and procedures for firing ammunition for training, target practice, and combat, 15 October 1983, p 7-3)	228
358	Angle court, circa 1927-1944 (War Department Document No. 918, Volume III, Manual of basic training and standards of proficiency for the national guard, 1927, p 436, FM 23-30, Grenades, 15 June 1942, p 20, FM 23-30, Hand and rifle grenades, rocket, AT, HE, 2.36-in., 14 February 1944, pp 30-34, Ibid., pp 31-34, FM 23-30, Grenades, January 1940, p 15, FM 23-30, Grenades, 15 June 1942, p 19)	229
359	Practice grenade course, crater court, Fort Bragg, NC, 1951 (Standard drawing 28-13-43 sheet 1 of 1, Practice grenade course layout and details, 21 November 1951)	230
360	Assault course layout, circa 1988 (FM 23-30, Hand and rifle grenades, rocket, AT, HE, 2.36-in., 14 February 1944, pp 39 – 41, FM 23-30, Hand and rifle grenades, 14 April 1949, pp 42-44)	231
361	Live grenade practice course, plan and profile, circa 1944-1949 (FM 23-30, Hand and rifle grenades, rocket, AT, HE, 2.36-in., 14 February 1944, pp 35, 36, FM 23-30, Hand and rifle grenades, 14 April 1949, pp 40, 41)	232
362	Live grenade course, plan and profile, Fort Bragg, NC, 1951 (Standard drawing 28-13-44 sheet 1 of 1, Live grenade course, plans and details, 21 November 1951)	233
363	Live grenade pit layout, circa 1988 (FM 23-30, Grenades and pyrotechnic signals, 27 December 1988, p 4-6)	234
364	Periscopes, throwing bays, and impact area at Remagen Hand Grenade Range at Fort Jackson, SC, 1 November 1961 (NARA College Park, RG 111-SC post-1955, box 371, photo SC592064)	234
365	Main grenade court, circa 1927-1949 (FM 23-30, Grenades, January 1940, p.16, FM 23-30, Hand and rifle grenades, rocket, AT, HE, 2.36-in., 14 February 1944, p 30, 31)	235
366	Throwing pits course, circa 1942 (FM 23-30, Grenades, 15 June 1942, p 23)	236
367	Practice grenade course, foxhole court, Fort Bragg, NC, 1951 (Standard drawing 28-13-43 sheet 1 of 1, Practice grenade course layout and details, 21 November 1951)	237
368	Trench court, circa 1927-1942 (War Department Document No. 918, Volume III, Manual of basic training and standards of proficiency for the National Guard, 1927, p 437, FM 23-30, Grenades, 15 June 1942, pp 22, 23)	238
369	Throwing live fragmentation hand grenades at the dummy targets at Fort Jackson, SC, 1 November 1943 (NARA College Park, RG 111-SC WWII, box 681, photo SC324452)	238
370	Practice grenade course, vertical target court, Fort Bragg, NC, circa 1927 to 1951 (Standard drawing 28-13-43 sheet 1 of 1, Practice grenade course layout and details, 21 November 1951)	239
371	Woods Court, circa 1945-1949 (change 2 to FM 23-30, Hand and rifle grenades, rocket, AT, HE, 2.36-in., 18 October 1945, pp 13, 14, FM 23-30, Hand and rifle grenades, 14 April 1949, pp 34-39)	240

Figure		Page
372	Practice grenade course, Woods Court, Fort Bragg, NC, 1951 (Standard drawing 28-13-43 sheet 1 of 1, Practice grenade course layout and details, 21 November 1951)	240
373	Wearing helmets and barricaded behind sand bags as a precaution against flying grenade splinters, a recruit assumes a perfect stance as he lets fly his live grenade at MCRD Parris Island, SC, 11 June 1946 (NARA College Park, RG 127-GC, box 35, photo A16066)	241
374	The straining on the faces of these recruits show that they want to give the grenade a prodigious heave to give them time to duck behind the concrete barrier before the missile explodes at MCRD Parris Island, SC, 28 April 1949 (NARA College Park, RG 127-GC, box 35, photo 19159)	241
375	A recruit lies in a grenade pit just after he has thrown a grenade, there is an instructor in each pit at MCRD Parris Island, SC, 6 December 1951 (NARA College Park, RG 127-GC-586, box 35, photo A60683)	242
376	Live bays on the Remagen Hand Grenade Range at Fort Jackson, SC, 25 March 1965 (NARA College Park, RG 111-SC post-1955, box 390, photo SC615409)	243
377	Grenade bursting in front of pits at MCRD Parris Island, SC, 6 December 1951 (NARA College Park, RG 127-GC-586, box 35, photo A60641)	244
378	Recruits lying behind wall waiting their turn to throw grenades at MCRD Parris Island, SC, 6 December 1951 (NARA College Park, RG 127-GC-586, box 35, photo A60662)	244
379	Live grenade course, section thru barrier and periscope detail, Fort Bragg, NC, 1951 (Standard drawing 28-13-44 sheet 1 of 1, Live grenade course, plans and details, 21 November 1951)	245
380	Aerial photograph of Millcreek Grenade Range, Fort Knox, KY, 1974 (Training 15, vol. 1, 3-T-321-4/AH-74 Stock Shot # 1075, Millcreek grenade range, photo by: Mr. Maurice Monday, 6 June 1974, photo sac. A-V Sys Br. TASO, DPT, Fort Knox, KY 40121)	246
381	Rifle grenade surface danger area, circa 1968-1973 (AR 385-63, Regulations for firing ammunition for training, target practice and combat, 17 June 1968, p 8-4, AR 385-63, Safety regulations for firing ammunition for training, target practice, and combat, 28 February 1973, p 8-5, AR 385-63/MCO P3570.1, Policies and procedures for firing ammunition for training, target practice, and combat, 22 February 1978, p 7-5, AR 385-63/MCO P3570.1A, Policies and procedures for firing ammunition for training, target practice, and combat, 15 October 1983, p 7-5)	248
382	Practice rifle grenade course, danger area, Fort Bragg, NC, 1952 (Standard drawing 28-13-45 sheet 1 of 2, Practice rifle grenade course plans, 30 January 1952)	249
383	Rifle grenade court, circa 1927-1942 (Basic Field Manual, Instruction with hand and rifle grenades, 14 July 1932, p 30)	250
384	Stationary target rifle grenade course, circa 1942-1949 (FM 23-30, Hand and rifle grenades, rocket, AT, HE, 2.36-in., 14 February 1944, pp 85, 90, FM 23-30, Hand and rifle grenades, 14 April 1949, pp 86, 90)	251
385	Practice rifle grenade course, silhouette court for high angle antipersonnel firing, Fort Bragg, NC, 1952 (Standard drawing 28-13-45 sheet 1 of 2, Practice rifle grenade course plans, 30 January 1952)	252
386	Practice rifle grenade course, vertical target court for flat trajectory antipersonnel firing, Fort Bragg, NC, 1952 (Standard drawing 28-13-45 sheet 1 of 2, Practice rifle grenade course plans, 30 January 1952)	253

Figure		Page
387	Moving target rifle grenade course, circa 1942-1964 (FM 23-30, Grenades, 15 June 1942, pp 54, 56, FM 23-30, Hand and rifle grenades, rocket, AT, HE, 2.36-in., 14 February 1944, pp 85, 89, 91, FM 23-30, Hand and rifle grenades, 14 April 1949, pp 88 – 91, AR 210-21, Training areas and facilities for ground troops, 18 December 1964, p 6)	254
388	Practice rifle grenade course, plan of range for stationary and moving target firing, Fort Bragg, NC, 1952 (Standard drawing 28-13-45 sheet 1 of 2, Practice rifle grenade course plans, 30 January 1952)	255
389	Employment of 37-mm antitank range for antitank practice rifle grenade field firing, circa 1942-1949 (FM 23-30, Grenades, 14 June 1942, p 57, FM 23-30, Hand and rifle grenades, rocket, AT, HE, 2.36-in., 14 February 1944, p 92, FM 23-30, Hand and rifle grenades, 14 April 1949, p 93)	257
390	Practice rifle grenade course, use of antitank range for antitank practice grenade field firing, Fort Bragg, NC, 1952 (Standard drawing 28-13-45 sheet 1 of 2, Practice rifle grenade course plans, 30 January 1952)	258
391	Practice rifle grenade course, foxhole details, Fort Bragg, NC, 1952 (Standard drawing 28-13-45 sheet 2 of 2, Practice rifle grenade course details, 30 January 1952)	258
392	Practice rifle grenade course, targets for stationary target range and vertical target court, Fort Bragg, NC, 1952 (Standard drawing 28-13-45 sheet 2 of 2, Practice rifle grenade course details, 30 January 1952)	260
393	Practice rifle grenade course, high angle antipersonnel target, Fort Bragg, NC, 1952 (Standard drawing 28-13-45 sheet 2 of 2, Practice rifle grenade course details, 30 January 1952)	261
394	Moving target, circa 1942-1949 (FM 23-30, Hand and rifle grenades, rocket, AT, HE, 2.36-in., 14 February 1944, p 54, FM 23-30, Hand and rifle grenades, rocket, AT, HE, 2.36-in., 14 February 1944, p 89, FM 23-30, Hand and rifle grenades, 14 April 1949, p 92)	261
395	Practice rifle grenade course, target sled details, Fort Bragg, NC, 1952 (Standard drawing 28-13-45 sheet 2 of 2, Practice rifle grenade course details, 30 January 1952)	262
396	Practice rifle grenade course, ground pulley details, Fort Bragg, NC, 1952 (Standard drawing 28-13-45 sheet 2 of 2, Practice rifle grenade course details, 30 January 1952)	262
397	Trainfire ranges layout plan, Fort Bragg, NC, 1960 (Standard drawing 28-13-115 sheet 1 of 24, Trainfire ranges, layout plan, 5 October 1960)	263
398	Aerial photograph of Clark Trainfire Range, Fort Knox, KY, 1960 (Training 4a, vol. 1, Stock Shot # 263, Fort Knox, KY, Clark Trainfire, Fort Knox, Ky, photo by: SFC John A. Gilstrap, 23 May 1960, USAARMC Photo Branch, Fort Knox, KY)	264
399	Aerial photograph of Captain O'Brien Trainfire Range, Fort Knox, KY, 1960 (Training 11, vol. 1, Stock Shot # 281, Fort Knox, KY, O'Brien Trainfire Range, Fort Knox, photo by: SFC John A. Gilstrap, 23 May 1960, USAARMC Photo Branch, Fort Knox, KY)	264
400	Rifle marksmanship course, trainfire I, 25m range (65 point), danger area, Fort Bragg, NC, 1958 (Standard drawing 28-13-105 sheet 2, Rifle marksmanship course, trainfire i, 25m range (65 point), plan, section and details, 5 August 1958)	265
401	Rifle marksmanship course, trainfire I, 25m range (65 point), plan, Fort Bragg, NC, 1958 (Standard drawing 28-13-105 sheet 2, Rifle marksmanship course, trainfire I, 25m range (65 point), plan, section and details, 5 August 1958)	266
402	Twenty-five meter Range at Fort Jackson, SC, 25 October 1965 (NARA College Park, RG 111-SC post-1955, box 396, photo SC624269)	267

Figure		Page
403	Trainfire ranges, target detection range plan, Fort Bragg, NC, 1960 (Standard drawing 28-13-115 sheet 12 of 24, Rifle marksmanship course, trainfire I, target detection range, plan, section and details, 5 October 1960)	268
404	Trainfire ranges, target detection range perspective, Fort Bragg, NC, 1960 (Standard drawing 28-13-115 sheet 12 of 24, Rifle marksmanship course, trainfire I, target detection range, plan, section and details, 5 October 1960)	268
405	Trainfire field firing range (35 points) danger area, Camp Gordon, GA, 1958 (Standard drawing 28-13-105 sheet 3, Field firing range (35 points), plan, section and details, 5 August 1958)	269
406	Trainfire field firing range (35 points) typical section, Camp Gordon, GA, 1958 (Standard drawing 28-13-105 sheet 3, Field firing range (35 points), plan, section and details, 5 August 1958)	270
407	Trainfire field firing range (35 points) plan legend, Camp Gordon, 1958 (Standard drawing 28-13-105 sheet 3, Field firing range (35 points), plan, section and details, 5 August 1958)	270
408	Trainfire I record firing range typical layout plan, Fort Knox, KY, 1960 (Standard drawing 28-13-02 sheet 12, Trainfire I ranges-1960 F.Y., Ditto Hill record firing range, typical layout plan and foxhole profile, 25 May 1960)	271
409	Trainfire ranges record firing range (16 point) plan legend, Fort Gordon, 1966 (Standard drawing 28-13-105 sheet 4, Rifle marksmanship course, trainfire I, record firing range (16 point) plan and section, 7 June 1966)	271
410	Trainfire ranges record fire No.4 (12 points), layout, Fort Bragg, NC, 1960 (Standard drawing 28-13-115 sheet 5 of 24, Record fire No.4 (12 Points), 5 October 1960)	272
411	Trainfire ranges record fire (12 points), legend, Fort Bragg, NC, 1960 (Standard drawing 28-13-115 sheet 2 of 24, Record fire No.1 (12 Points), 5 October 1960)	272
412	Trainfire ranges record firing range (16 Point) danger area, Fort Gordon, 1966 (Standard drawing 28-13-105 sheet 4, Rifle marksmanship course, trainfire I, Record firing range (16 point) plan and section, 7 June 1966)	273
413	Trainfire ranges record firing range (16 point) plan, Fort Gordon, 1966 (Standard drawing 28-13-105 sheet 4, Rifle marksmanship course, trainfire I, Record firing range (16 point) plan and section, 7 June 1966)	273
414	Rifle marksmanship course, trainfire I, record firing range (16 point), perspective view of layout plan, Fort Bragg, NC, 1958 (Standard drawing 28-13-105 sheet 4, Rifle marksmanship course, trainfire I, record firing range (16 point), plan and section, 5 August 1958)	274
415	Automated record fire range location map, Fort Knox, KY, 1985 (FK-360-2, Automated record fire range, infantry remote electronic target system, PN 308 FY 85, location map, 1 May 1985)	274
416	Automated record fire range site layout, Fort Knox, KY, 1985 (FK-360-7, Automated record fire range, infantry remote electronic target system, PN 308 FY 85, site layout, 1 May 1985)	275
417	Automated record fire range site layout legend, Fort Knox, KY, 1985 (FK-360-7, Automated record fire range, infantry remote electronic target system, PN 308 FY 85, site layout, 1 May 1985)	275
418	Automated record fire range site layout, Fort Knox, KY, 1985 (FK-360-8, Automated record fire range, infantry remote electronic target system, PN 308 FY 85, site layout, 1 May 1985)	276
419	Automated record fire range site layout, Fort Knox, KY, 1985 (FK-360-9, Automated record fire range, infantry remote electronic target system, PN 308 FY 85, site layout, 1 May 1985)	276

Figure		Page
420	Automated record fire range site layout, Fort Knox, KY, 1985 (FK-360-10, Automated record fire range, infantry remote electronic target system, PN 308 FY 85, Site layout, 1 May 1985)	277
421	Trainfire ranges record fire No. 1 (12 points), foxholes, Fort Bragg, NC, 1960 (Standard drawing 28-13-115 sheet 2 of 24, Record fire No.1 (12 points), 5 October 1960)	277
422	Trainfire I record firing range typical layout plan, Fort Knox, KY, 1960 (Standard drawing 28-13-02 sheet 12, Trainfire I ranges-1960 F.Y., Ditto Hill record firing range, typical layout plan and foxhole profile, 25 May 1960)	278
423	Range 24—25m range at Fort Leonard Wood, MO, 27 April 1966 (NARA College Park, RG 111-CRB box 86, photo SC35770)	279
424	Trainfire ranges, record fire range foxhole section, Fort Bragg, NC, 1960 (Standard drawing 28-13-115 sheet 13 of 24, Trainfire ranges details, 5 October 1960)	279
425	Trainfire ranges, record fire range foxhole plan, Fort Bragg, NC, 1960 (Standard drawing 28-13-115 sheet 13 of 24, Trainfire ranges details, 5 October 1960)	280
426	Trainfire ranges field fire No.2 (35 Points), Fort Bragg, NC, 1960 (Standard drawing 28-13-115 sheet 7 of 24, Field fire No.2 (35 Points), 5 October 1960)	280
427	Trainfire ranges field fire No.3 (35 Points), Fort Bragg, NC, 1960 (Standard drawing 28-13-115 sheet 8 of 24, Field fire No.3 (35 Points), 5 October 1960)	281
428	Trainfire ranges, target detection range observation line plan, Fort Bragg, NC, 1960 (Standard drawing 28-13-115 sheet 12 of 24, Rifle marksmanship course, trainfire I, target detection range, plan, section and details, 5 October 1960)	281
429	Trainfire ranges, target detection range sighting device, Fort Bragg, NC, 1960 (Standard drawing 28-13-115 sheet 12 of 24, Rifle marksmanship course, trainfire I, target detection range, plan, section and details, 5 October 1960)	282
430	Trainfire ranges, target detection range sighting device, Fort Bragg, NC, 1960 (Standard drawing 28-13-115 sheet 12 of 24, Rifle marksmanship course, trainfire I, target detection range, plan, section and details, 5 October 1960)	282
431	Trainfire ranges record fire No.1 (12 points), targets, Fort Bragg, NC, 1960 (Standard drawing 28-13-115 sheet 2 of 24, Record fire No.1 (12 Points), 5 October 1960)	283
432	Trainfire ranges, target holder details, Fort Bragg, NC, 1960 (Standard drawing 28-13-115 sheet 13 of 24, Trainfire ranges details, 5 October 1960)	283
433	Trainfire ranges, target detection range detection panels, Fort Bragg, NC, 1960 (Standard drawing 28-13-115 sheet 12 of 24, Rifle marksmanship course, trainfire I, target detection range, plan, section and details, 5 October 1960)	284
434	Type F and E silhouette targets, Fort Bragg, NC, 1951 (Standard drawing 28-13-21 sheet 1 of 1, Range, moving vehicle, submachine gun, plan and details, 21 November 1951)	284
435	Trainfire ranges, concrete pop-up target emplacement plan, Fort Bragg, NC, 1960 (Standard drawing 28-13-115 sheet 13 of 24, Trainfire ranges details, 5 October 1960)	285
436	Trainfire ranges, concrete pop-up target emplacement section, Fort Bragg, NC, 1960 (Standard drawing 28-13-115 sheet 13 of 24, Trainfire ranges details, 5 October 1960)	285
437	Electrical targets at Camp Lejeune, NC, 2 May 1958 (NARA College Park, RG 127-GC, box 22, photo 340848)	286
438	Typical rifle range, m31a1 target device, plan of pop-up target enclosure, Fort Bragg, NC, 1963 (Standard drawing No. 28-13-09 drawing 5 of 7, Range, rifle, known distance, layout and details for installation of M31A1 target device, 6 November 1963)	287

Figure		Page
439	Typical rifle range, M31A1 target device, section through target butt and pop-up target enclosure, Fort Bragg, NC, 1963 (Standard drawing No. 28-13-09 drawing 5 of 7, Range, rifle, known distance, layout and details for installation of M31A1 target device, 6 November 1963)	287
440	Trainfire ranges record fire No.4 (12 points), typical berm section, Fort Bragg, NC, 1960 (Standard drawing 28-13-115 sheet 5 of 24, Record fire No.4 (12 Points), 5 October 1960)	288
441	ATC facilities, record firing range (16 point), range "K," Fort Bragg, NC, 1966 (Standard drawing 28-13-117 sheet 5, Construction of ranges, phase 1, U.S. ATC facilities, Fort Bragg, NC, range "K," record firing range (16 Point), 28 June 1966)	288
442	Rifle marksmanship course, trainfire I, record firing range (16 point), layout plan, Fort Bragg, NC, 1958 (Standard drawing 28-13-105 sheet 4, Rifle marksmanship course, trainfire I, record firing range (16 point), plan and section, 5 August 1958)	288
443	Trainfire I record firing range foxhole profile, Fort Knox, KY, 1960 (Standard drawing 28-13-02 sheet 12, Trainfire I Ranges-1960 F.Y., Ditto Hill Record Firing Range, Typical layout plan and foxhole profile, 25 May 1960)	289
444	Trainfire I field firing range cross sections, Fort Knox, KY, 1960 (Standard drawing 28-13-02 sheet 4,Trainfire I ranges-1960 F.Y., addition to Dripping Springs-Field Firing Range, Cross Sections, 25 May 1960)	289
445	Trainfire ranges record fire No.3 (12 Points), Buildings and foxholes, Fort Bragg, NC, 1960 (Standard drawing 28-13-115 sheet 4 of 24, Record fire No.3 (12 points), 5 October 1960)	290
446	Rifle marksmanship course, trainfire I, record firing range (16 point), layout plan, Fort Bragg, NC, 1958 (Standard drawing 28-13-105 sheet 4, Rifle marksmanship course, trainfire I, record firing range (16 point), plan and section, 5 August 1958)	290
447	25m range "R" electrical layout, Fort Bragg, NC, 1966 (Standard drawing 28-13-117 sheet 28, Electrical distribution, 28 June 1966)	291
448	Rifle marksmanship course, trainfire I, control tower (closed bottom), section, Fort Bragg, NC, 1958 (Standard drawing 28-13-105 sheet 7, Rifle marksmanship course, trainfire I, control tower (closed bottom), plans, elevations, section and details, 5 August 1958)	292
449	Rifle marksmanship course, trainfire I, control tower (closed bottom), elevations, Fort Bragg, NC, 1958 (Standard drawing 28-13-105 sheet 7, Rifle marksmanship course, trainfire I, control tower (closed bottom), plans, elevations, section and details, 5 August 1958)	293
450	Rifle marksmanship course, trainfire I, control tower (closed bottom), floor plans, Fort Bragg, NC, 1958 (Standard drawing 28-13-105 sheet 7, Rifle marksmanship course, trainfire I, control tower (closed bottom), plans, elevations, section and details, 5 August 1958)	293
451	ATC facilities, control towers, alternate details, Fort Bragg, NC, 1966 (Standard drawing 28-13-117 sheet 23, Construction of ranges, phase 1, U.S. ATC facilities, Fort Bragg, NC, option details, range buildings, 28 June 1966)	293
452	Rifle marksmanship course, trainfire i, control tower, section A, Fort Bragg, NC, 1958 (Standard drawing 28-13-105 sheet 6, Rifle marksmanship course, trainfire 1, control tower (open bottom), plans, elevations, sections and details, 5 August 1958)	294
453	Rifle marksmanship course, trainfire I, control tower, elevations, Fort Bragg, NC, 1958 (Standard drawing 28-13-105 sheet 6, Rifle marksmanship course, trainfire 1, control tower (open bottom), plans, elevations, sections and details, 5 August 1958)	295

Figure		Page
454	Rifle marksmanship course, trainfire I, control tower, foundation and floor plans, Fort Bragg, NC, 1958 (Standard drawing 28-13-105 sheet 6, Rifle marksmanship course, trainfire 1, control tower (open bottom), plans, elevations, sections and details, 5 August 1958)	295
455	Rifle marksmanship course, trainfire I, target house, elevations, Fort Bragg, NC, 1958 (Standard drawing 28-13-105 sheet 8, Rifle marksmanship course, trainfire I, target house, elevations, plans and details, 5 August 1958)	295
456	Rifle marksmanship course, trainfire I, target house, section and plan, Fort Bragg, NC, 1958 (Standard drawing 28-13-105 sheet 8, Rifle marksmanship course, trainfire i, target house, elevations, plans and details, 5 August 1958)	296
457	ATC facilities, target house, alternate details, Fort Bragg, NC, 1966 (Standard drawing 28-13-117 sheet 23, Construction of ranges, phase 1, U.S. ATC facilities, Fort Bragg, NC, option details, range buildings, 28 June 1966)	296
458	Rifle marksmanship course, trainfire i, latrine, elevations, Fort Bragg, NC, 1958 (Standard drawing 28-13-105 sheet 9, Rifle marksmanship course, trainfire I, latrine, plans, elevations, section and details, 5 August 1958, "Typical elevation with plywood siding" Standard drawing 28-13-117 sheet 23, Construction of ranges, phase 1, U.S. ATC facilities, Fort Bragg, NC, option details, range buildings, 28 June 1966)	297
459	Rifle marksmanship course, trainfire I, latrine, section and plan, Fort Bragg, NC, 1958 (Standard drawing 28-13-105 sheet 9, Rifle marksmanship course, trainfire I, latrine, plans, elevations, section and details, 5 August 1958)	298
460	Rifle marksmanship course, trainfire I, latrine, interior elevation and pit revetment, Fort Bragg, NC, 1958 (Standard drawing 28-13-105 sheet 9, Rifle marksmanship course, trainfire i, latrine, plans, elevations, section and details, 5 August 1958)	298
461	A 25m Range, Fort Bragg, NC, 17 May 2006	306
462	A 25m Range, Fort Bragg, NC, 17 May 2006	306
463	Trainfire Range, Fort Bragg, NC, 17 May 2006	307
464	Trainfire Range, Fort Bragg, NC, 17 May 2006	307
465	Transition Range, Fort Bragg, NC, 17 May 2006	308
466	Transition Range, Fort Bragg, NC, 17 May 2006	308
467	Transition Range, Fort Bragg,, NC, 17 May 2006	308
468	Cleared firing positions on 25m range, Fort Bragg, NC, 17 May 2006	309
469	Sandbag support on firing line of trainfire range, Fort Bragg, NC, 17 May 2006	17
470	Foxholes on firing line of 25m range, Fort Bragg, NC, 17 May 2006	310
471	Abandoned known distance range (overgrown foxholes on firing lines), Fort Bragg, NC, 17 May 2006	310
472	Covered firing line on Range 2, Fort Gordon, January 2004	311
473	Grenade firing pit, Fort Bragg, NC, 17 May 2006	19
474	Grenade firing trench, Fort Bragg, NC, 17 May 2006	312
475	Machine gun firing platform, Fort Bragg, NC, 17 May 2006	312
476	Transition range firing line with window, foxhole, door, rooftop and other simulated firing positions, Fort Bragg, NC, 17 May 2006	313
477	Stationary target frames on 25m range, Fort Bragg, NC, 17 May 2006	21
478	Stationary silhouette targets on grenade range, Fort Bragg, NC, 17 May 2006	314
479	Raised panel targets with pulleys, Fort Bragg, NC, 17 May 2006	315

Figure		Page
480	Raised panel targets with weights, Fort Bragg, NC, 17 May 2006	315
481	Firing line view of raised panel targets (in lowered positions), Fort Bragg, NC, 17 May 2006	315
482	Remnant of pop-up targets, Fort Bragg, NC, 17 May 2006	23
483	Remnant of pop-up targets, Fort Bragg, NC, 17 May 2006	23
484	Transition range target line with pop-up targets, Fort Bragg, NC, 17 May 2006	317
485	Transition range pop-up target, Fort Bragg, NC, 17 May 2006	317
486	Transition range pop-up target, Fort Bragg, NC, 17 May 2006	317
487	Pop-up target, Fort Bragg, NC, 17 May 2006	25
488	Pop-up targets, Fort Bragg, NC, 17 May 2006	25
489	Silhouette target mounted to a dolly on a moving target track, Fort Bragg, NC, 17 May 2006	319
490	Abandoned target track, Fort Bragg, NC, 17 May 2006	319
491	Cable pulley system on moving target track, Fort Bragg, NC, 17 May 2006	27
492	Cable pulley system on moving target track, Fort Bragg, NC, 17 May 2006	27
493	Target cars mounted on dollies and pulled by a cable system on a moving target track, Fort Bragg, NC, 17 May 2006	321
494	Moving target track operator and storage building, Fort Bragg, NC, 17 May 2006	321
495	Embankments behind target lines, Fort Bragg, NC, 17 May 2006	321
496	Embankments in front of pop-up targets, Fort Bragg, NC, 17 May 2006	322
497	Walled embankments at grenade range firing lines, Fort Bragg, NC, 17 May 2006	322
498	Grenade firing trench, Fort Bragg, NC, 17 May 2006	29
499	Grenade range protection walls, Fort Bragg, NC, 17 May 2006	324
500	Grenade range protection walls, Fort Bragg, NC, 17 May 2006	324
501	Grenade area on old Range 30, Fort Gordon, July 2004	324
502	Remnants of WWII target butt, Fort Gordon, January 2004	31
503	Remnants of WWII known distance range target butt, Fort Jackson (SCARNG), June 2004	31
504	Remnants of WWII known distance range target butt, Fort Jackson (SCARNG), June 2004	326
505	Remnants of WWII known distance range target butt, Fort Jackson (SCARNG), June 2004	33
506	Remnants of WWII known distance range target pulley, Fort Jackson (SCARNG), June 2004	327
507	Remnants of WWII Observation Tower Bldg R122, Fort Gordon, GA, January 2004	327
508	WWII Observation Tower Building 9805, Fort Bliss, TX, November 2005	328
509	Remnants of WWII Observation Tower Building R122, Fort Gordon, GA, January 2004	328
510	WWII Observation Tower Building 9789, Fort Knox, KY, November 2005	329
511	Remnants of WWII Observation Tower Building 9606, Fort Knox, KY, November 2005	329
512	Remnants of 1950's Observation Tower, Fort Jackson, SC (SCARNG), June 2004	330
513	Remnants of 1950's observation tower on Combat II Range, Fort Jackson, SC, June 2004	330
514	Observation tower, Fort Bragg, NC, 17 May 2006	331

Figure		Page
515	1966 Observation tower near Old Range #19, Fort Gordon, GA, January 2004	331
516	Remnants of 1966 Observation tower (near Building 421), Fort Gordon, GA, July 2004	332
517	1987 Range Bldg 486, Fort Gordon, GA, January 2004	332
518	1995 Range 10 observation tower and grandstand, Fort Jackson, SC, July 2003	332
519	Control tower and range buildings, Fort Bragg, NC, 17 May 2006	333
520	Observation tower, Fort Bragg, NC, 17 May 2006	333
521	25m trainfire range observation tower, Fort Bragg, NC, 17 May 2006	334
522	Observation tower, Fort Bragg, NC, 17 May 2006	334
523	Range firing tower, Fort Bragg, NC, 17 May 2006	334
524	Abandoned known distance range (foundations of observation tower and range building), Fort Bragg, NC, 17 May 2006	335
525	WWII Range Target Storage Bldg 9898, Fort Bliss, TX, November 2005	335
526	WWII Range Target Storage Bldg 9898, Fort Bliss, TX, November 2005	336
527	Remains of WWII known distance range target butt, latrine, and target storage, Fort Jackson, SC (SCARNG), June 2004	336
528	Remains of WWII Range Target Shed Bldg R161, Fort Gordon, GA, January 2004	336
529	1966 Bldg 421 range target shed, Fort Gordon, GA, January 2004	337
530	Range storage building, Fort Bragg, NC, 17 May 2006	337
531	Range storage building, Fort Bragg, NC, 17 May 2006	337
532	Range storage building, Fort Bragg, NC, 17 May 2006	338
533	Range storage building, Fort Bragg, NC, 17 May 2006	338
534	Range storage building, Fort Bragg, NC, 17 May 2006	338
535	1987 Building 485 range storage, Fort Gordon, January 2004	339
536	Range ammunition storage building, Fort Bragg, NC, 17 May 2006	339
537	Range ammunition storage building, Fort Bragg, NC, 17 May 2006	339
538	Remnants of WWII Latrine Bldg R152, Fort Gordon, GA, January 2004	340
539	Remnants of WWII Latrine Bldg 9347, Fort Knox, KY, November 2005	340
540	WWII Latrine Bldg 9606, Fort Knox, KY, November 2005	340
541	Remnants of WWII Latrine Bldg 9606, Fort Jackson, SC, November 2003	341
542	remnants of wwii known distance range target butt, latrine, and target storage, Fort Jackson, SC (SCARNG), June 2004	341
543	Range latrine, Fort Bragg, NC, 17 May 2006	342
544	Range latrine, Fort Bragg, NC, 17 May 2006	342
545	Range latrine, Fort Bragg, NC, 17 May 2006	342
546	Range bleachers, Fort Bragg, NC, 17 May 2006	343
547	1995 Range 10 observation tower and grandstand, Fort Jackson, SC, July 2003	343
548	Range bleachers, Fort Bragg, NC, 17 May 2006	343
549	1966 Bldg 427 Range Mess Hall, Fort Gordon, GA, January 2004	344
550	1966 Bldg 427 Range Mess Hall Interior, Fort Gordon, GA, January 2004	344
551	Range Mess Facilities 0-7906, Fort Bragg, NC, 17 May 2006	345
552	1995 Range 10 Weapons cleaning point, Fort Jackson, July 2003	345

Figure		Page
553	Weapons cleaning point, Fort Bragg, NC, 17 May 2006	345
554	Range firing tower, Fort Bragg, NC, 17 May 2006	346

Preface

This study was conducted for the Legacy Resource Management Program, Cultural Resources Management, under project "Activity A1450-MIPR to ERDC PN05-265." Funding was provided by Military Interdepartmental Purchase Request (MIPR) W31RYO51541162. The Legacy Resource Management Program technical monitor was Hillori Schenker, Cultural Resources Specialist.

The work was performed by the Land and Heritage Conservation Branch (CN-C) of the Installations Division (CN), Construction Engineering Research Laboratory (CERL). The CERL Project Manager was Adam Smith. Daniel Archibald, was primary compiler of the historical information; Sunny Adams was assistant architectural historian; Manroop Chawla was environmentalist, and Daniel Smith (IMCOM) was the military training history expert. Special thanks are owed to those that assisted with the development of this historic context: Holly Reed, Teresa Roy, and Donna Larker in the Still Pictures Room at the National Archives in College Park, Maryland; Andrew Knight, Priscilla Dyson, and Ivy Yarbough in the Cartographic and Architectural Record Room at the National Archives in College Park, Maryland; Pat Lacey, ERDC-CERL Librarian; Michelle Michael at Fort Bragg, NC; John Doss at Fort Bragg, NC; Laurie Rush at Fort Drum, NY; Ruth Lewis at Fort Gordon, GA; Pam Anderson at Naval Base, Norfolk, VA; Jim Dolph at Portsmouth Navy Yard; and Brian Lione, former Deputy Federal Preservation Officer, Office of the Secretary of Defense. Dr. Christopher White is Chief, CN-C, and Dr. John T. Bandy is Chief, CN. The Director of CERL is Dr. Ilker R. Adiguzel.

CERL is an element of the U.S. Army Engineer Research and Development Center (ERDC), U.S. Army Corps of Engineers. The Commander and Executive Director of ERDC is COL Gary E. Johnston, and the Director of ERDC is Dr. Jeffery p Holland.

Unit Conversion Factors

Multiply	By	To Obtain
acres	4,046.873	square meters
degrees Fahrenheit	(F-32)/1.8	degrees Celsius
feet	0.3048	meters
gallons (U.S. liquid)	3.785412 E-03	cubic meters
horsepower (550 foot-pounds force per second)	745.6999	watts
inches	0.0254	meters
miles (U.S. statute)	1,609.347	meters
square feet	0.09290304	square meters
square miles	2.589998 E+06	square meters
square yards	0.8361274	square meters
yards	0.9144	meters

1 Introduction

Background

Through the years, laws have been enacted to preserve our national cultural heritage. The Antiquities Act of 1906, which was the first major Federal preservation legislation to be enacted, was instrumental in securing protection for archeological resources on Federal property. The benefits derived from this Act and subsequent legislation precipitated an expanded and broader need for the preservation of historic cultural resources. This growing awareness was codified in the most sweeping legislation to date, the National Historic Preservation Act of 1966 (NHPA).

The NHPA was created to provide guidelines and requirements aimed at preserving tangible elements of our past primarily through the creation of the National Register of Historic Places (NRHP). Contained within this piece of legislation (Sections 110 and 106) are requirements for Federal agencies to address their cultural resources, defined as any prehistoric or historic district, site, building, structure, or object. Section 110 requires Federal agencies to inventory and evaluate their cultural resources. Section 106 requires the determination of effect of Federal undertakings on properties listed on, deemed eligible for, or potentially eligible for the NRHP, and requires Federal agencies to take into account the effect of a project on a property and to afford the State Historic Preservation Officer/Tribal Historic Preservation Officer (SHPO/THPO) a reasonable opportunity to comment on the undertaking.

According to National Register Bulletin #15, "How to Apply the National Register of Criteria for Evaluation," and National Register Bulletin #16a, "How to Complete the National Register Registration Form define historic contexts," for a building, structure, object, or a district to be eligible for the National Register, it must:

> represent a significant part of the history, architecture, archeology, engineering, or culture of an area, and it must have the characteristics that make it a good representative of properties associated with that aspect of the past. The significance of a historic property can be judged and explained only when it is evaluated within its historic context. ... Historic contexts are those patterns or trends in history by which a specific occurrence, property, or site is understood and its meaning (and ultimately its

significance) within history or prehistory is made clear.

A historic context is necessary to help researchers and persons involved in inventorying buildings for eligibility to the National Register, address these five factors:
1. The facet of prehistory or history of the local area, State, or the nation that the property represents
2. Whether that facet of prehistory or history is significant
3. Whether it is a type of property that has relevance and importance in illustrating the historic context
4. How the property illustrates that history
5. Whether the property possesses the physical features necessary to convey the aspect of prehistory or history with which it is associated.
National Register Bulletin #15

This project work was undertaken to develop a historic context for the development of military training lands used by the U.S. Department of Defense (DOD) and its forerunners.

Objectives

The initial objective of this project was to develop a historic context for the development of military training lands used by the DOD and its forerunners.

Approach

This work was performed in four steps:
1. A literature review was done in the area of military training.
2. Original photographs and training plans were gathered from a variety of archival centers.
3. A site visit was made to a large-scale training installation to photograph extant training facilities.
4. Data was collected and analyzed, and conclusions were drawn.

Literature review

The research team used secondary literature to determine the general history of military training throughout the development of War Department and the Navy Department (and subsequently the DOD—Army, Navy, and Air Force). The military literature review consisted of reading the various training manuals pushed out by those departments and a variety of military training histories published by and for those departments.

Archival research

The research team then located primary research materials and additional secondary materials to establish a strategy to best use these resources. The research team conducted four visits to the National Archives in Washington, DC and the National Archives at College Park, MD. They occurred during the weeks of 6 February 2006, 27 February 2006, 17 April 2006, and 22 May 2006. Other archival depositaries visited were the Library of Congress, 27 February 2006; the Naval Photo Library at the Washington Navy Yard, 17 April 2006; the History Office at the Corps of Engineers, Alexandria, VA, 17 April 2006; and a variety of installation museums, cultural resources offices, and archives across the country.

Site visits

Two members of the research conducted a site visit to Fort Bragg, NC. Fort Bragg was chosen for the site visit because it had one of the largest groupings of different training lands in the DOD; the complexity of its training lands; and the level of historical background that Fort Bragg had on its training lands.

Analysis

After the initial research was complete, the team analyzed the gathered information. The researchers outlined the historical context for military training, identified changes in history and use over time, identified important chronological periods, established a geographical context, and identified historical themes. The analysis resulted in an outline of military training divided into eight significant periods:

- Pre-Civil War (up to 1861)
- Civil War (1861-1865)
- National Expansion (1865-1916)
- World War I (1917-1920)
- Interwar (1921-1940)
- World War II (1941-1945)
- Early Cold War (1946-1955)
- Late Cold War (1956-1989).

Scope

Military training that occurred inside buildings and the Cold War missile programs are not part of this historic context.

The complexity of military training across the services required four historic contexts to be developed, each geared to a particular type of training:

5. Small arms ranges
6. Large arms ranges
7. Training villages, mock sites, and large-scale operation areas
8. Miscellaneous training sites.

This report details the history of small arms ranges.

Mode of technology transfer

This report will be made accessible through the World Wide Web (WWW) at URL: http://www.cecer.army.mil

2 Small Arms Ranges

Small arms ranges have long been used for a variety of marksmanship training activities in the U.S. military. At these ranges, soldiers learned how to fire small arms weapons from stationary and moving positions at stationary and moving targets. "Training included munitions handling, first echelon (Operator) field maintenance, weapons firing, and range clearance" ("RO-1" 2). Pistols, rifles, shotguns, machine guns, and grenades were the principle weapons fired.

A range typically had a set of firing points laid out on a firing line, firing lanes that soldiers traveled down as they fired, or sections of a course or road on which firing was completed. Firing points were sometimes fitted with foxholes, trenches, or sandbags for positional support. A range also typically had stationary targets or moving target systems (cables, pulleys, tracks, pop-up targets, miniature airplanes, etc). Ranges may have had embankments or walls built up behind targets (to catch ammunition), in front of targets or target tracks (for concealment and protection), at firing lines (for firing support or to stabilize firing positions), or between ranges (to protect from adjacent fire). Ranges also had trenches or foxholes on some firing lines (for firing support) and on some target lines (for partial concealment). Ranges often had a control or observation tower (for fire directing officers), bleachers (for observers), latrines, target storage houses, ammunition storage buildings, and a variety of other buildings to support general range functions. A range may also have been part of a larger installation range complex that contained these buildings.

"Although there was great variation in small arms ranges, they were divided into four basic types: ranges with fixed firing points and fixed targets, ranges with fixed firing points and moving targets, ranges with moving firing points and fixed targets, and ranges with moving firing points and moving targets" ("RO-1" 2). This report lists ranges according to these types, with grenade and train fire ranges broken out into separate sections for clarity. Some general information that applied to many small arms ranges is listed below, including: training procedures, weapons, safety fans and distances, multiple range layouts, firing lines, targets, embankments/trenches/etc., and buildings. Following this general information, ranges are discussed individually in sections according to range type.

The first paragraphs under each range name heading summarizes important information about the range, including: who was trained to do what, what the typical range features were, how training was typically completed, and what weapons were used in training. Following this summary is more information on layouts, firing lines, targets, embankments/trenches/etc., and buildings associated with the range. Construction drawings and historic photographs are shown where available. Plan drawings are colored according to the key below (Figure 1) to more clearly show a range's typical features. Present-day photographs and evaluation material follow the list of ranges.

Figure 1. Color key for plan drawings.

General information

Training procedures

"Upon arrival at the Small Arms Range, personnel received weapons delivered from the ammunition supply point. Normally near the firing point, ammunition was removed from its shipping/packing case and issued to troops at a breakdown site. Ammunition at the firing point was placed out of range of any weapon backblast and stored to minimize accidental ignition, explosion, or detonation. Ammunition was issued to troops only on the ready or firing line" ("RO-1" 9).

"After being issued ammunition, an individual or firing crew moved to the firing position. The firing of small arms began only on the order of the officer in charge. After firing, the gunner removed the magazine, unloaded the gun, and reported that the gun was clear. The gunner inspected the weapon and performed after firing first echelon (Operator) maintenance including cleaning and lubricating. If the crew was returning to garrison immediately after firing, first echelon (Operator) maintenance may have been postponed until they arrived at their destination. Unused ammunition was repackaged and returned to the appropriate storage facility." Duds were either destroyed at the site or taken to an appropriate disposal facility. Trash was gathered by troops and taken to a landfill, or buried in a foxhole if they were on their way to do more training" ("RO-1" 10-12).

Weapons

"From 1917 through 1986, small arms weapon systems included pistols (.22-caliber, .32-caliber, .38-caliber, and .45-caliber), rifles and carbines (.22-caliber, .30-caliber, 7.62-mm, and 5.56-mm), automatic rifles (.30-caliber), submachine guns (.45-caliber), machine guns (.30-caliber, .50-caliber, and 7.62-mm), and shotguns (12-gauge). The types of small arms used by the military and their respective periods of use are presented in Table 1 below" ("RO-1" 2).

Table 1. Small Arms Types and Use Periods ("RO-1" 2 (Table 1); War Department Document No. 72, Firing Regulations for Small Arms, 7 February 1898; TM 9-2200, Small Arms, Light Field Mortars and 20mm Aircraft Guns, 11 October 1943; TM 9-2200, Small Arms Material and Associated Equipment, 14 April 1949; TM 9-2200, Small Arms Material and Associated Equipment, 9 October 1956; TM 9-500, Data Sheets for Ordnance Type Material, 11September 1962).

Weapon Type	Earliest Mention	Latest Mention
Rifle, .22-Caliber, M1, M2, M1922; Winchester Model 75; Remington Model 513T; Stevens Model 416-2	1917	Present
Carbine, .30-Caliber, M1, M1A1	1939	1970
Carbine, .30-Caliber M2, M3	1939	1970
Rifle, .30-Caliber, M1903 (all models)	1917	1970
Rifle, .30-Caliber, M1917	1917	1945
Rifle, .30-Caliber, M1, M1C, M1D	1930	1960
Rifle, 7.62-mm, M14	1950	1970
Rifle, 5.56-mm, M16	1965	Present
Pistol, .22-Caliber, Colt Ace; Colt Woodsman Match; Colt Woodsman Standard; Highstandard Model B or H-D	1917	Present
Pistol, Automatic, Colt, .32-Caliber	Unknown	Unknown
Pistol, Automatic, Colt .38-Caliber	Unknown	Unknown
Revolver, .38-Caliber, Colt Commando; Colt Junior Commando; Smith & Wesson Police Special; Smith & Wesson Police Regular	1917	Present
Revolver, .45-Caliber, M1917	1917	1945
Pistol, .45-Caliber, M1911	1917	Present

Weapon Type	Earliest Mention	Latest Mention
Automatic Rifle, .30-Caliber, M1918	1917	1960
Shotgun, 12-gauge, Stevens Model 620; Stevens Model 520-30; Winchester Model 12; Winchester Model M1897; Ithaca Model 37; Savage Model 720; Remington Model 31; Remington Model 11; Remington Sportsman;	1917	Present
Gun, Submachine, .45-Caliber M3	1939	1960
Gun, Submachine, .45-Caliber M1, M1928	1917	1955
Gun, Machine, .30-Caliber, M1917	1917	1960
Gun, Machine, .30-Caliber, M1919	1917	1960
Gun, Machine, 7.62-mm, M60	1960	Present
Gun, Machine, .50-Caliber, M2	1917	Present
Note: 'Present' means the weapon was being used in 1986.		

Safety fans and distances

"The range fans depicted in Figures 2, 3, and 4 below apply to all small arms ranges, even though firing small arms at aerial targets was different than firing at ground targets because it was done at a higher angle. When applied to a specific range, range fans were modified to account for the length of the firing line and the width of the target area. This was done by dividing the range fan drawing in half and anchoring the base at the left and right ends of the firing line, then adjusting the line of fire line between the left and right outer lateral limits. For ranges with moving firing points, the range fan was established for each firing point, and the danger area then became the outside boundary of all individual range fans. The A and B distances for small arms ranges was 100 yards" ("RO-1" 16).

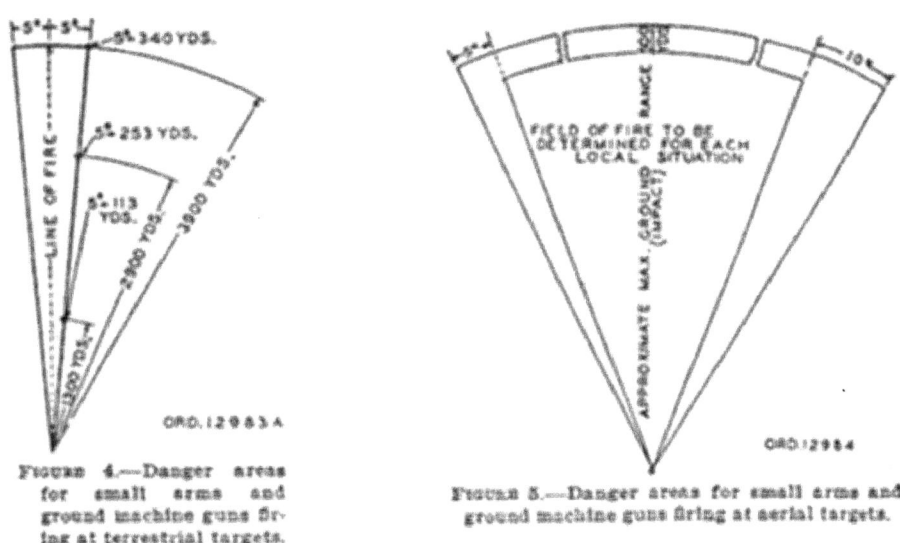

Figure 2. Small arms range fans, circa 1942 (AR 750-10, Range regulations for firing ammunition for training and target practice, 14 February 1942, pp 7-9).

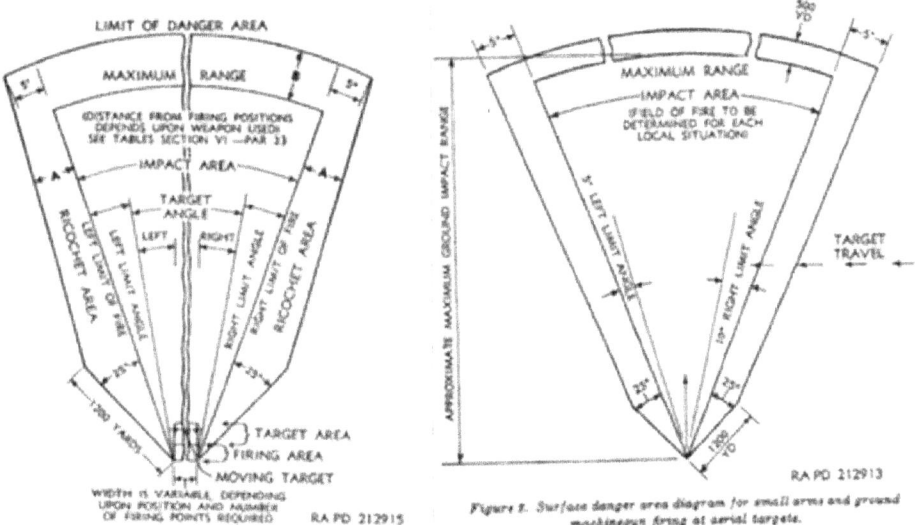

Figure 3. Small arms range fans, circa 1955 (AR 385-63, Regulations for firing ammunition for training, target practice, and combat, 8 December 1955, pp 12, 13).

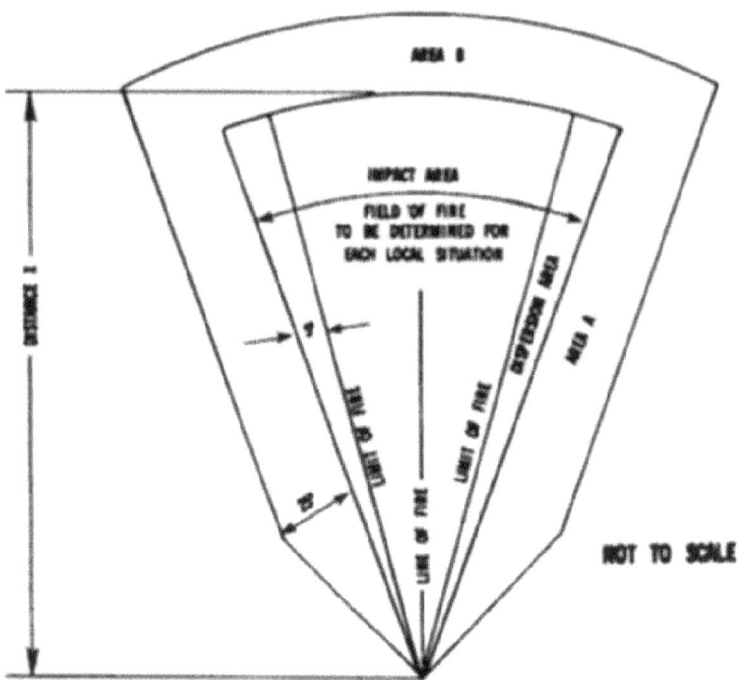

Figure 4. Small arms range fan, circa 1983 (AR 385-63, Regulations for firing ammunition for training, target practice, and combat, 15 October 1983, Chapter 6).

"The maximum ranges for various types of small arms ammunition are presented in Table 2 below. The maximum range is used to determine the X-distance or maximum range distance for standard range fans presented above. Refer to the appropriate regulation for a complete listing of ammunition covered" ("RO-1" 18).

Table 2. Small Arms Ammunition Maximum Ranges 1942, 1955, 1983 ("RO-1" 18 (Table 2); AR 750-10, Range Regulations for Firing Ammunition for Training and Target Practice, 14 February 1942, p 23; AR 385-63, Policies and Procedures for Firing Ammunition for Training, Target Practice, and Combat, 5 December 1955, pp 76-77, AR 385-63, Policies and Procedures for Firing Ammunition for Training, Target Practice, and Combat, 15 November 1983, p 6-2).

Ammunition	1942 Maximum Range (in yds)	1955 Maximum Range (in yds)	1983 Maximum Range (in meters/yds)
Caliber .22 Long Rifle	1,500	1,500	1,400/1,532
5.56-mm Ball, M193	not covered	not covered	3,100/3,391
7.62-mm Ball M80	not covered	not covered	4,100/4,485
Caliber .30 Rifle and machine gun, Ball, M2	4,000	3,500	3,100/3,391

Ammunition	1942 Maximum Range (in yds)	1955 Maximum Range (in yds)	1983 Maximum Range (in meters/yds)
Caliber .30 Rifle and machine gun, AP, M2	4,000	3,160	4,400/4,814
Caliber .30 Carbine, Ball, M1	not covered	2,200	2,300/2,516
Caliber .45, Pistol, Ball, M1911	1,600	1,640	1,300/1,422
Caliber .45, Submachine Gun	1,600	1,760	1,400/1,532
Caliber .50 machine gun, Ball, M33	not covered	6,380	6,500/7,111
Caliber .50 machine gun, AP, M2	7,600	7,280	6,100/6,673
Shotgun, 12-gauge (00 Buckshot)	not covered	600	600/656
Caliber .38 Revolver, Ball, M41	not covered	1,600	1,600/1,750

Multiple range layouts

"Small arms ranges could have been stand-alone ranges or shared with other ranges on a Ground Forces Training Center. The integration of a typical small arms range into a range layout for a division-training site that accommodated numerous weapons is depicted in Figures 5 and 6 below. The range layouts show stand-alone small arms ranges with overlapping cell boundaries covering several other ranges" ("RO-1" 15).

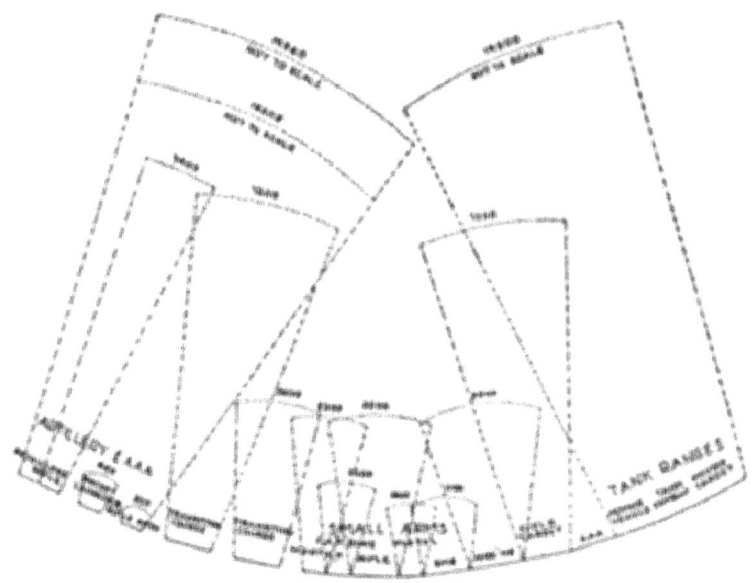

Figure 5. Typical division range layout, circa 1950 (SR 210-20-20, Installations training areas and facilities for ground troops, 23 May 1950, p 4).

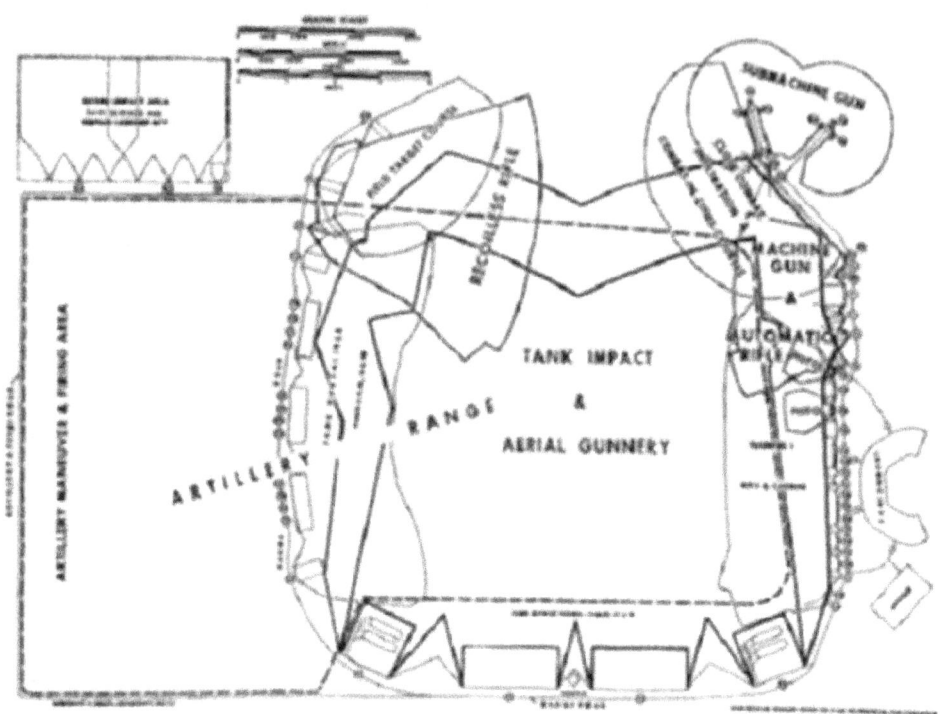

Figure 6. Typical division range layout, circa 1964 (AR 210-21, Training areas and facilities for ground troops, 18 December 1964, p.10).

"The division training range layouts and the small arms range locations depicted in Figures 5 and 6 are examples of the many variations of division and small arms ranges that were built. The figure does not represent the standard range configuration, nor is it representative of the entire period of use. Division training ranges varied or evolved based on unit training needs, terrain considerations, and the evolution of the different types of small arms. At smaller installations (i.e., installations with inadequate space for all the required ranges), a small arms range may have been the only area available to train with other weapons, such as mortars and rifle grenades" ("RO-1" 16). Figure 7 below shows an example of how spaces for small arms ranges were combined at the Murch Range at Fort Bragg.

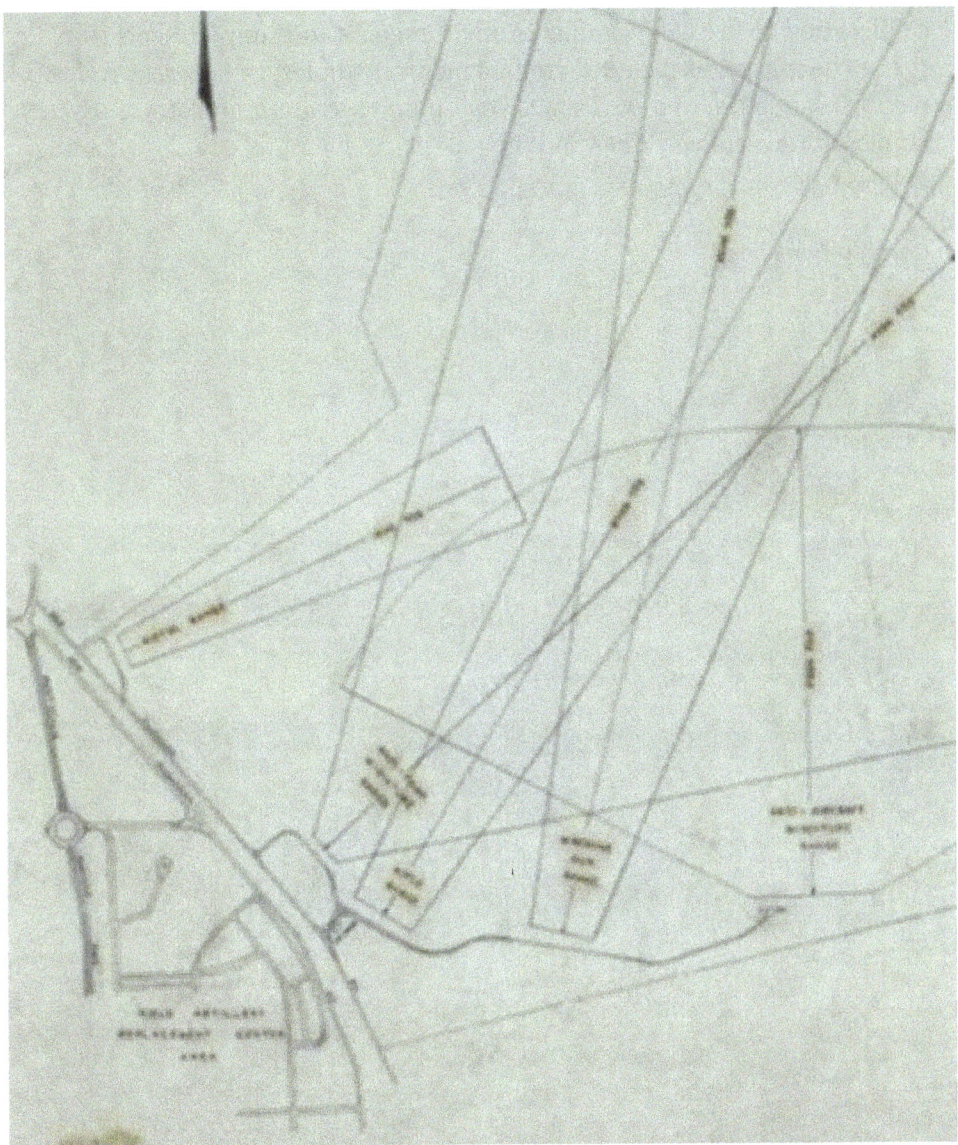

Figure 7. Firing ranges, general layout plan, Murch Range Fort Bragg, NC, 1944 (Record drawing FBR 150, "General layout, firing ranges," 15 March 1944).

Firing lines

A range typically had a set of firing points laid out on a firing line, firing lanes that soldiers traveled down as they fired, or sections of a course or road on which firing was completed. Some firing lines were covered and some had benches, stands, aiming devices, stakes, and other accessories. Firing was often completed from basic standing, kneeling, sitting, or prone (laying) positions. Some firing lines also had a variety of features to train soldiers how to fire from simulated environments and supported positions. These features included embankments, foxholes, trenches, swivel mounts, sandbags, window frames, logs, stumps, craters, and others. Examples of several firing lines and their associated features are shown in the images below.

Covered firing lines

Figure 8. Rifle range firing line at Finger Bay, Adak, AK, 23 June 1945 (NARA College Park, RG 80-G, box 1285, photo 342163).

Benches

Figure 9. Platoon 381 on the pistol range at Camp Matthews, MCRD San Diego, CA, 29 November 1951 (NARA College Park, RG 127-GC, box 5, photo A77625).

Stands

Figure 10. Firing position on pistol range during training at Amphibious Training Base Fort Pierce, FL, 23 November 1943 (NARA College Park, RG 80-G, box 861, photo 264381).

Aiming devices

Figure 11. Recruit using aiming device at MCRD Parris Island, SC, 1970 (NARA College Park, RG 127-GG-919, box 32, photo A601880).

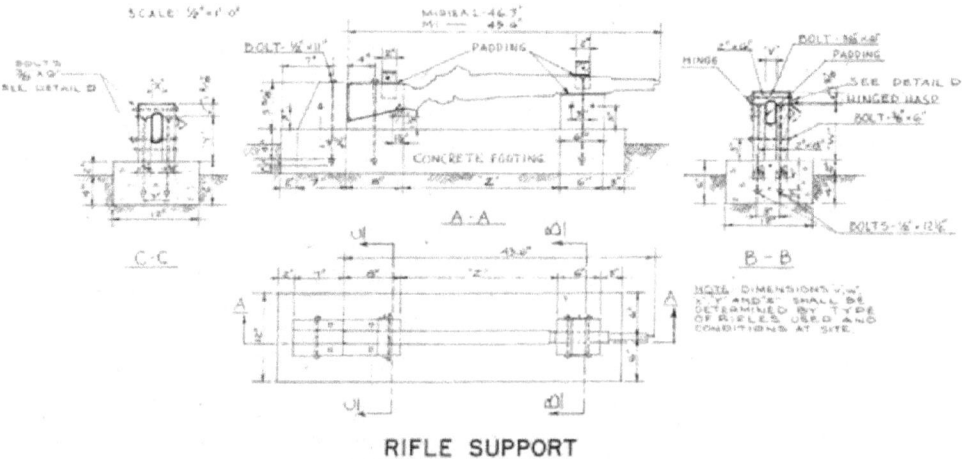

Figure 12. Small arms fire course, rifle support, Fort Bragg, NC, 1952 (Standard drawing 28-13-37 sheet 2 of 2, "Small arms fire course, details," 20 June 1952).

Embankments at firing lines

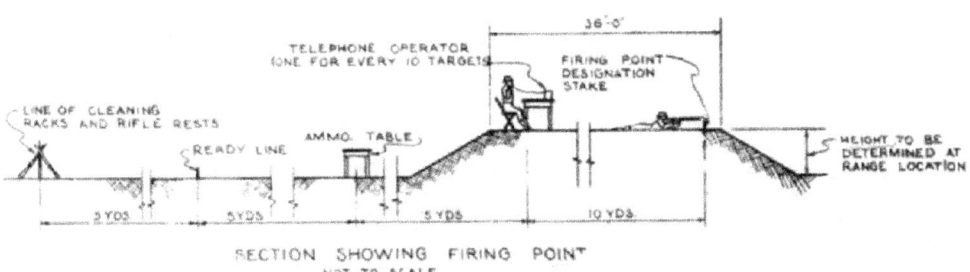

Figure 13. Typical rifle range section showing firing point, Fort Bragg, NC, 1952 (Standard drawing No. 28-13-09 drawing 1 of 7, "Range, rifle, known distance, plans and details," 5 January 1952).

Foxholes

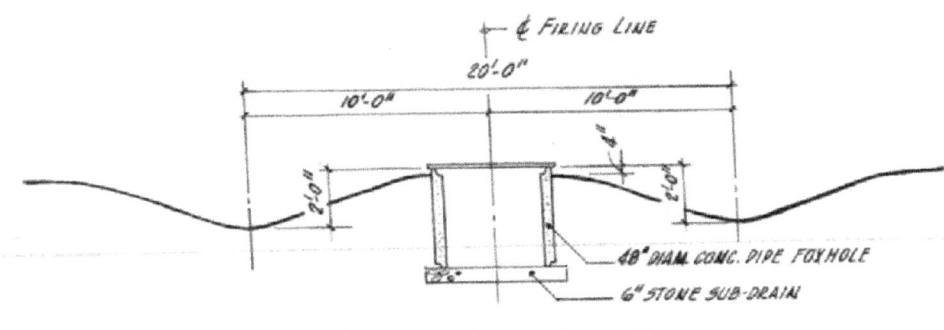

Figure 14. ATC facilities, night firing range "N", firing line, Fort Bragg, NC, 1966 (Standard drawing 28-13-117 sheet 9, "Construction of ranges, phase 1, U.S. ATC Facilities, Fort Bragg, NC, range "N", night firing range," 28 June 1966).

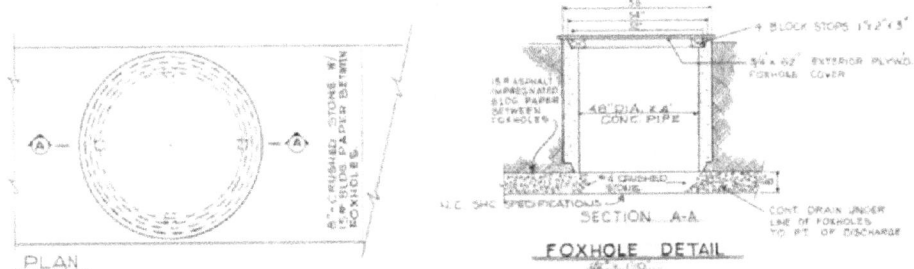

Figure 15. ATC facilities, range details, foxhole detail, Fort Bragg, NC, 1966 (Standard drawing 28-13-117 sheet 14, Construction of ranges, phase 1, U.S. ATC Facilities, Fort Bragg, NC, range details, 28 June 1966).

Trenches

Figure 16. Trench firing line at Camp Wheeler, GA, 1918 (New York Public Library, digital No. 117146).

Figure 17. Throwing live fragmentation hand grenades at the dummy targets at Fort Jackson, SC, 1 November 1943 (NARA College Park, RG 111-SC WWII, box 681, photo SC324452).

Swivel mounted firing positions

Figure 18. In their first contact with guns (that work by air power) that keep on firing, students pepper plane silhouettes (that are mounted on a track developed for shooting galleries) at Tyndall Army Air Field (Panama City), FL, 29 December 1942 (NARA College Park, RG 342-FH, box 2202, photo 4A-17270).

Figure 19. Students fire swivel mount shot guns at Laredo Army Air Field, TX, 16 July 1943 (NARA College Park, RG 342-FH, box 2202, photo 4A-17294).

Sandbags

Figure 20. Marines operating machine gun at MCAS Cherry Point, NC, 21 August 1943 (NARA College Park, RG 80-G, box 1360, photo 358872).

Window frames

Figure 21. A Drill Instructor teaches recruits the proper method of firing from inside a building at MCRD San Diego (Camp Pendleton), CA, December 1972 (NARA College Park, RG 127-GG-936, photo 230641).

Targets

A range typically had stationary targets or moving target systems. Stationary targets included wood framed panel targets, silhouettes, paper targets, cloth circle outlines, trenches, objects in the landscape, and others. Moving target systems included target sleds, target cars, cables, pulleys, dolly mounted targets on tracks, pop-up targets, miniature airplanes, clay pigeons, and others. "Initially, stationary and moving targets for small arms ranges consisted of wooden frames with cardboard, paper, or fabric targets attached. The materials used included target paper, wood, target cloth, steel cable, rope, concrete, and sheet metal" ("RO-1" 60). "More modern targets were also constructed of old wood and paper, but plastic targets were also used. These plastic targets were used to provide more accurate depictions of opposing force weapons and personnel. Plastic was also used to construct pop-up and moving targets. The plastic target had the benefit of allowing more use before replacement than the older paper targets. On the newer pop-up and moving targets, the line and pulley systems were replaced by electronic controls. These controls allowed safer operation with fewer maintenance problems" ("RO-1" 66). A variety of small arms range targets are shown in the images and drawings below.

Stationary panel targets

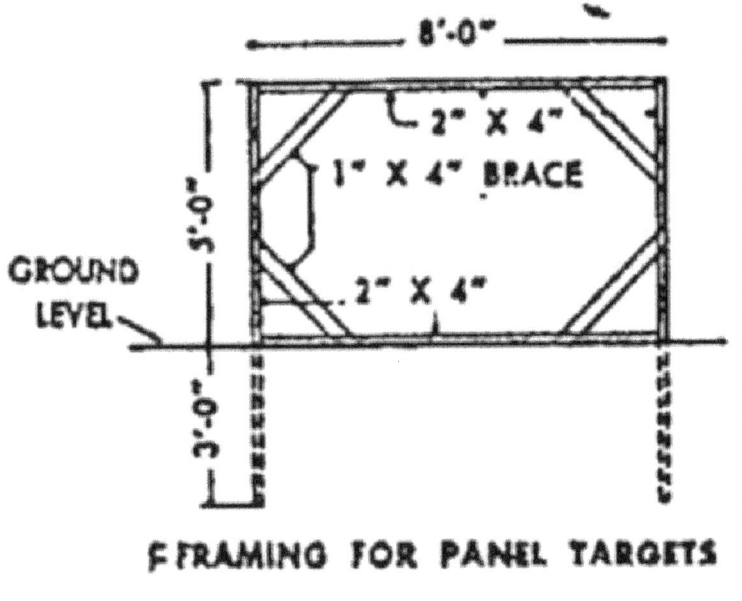

Figure 22. Panel target, circa 1942 (Standard drawing No. 1600-135, Moving vehicle ranges, 23 October 1942).

Figure 23. CMTC students on the pistol range at Camp Vail, NJ, 1923 (NARA College Park, RG 111-SC WWI, box 700, photo 94912).

Figure 24. CMTC students on machine gun range at Camp Del Monte, CA, 13 August 1925 (NARA College Park, RG 111-SC WWI, box 700, photo 94896).

Raised panel targets

Figure 25. Target Line at Sioux Falls Army Air Field, SD, 1942 (NARA College Park, RG 342-FH, box 2202, photo 4A-17256).

Figure 26. The circular slow-fire targets are being ripped full of holes, Marines observe each shot and mark its value with colored discs, at the bottom of each frame is a rapid fire target, these will soon be hoisted into place at MCB San Diego, CA, February 1944 (NARA College Park, RG 127-GC, box 5, photo 35366).

Figure 27. Here is the rapid-fire target complete with markers, the first one shows the results of aiming too low, those firing on targets 2 and 3 registered better patterns at MCB San Diego, CA, February 1944 (NARA College Park, RG 127-GC, box 5, photo 35364).

Figure 28. Butts at rifle range at Camp Matthews, MCRD San Diego, CA, 6 February 1946 (NARA College Park, RG 127-GC, box 5, photo 401016).

Figure 29. Recruits watch their targets during firing practice at MCRD Parris Island, SC, 6 December 1951 (NARA College Park, RG 127-GC, box 34, photo A60616).

Figure 30. Recruits watching their targets for hits at MCRD Parris Island, SC, 6 December 1951 (NARA College Park, RG 127-GC, box 34, photo A60628).

Figure 31. Platoon 23 works in the butts while Platoons 532 and 533 fire on their record day at Camp Matthews, MCRD San Diego, CA, 14 February 1952 (NARA College Park, RG 127-GC, box 5, photo A219158).

Figure 32. "Baker" targets cover the landscape near "Echo" range as they are laid out to dry at Camp Matthews, MCRD San Diego, CA, 4 May 1959 (NARA College Park, RG 127-GG-2019, box 40, photo A229012).

Figure 33. Production-line teamwork is employed by a detail of Marines to build some 2000 targets at Camp Matthews, MCRD San Diego, CA, 4 May 1959 (NARA College Park, RG 127-GG-2019, box 40, photo A229014).

Figure 34. Target handlers work in the butts marking and discing targets at Camp Matthews, MCRD San Diego, CA, 4 May 1959 (NARA College Park, RG 127-GG-2019, box 40, photo A229013).

Figure 35. Butt of X Course target in the air at MCB Camp Lejeune, NC, 1966 (NARA College Park, RG 127-GG-2019, box 40, photo A342975).

Paper targets

"An example of a stationary paper target (Figure 36) is one representing a soldier in the prone position. This target was a square 6 ft in height with a black silhouette in the middle representing a prone soldier. The silhouette was printed with black ink on buff manila target paper. The targets were packed 50 to a roll" ("RO-1" 60).

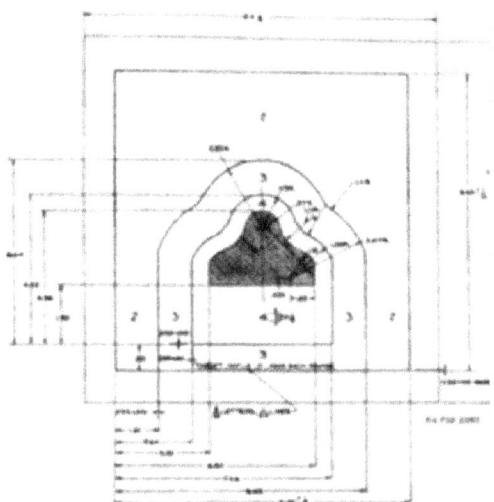

Figure 36. Sustained fire stationary paper target, circa 1941 (TM 9-855, Targets, target material, and rifle range construction, 19 June 1941, pp 98).

Figure 37. Perhaps the most important day in the recruit's boot training is qualification day, he will not only fire the rifle for an official score to be entered in his record book, he will also assist in operating some part of the firing line like phoning to the butts or keeping score at Camp Matthews, MCRD San Diego, CA, 6 June 1955 (NARA College Park, RG 127-GC, box 5, photo A227310).

Silhouette targets

"Soldiers were depicted in various positions by drab pasteboard targets attached to a wooden stave. These positions included prone, kneeling, and standing (Figure 38)" ("RO-1" 62).

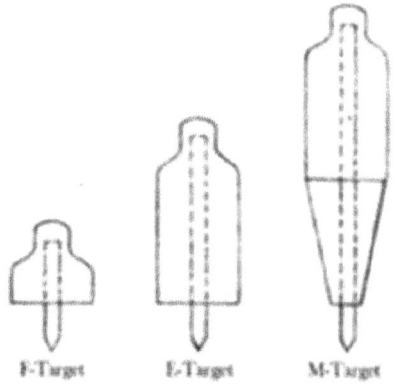

Figure 38. Silhouette targets for prone (F), kneeling (E), and standing (M) positions, circa 1943 (FM 23-41, Basic field manual, Submachinegun caliber .45, M3, 30 October 1943, p 60).

Sled targets

"An example of a moving wood and fabric target is depicted in Figure 39 below. A sled was constructed using wood, sheet metal, and target cloth. This target was constructed to represent a moving vehicle and was towed by a 1 1/2-ton truck using pulleys and ropes. The target could also be towed along a narrow gage railroad track" ("RO-1" 63).

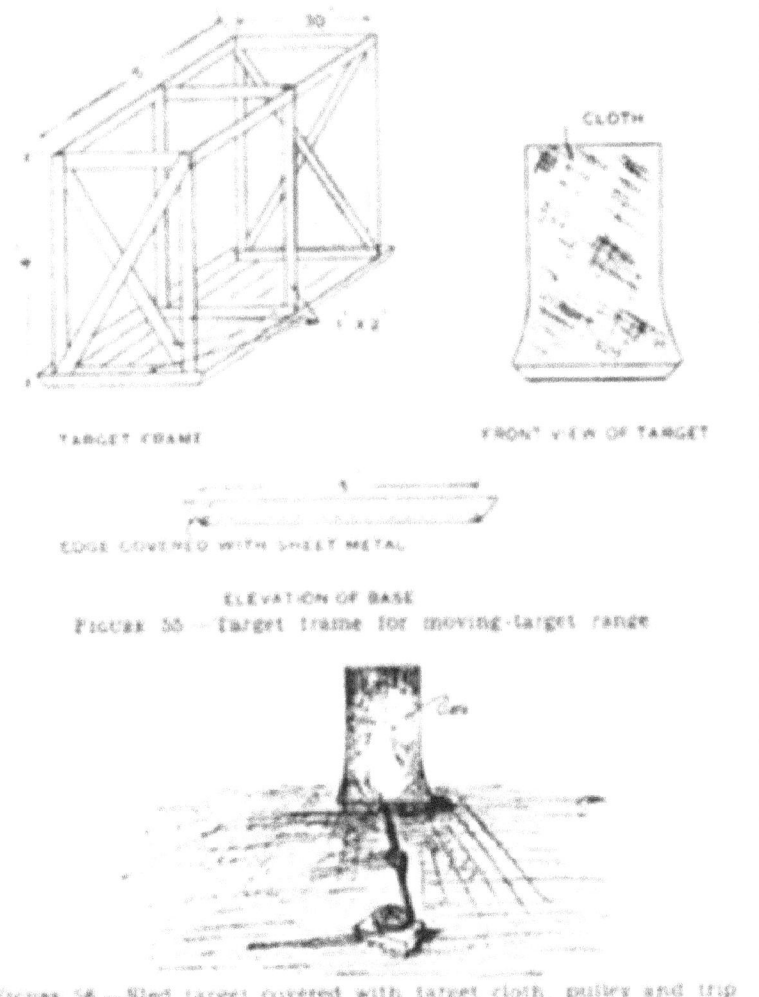

Figure 39. Moving target, circa 1940 to 1943 (FM 23-5, Basic field manual, U.S. rifle, caliber .30, M1, 30 July 1943, pp 194, 195, FM 23-7. Basic field manual, U.S. carbine, caliber .30, M1, 20 May 1942, pp 95, 96, FM 23-60, Basic field manual browning machine gun, caliber .50, HB, M2 ground, 25 September 1940, p 134).

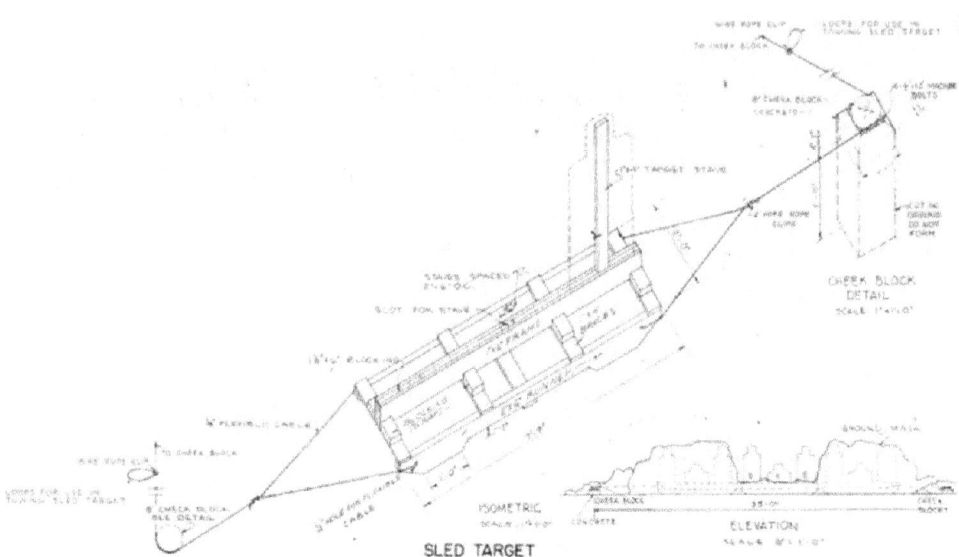

Figure 40. Dismounted submachine gun practice course, sled target, Fort Bragg, NC, 1951 (Standard drawing 28-13-13, sheet 1 of 2, Range, submachine gun, dismounted practice course, plans and details, 7 December 1951).

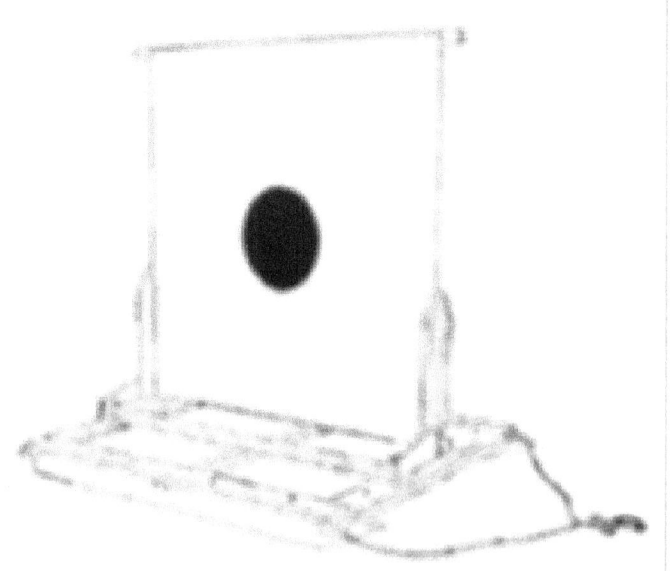

Figure 41. Moving target, circa 1942-1949 (FM 23-30, Hand and rifle grenades, rocket, AT, HE, 2.36-in., 14 February 1944, p 54, FM 23-30, Hand and rifle grenades, rocket, AT, HE, 2.36-in., 14 February 1944, p 89, FM 23-30, Hand and rifle grenades, 14 April 1949, p 92).

Dolly mounted track type targets

"Another type of moving target (Figure 42) consisted of a steel cable fastened to a target carriage and pulled by means of the cable around a pulley and cylindrical drum. The shaft of the drum was attached through a transmission and clutch to a motor. The drum was rotated by the motor in both the forward and reverse direction. The target carriage was equipped with flanged wheels and ran on narrow gage tracks" ("RO-1" 64).

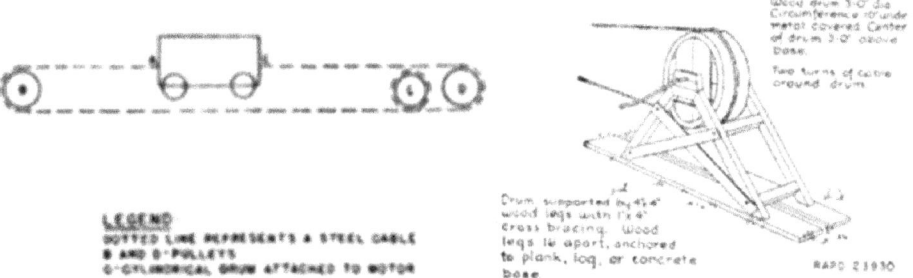

Figure 42. Motorized movable target plan with drum detail, circa 1943 (FM 23-41, Basic field manual, Submachine gun caliber .45, M3, 30 October 1943, pp 68-70, TM 9-855, Targets, target material, and rifle range construction, changes No. 1, 5 August 1942, p 5).

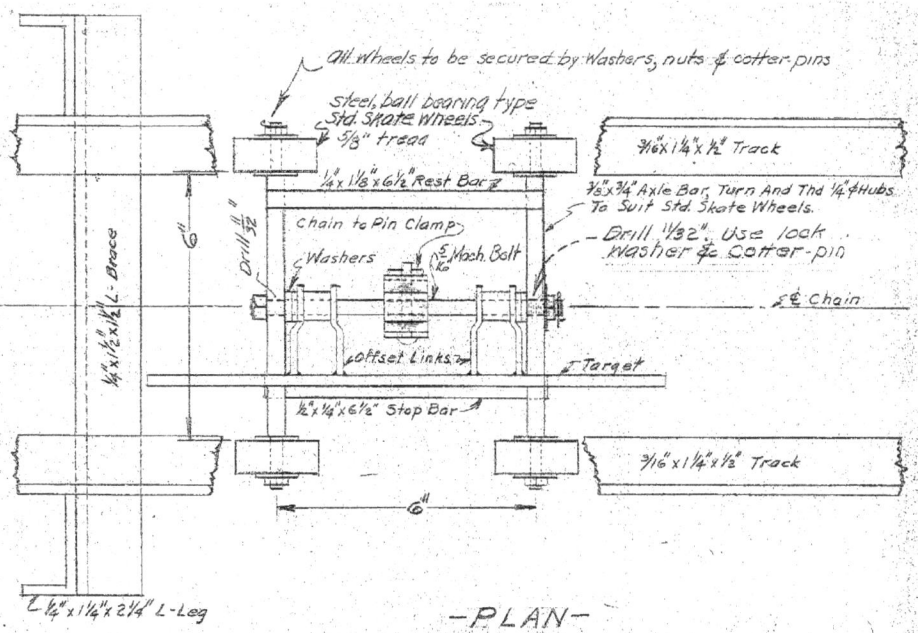

Figure 43. Moving target assembly, Fort Bragg, NC, June 1949 (Standard drawing 28-09-01 sheet 7 of 8, Training aids, moving target assembly, 6 June 1949).

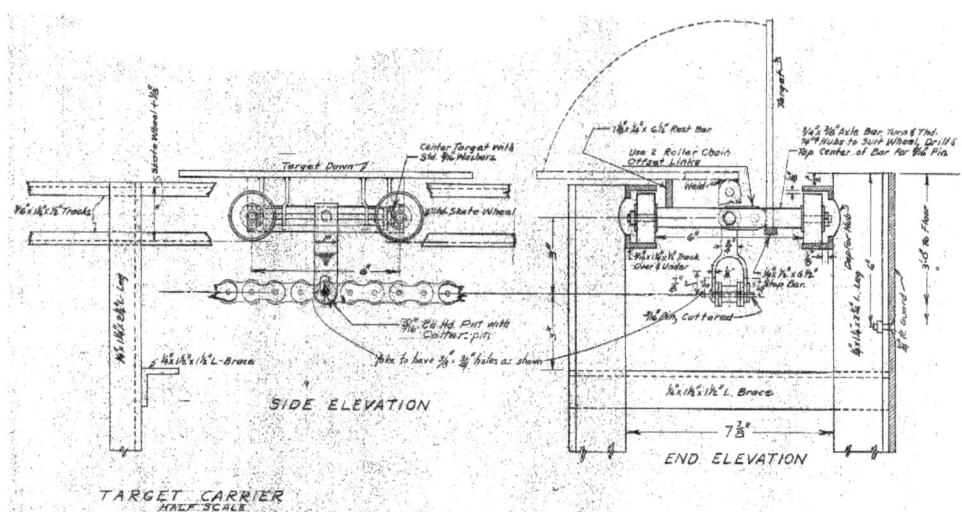

Figure 44. Moving target assembly, Fort Bragg, NC, June 1949 (Standard drawing 28-09-01 sheet 7 of 8, Training aids, moving target assembly, 6 June 1949).

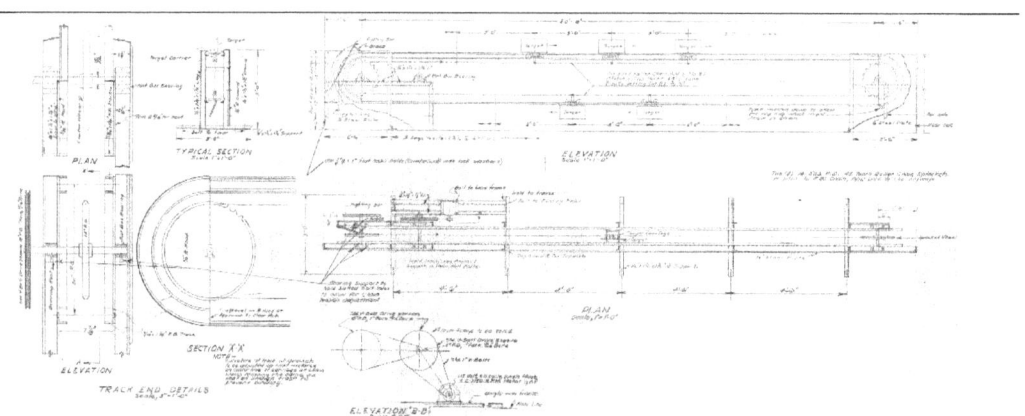

Figure 45. Moving target assembly, Fort Bragg, NC, June 1949 (Standard drawing 28-09-01 sheet 7 of 8, Training aids, moving target assembly, 6 June 1949).

Figure 46. Silhouette targets, moving target assembly, Fort Bragg, NC, June 1949 (Standard drawing 28-09-01 sheet 8 of 8, Training aids, moving target assembly, 6 June 1949).

Figure 47. Looking down on target track on submarine target range at Camp Wissahickon, NJ, undated (circa 1918) (NARA College Park, RG 24-TC, box 1, Folder K).

Figure 48. Motor driven buda target car on the aviation free gunnery range at NAS Kaneohe Bay, HI, March 1944 (NARA College Park, RG 80-G, box 1705, photo 419200).

Aerial pulley targets

"Aerial targets were also used to train riflemen to protect themselves against low-flying aircraft and descending parachutes. Aerial targets were classified as overhead, non-overhead, direct diving, and direct climbing. Overhead targets were those that passed directly over or nearly over the rifleman. Non-overhead targets were those that did not pass directly over the rifleman. Direct diving targets were those that dove directly toward a rifleman. Direct climbing targets were those that climbed directly away from the rifleman. Many of these targets were printed images on buff manila paper and apparently mounted on wooden poles. They appear to have been manipulated using ropes and pulleys" ("RO-1" 65).

Figure 49. Overhead rifle target, circa 1943 (FM 23-5, Basic field manual, U.S. rifle, .30 caliber, M1, 30 July 1943, p 204).

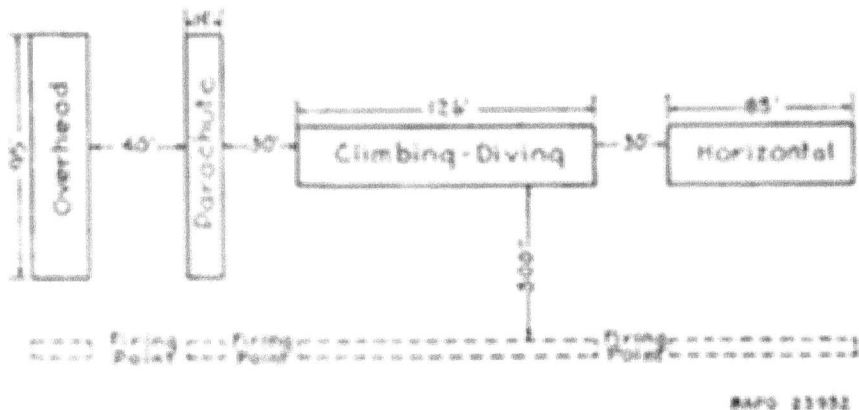

Figure 50. Aerial rifle target layout, circa 1942 (TM 9-855, Targets, target material, and rifle range construction, changes No. 1, 5 August 1942, p 3).

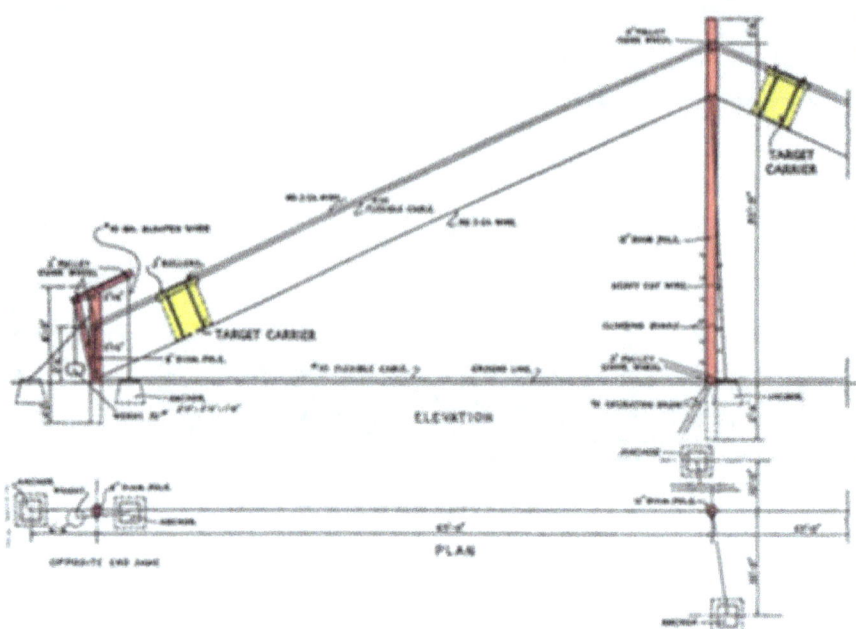

Figure 51. Miniature anti-aircraft range, target pulley system plan and elevation, circa 1942-1951 (Standard drawing No. 1600-130/131, Training aids, A. A. Range miniature, 14 October 1942, TM 9-855, Targets, target material, and training course layouts, 1 November 1951, pp 28, 29).

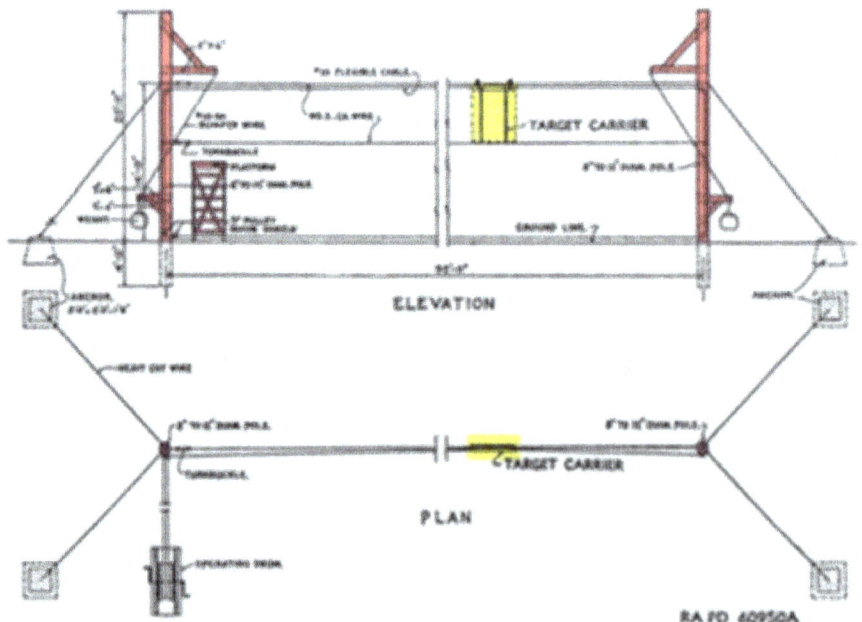

Figure 52. Miniature anti-aircraft range, target pulley system plan and elevation, circa 1942-1951 (Standard drawing No. 1600-130/131, Training aids, A. A. range miniature, 14 October 1942, TM 9-855, Targets, target material, and training course layouts, 1 November 1951, pp 28, 29).

Clay pigeons

Clay pigeons were used to simulate incoming and outgoing aerial targets.

Figure 53. High tower and turret range NAGS shot gun range at NAS Jacksonville, FL, 27 February 1944 (NARA College Park, RG 80-G, box 740, photo 241829).

Pop-up targets

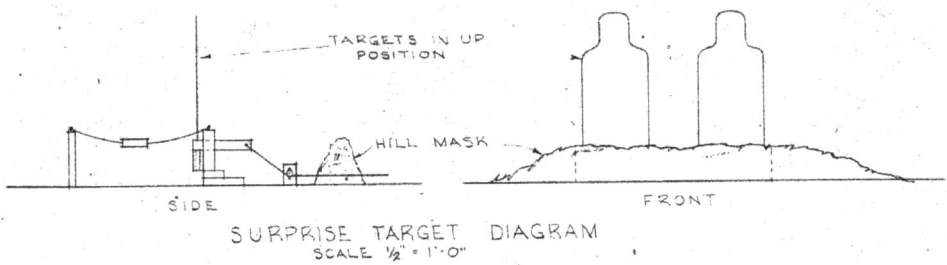

Figure 54. Dismounted submachine gun practice course, surprise target diagram, Fort Bragg, NC, 1951 (Standard drawing 28-13-13 sheet 2 of 2, Range, submachine gun, dismounted practice course, details, 7 December 1951).

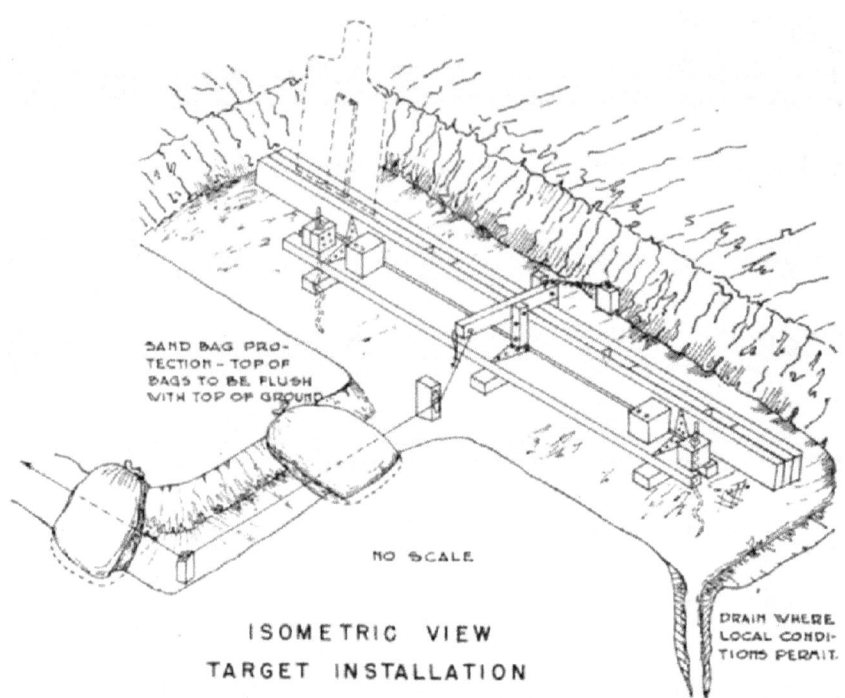

Figure 55. Small arms range, field target, isometric view of target installation, Fort Bragg, NC, 1951 (Standard drawing 28-13-18 sheet 1 of 1, Range, field target, small arms, plan and details, 21 November 1951).

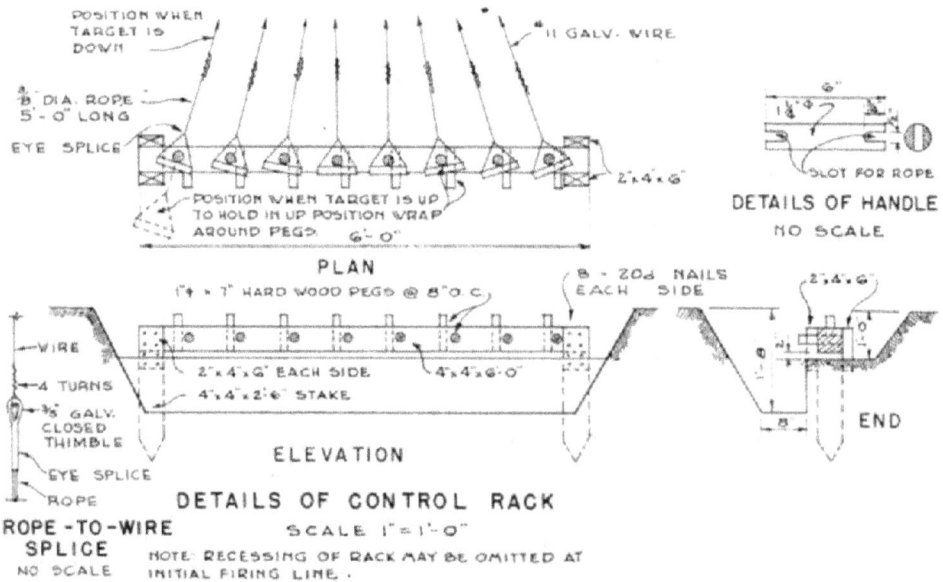

Figure 56. Small arms range, field target, details of control rack, Fort Bragg, NC, 1951 (Standard drawing 28-13-18 sheet 1 of 1, Range, field target, small arms, plan and details, 21 November 1951).

Figure 57. Electrical targets at Camp Lejeune, NC, 2 May 1958 (NARA College Park, RG 127-GC, box 22, photo 340848).

Radio controlled airplane targets

Figure 58. OQ-14/TDD-3/TDD-4 Radioplane, circa 1944 (TR 140-5, http://www.designation-systems.info/dursm/app1/oq-14.html).

Figure 59. Radio controlled airplane target OQ-2A taking off from catapult at the AA Range at Fort Jackson, SC, 1 November 1943 (NARA College Park, RG 111-SC WWII, box 415, photo SC246692).

Embankments/trenches/etc.

Ranges may have had embankments or walls built up behind targets (to catch ammunition), in front of targets or target tracks (for concealment and protection), at firing lines (for firing support or to stabilize firing positions), or between ranges (to protect from adjacent fire). Ranges also had trenches or foxholes on some firing lines (for firing support) and on some target lines (for partial concealment). Some earlier ranges had trenches for training in trench warfare. Earth was otherwise removed from the ground in some places and piled on the ground in others for a variety of uses such as earthen walls, target control pits, or demolition pits used to dispose of hazardous ordnance. Examples are shown below.

Embankments behind target lines

Figure 60. Small arms target practice for cadets at NAS Corpus Christi, TX, 23 July 1941 (NARA College Park, RG 80-G, box 1978, photo 463689).

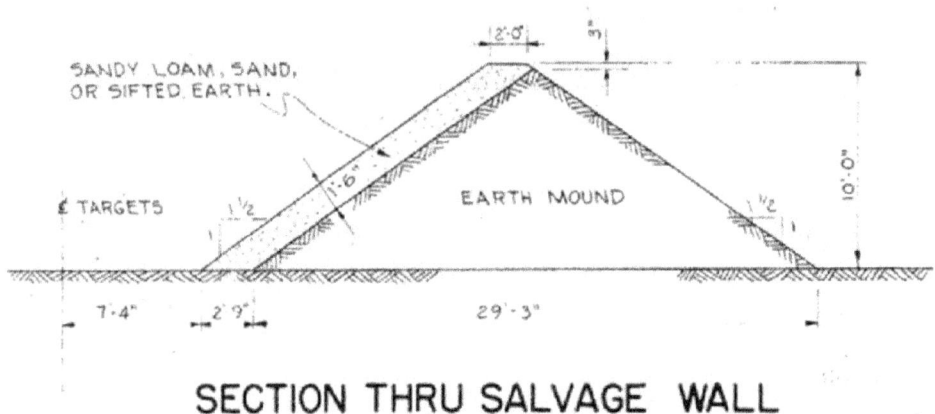

Figure 61. A 1000-in. fixed target track type range, section through salvage wall, Fort Bragg, NC, 1952 (Standard drawing 28-13-06 sheet 1 of 2, Range, 1000-in. fixed target, track type, plans and details, 20 June 1952).

Figure 62. Horses digging target trench, Fort Knox, KY, undated (scans from Knox, training 5, vol. 5).

Figure 63. Horses digging target trench, Fort Knox, KY, undated (scans from Knox, training 6, vol. 5).

Figure 64. Target complex built on top of trench, Fort Knox, KY, undated (scans from Knox, training 7, vol. 5).

Figure 65. Looking up out of target trench (rifle range pits), Fort Knox, KY, undated (scans from Knox, training 8, vol. 5, #358, Kirkpatrick photo).

Target butts

Figure 66. Target line at the USMC Winthrop rifle range, Indian Head, MD, 1915 (NARA College Park, RG 127-G-100b, box 23, photo 521528).

Timber wall target butts

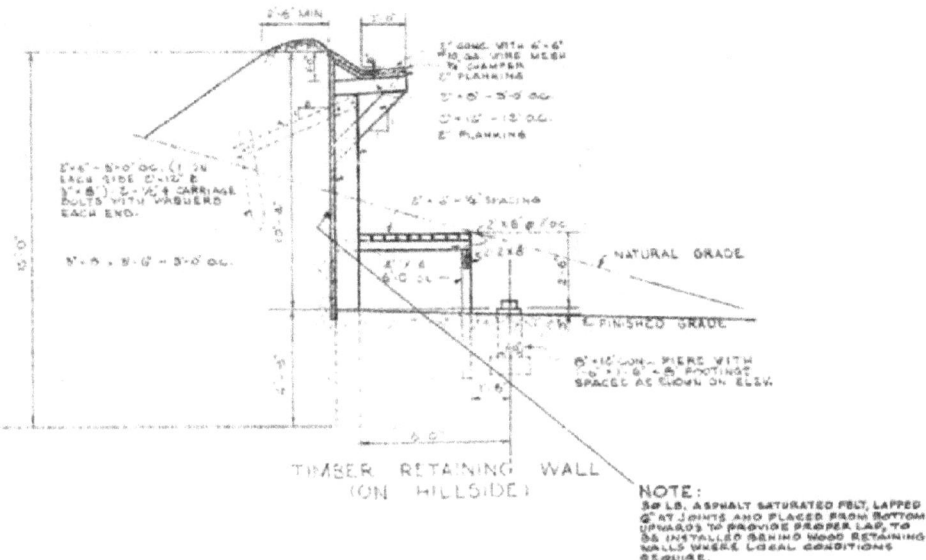

Concrete wall target butts

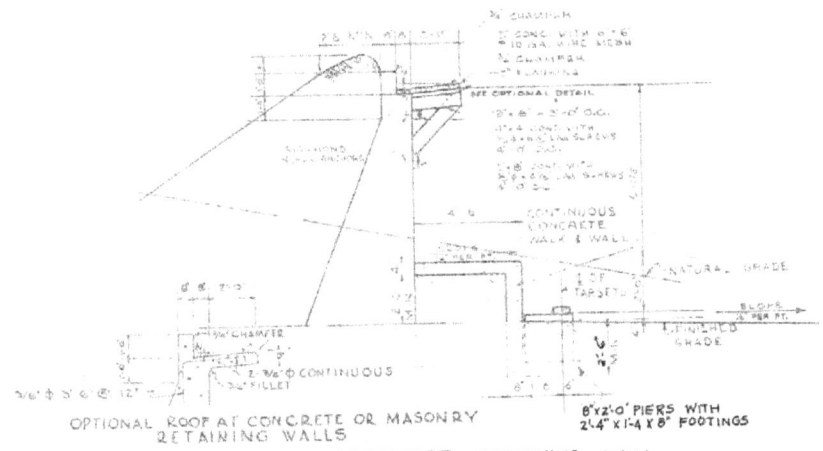

Figure 67. Typical rifle range target butts concrete retaining wall, Fort Bragg, NC, 1952 (Standard drawing No. 28-13-09 drawing 2 of 7, Range, rifle, known distance, details, 5 January 1952).

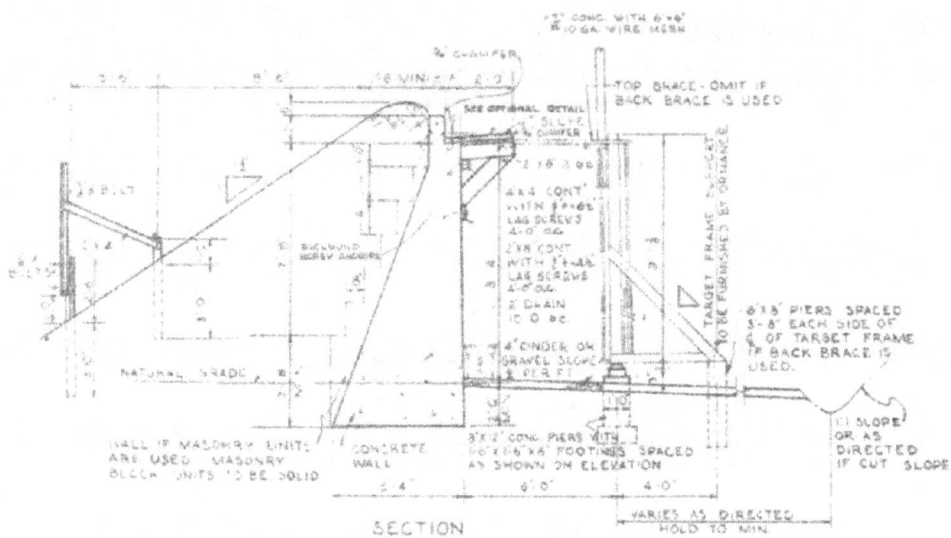

Figure 68. Typical rifle range target butts section detail, Fort Bragg, NC, 1952 (Standard drawing No. 28-13-09 drawing 2 of 7, Range, rifle, known distance, details, 5 January 1952).

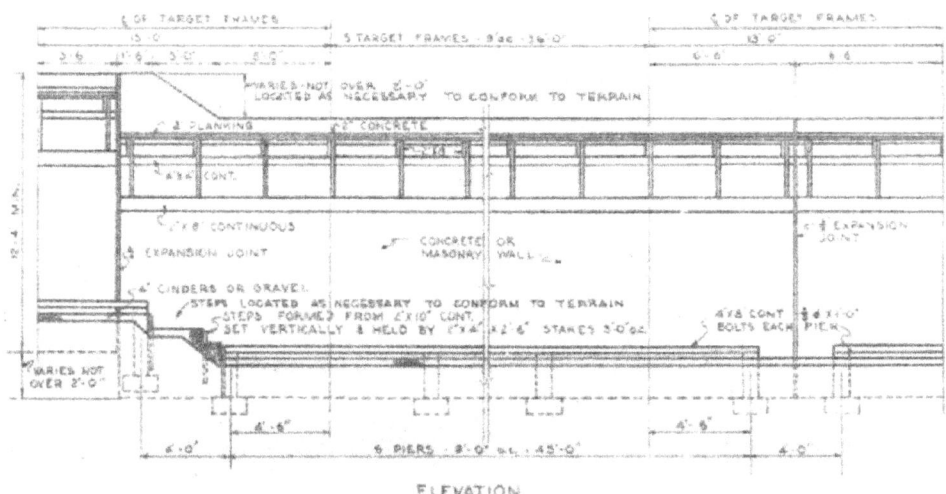

Figure 69. Typical rifle range target butts section elevation, Fort Bragg, NC, 1952 (Standard drawing No. 28-13-09 drawing 2 of 7, Range, rifle, known distance, details, 5 January 1952).

Figure 70. Typical rifle range target butts timber wall (on hillside), Fort Bragg, NC, 1952 (Standard drawing No. 28-13-09 drawing 2 of 7, Range, rifle, known distance, details, 5 January 1952).

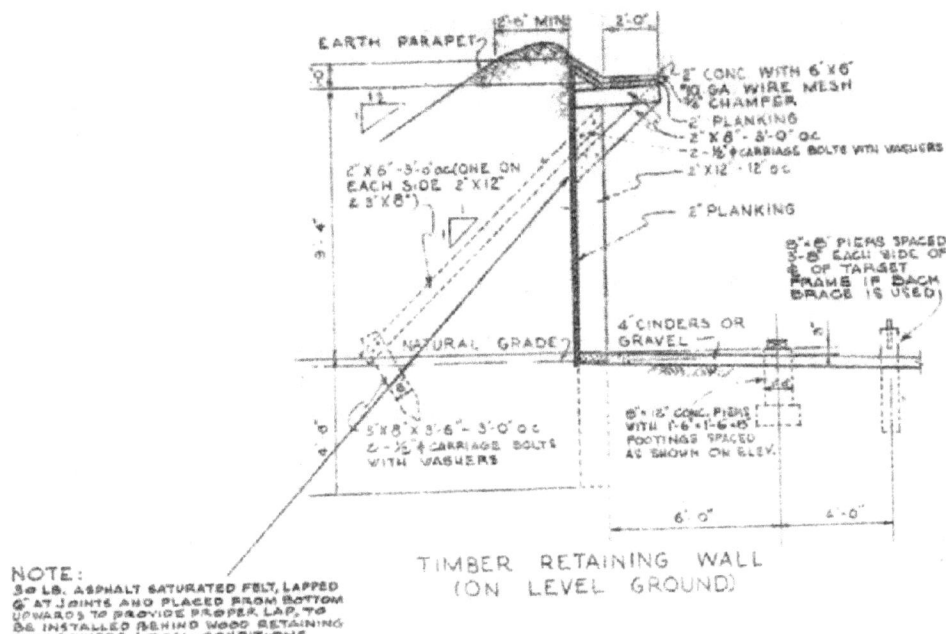

Figure 71. Typical rifle range target butts timber wall (on level ground), Fort Bragg, NC, 1952 (Standard drawing No. 28-13-09 drawing 2 of 7, Range, rifle, known distance, details, 5 January 1952).

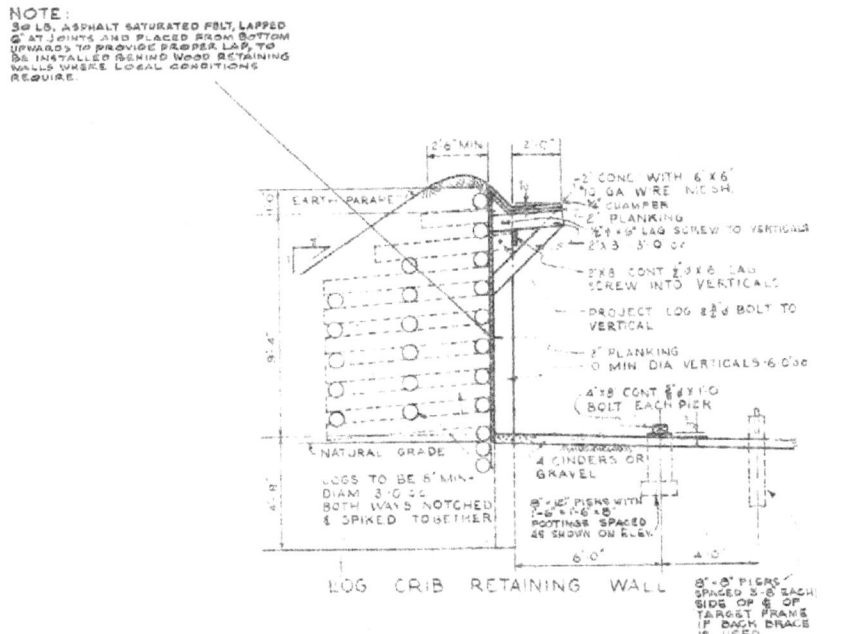

Figure 72. Typical rifle range target butts log crib retaining wall, Fort Bragg, NC, 1952 (Standard drawing No. 28-13-09 drawing 2 of 7, Range, rifle, known distance, details, 5 January 1952).

Embankments in front of targets or target tracks

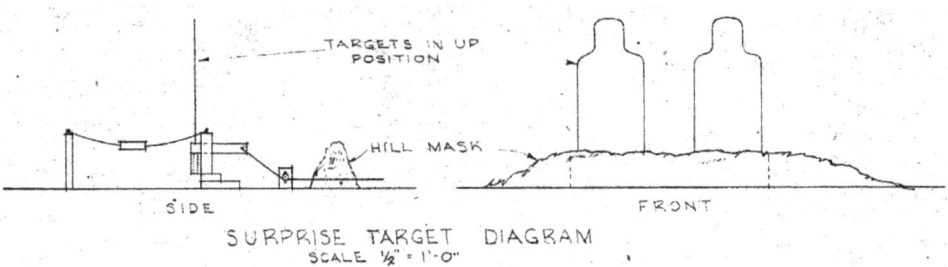

Figure 73. Dismounted submachine gun practice course, surprise target diagram, Fort Bragg, NC, 1951 (Standard drawing 28-13-13 sheet 2 of 2, Range, submachine gun, dismounted practice course, details, 7 December 1951).

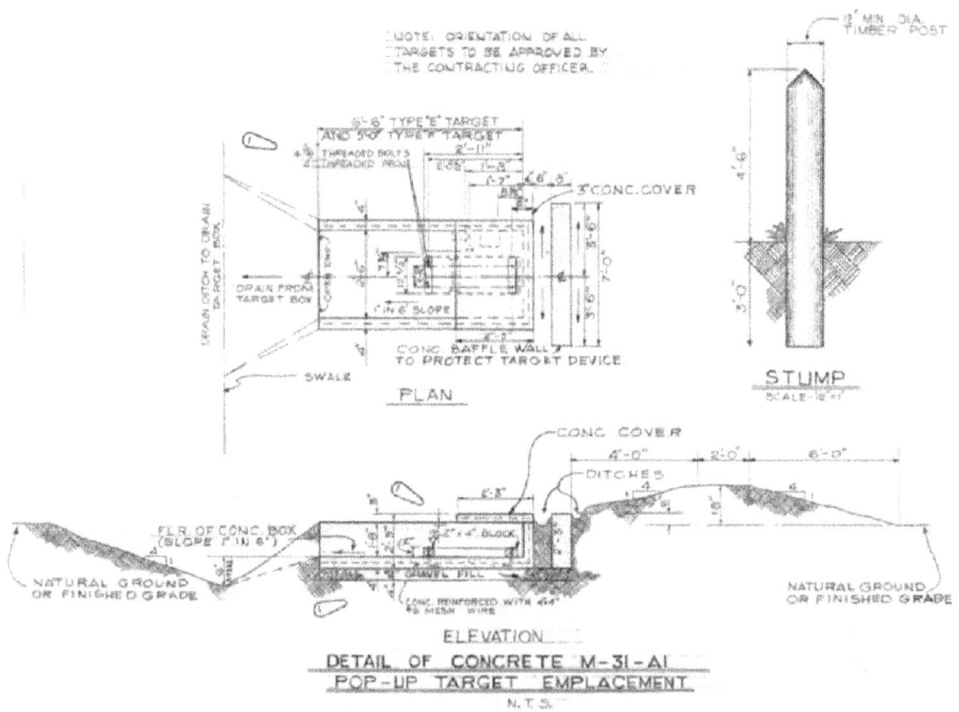

Figure 74. ATC facilities, range details, detail of concrete M-31-A1 pop-up target emplacement, Fort Bragg, NC, 1966 (Standard drawing 28-13-117 sheet 14, Construction of ranges, phase 1, U.S. ATC Facilities, Fort Bragg, NC, range details, 28 June 1966).

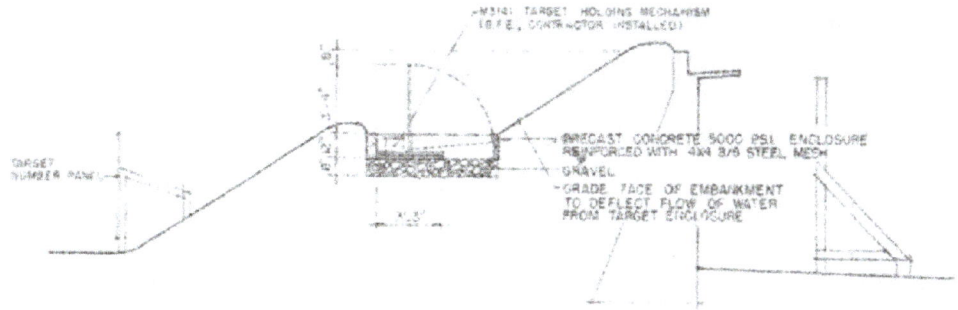

Figure 75. Typical rifle range, M31A1 target device, section through target butt and pop-up target enclosure, Fort Bragg, NC, 1963 (Standard drawing No. 28-13-09 drawing 5 of 7, Range, rifle, known distance, layout, and details for installation of M31A1 target device, 6 November 1963).

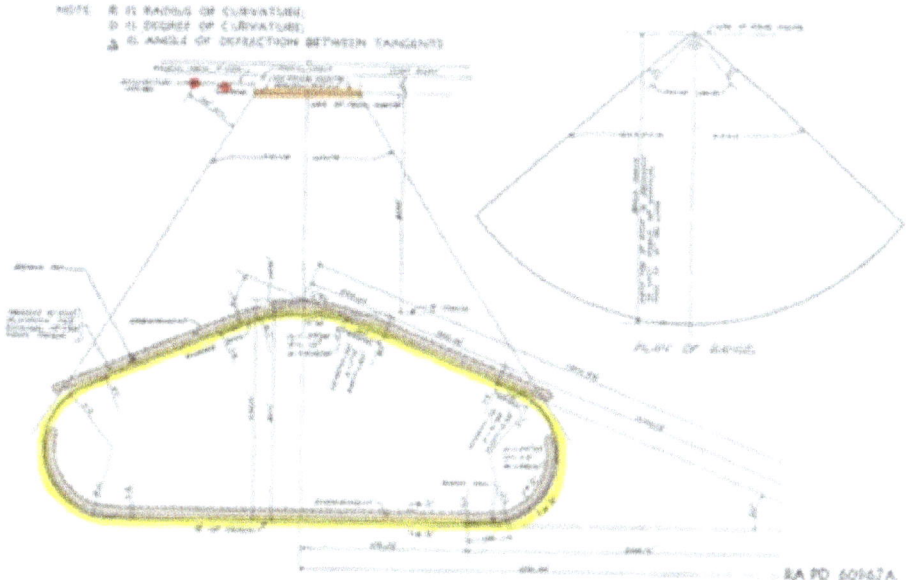

Figure 76. Moving target range, circa 1951 (TM 9-855, Targets, target material, and training course layouts, 1 November 1951, p 35).

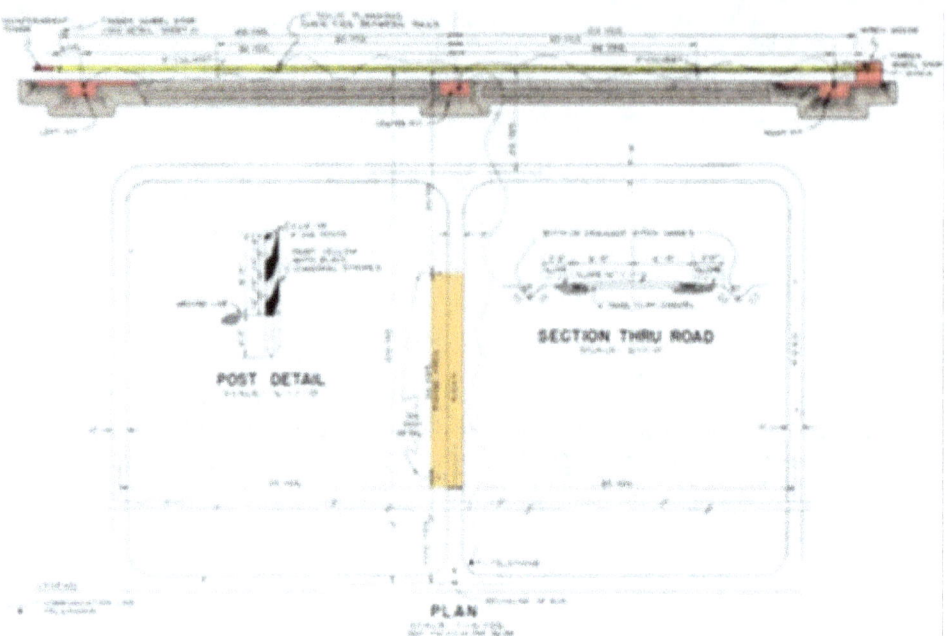

Figure 77. Moving target submachine gun range, plan, Fort Bragg, NC, 1952 (Standard drawing 28-13-14 sheet 1 of 3, Range, moving target, submachine gun, plans and details, 20 June 1952).

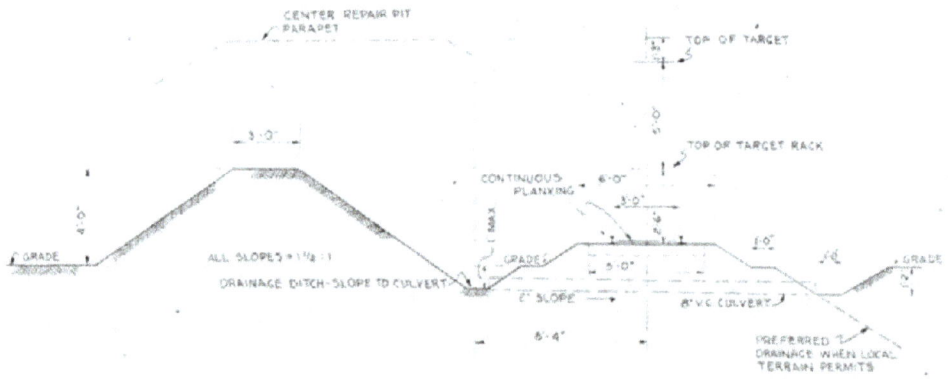

Figure 78. Moving target submachine gun range, section thru firing area, Fort Bragg, NC, 1952 (Standard drawing 28-13-14 sheet 2 of 3, Range, moving target, submachine gun, repair pit and parapet details, 20 June 1952).

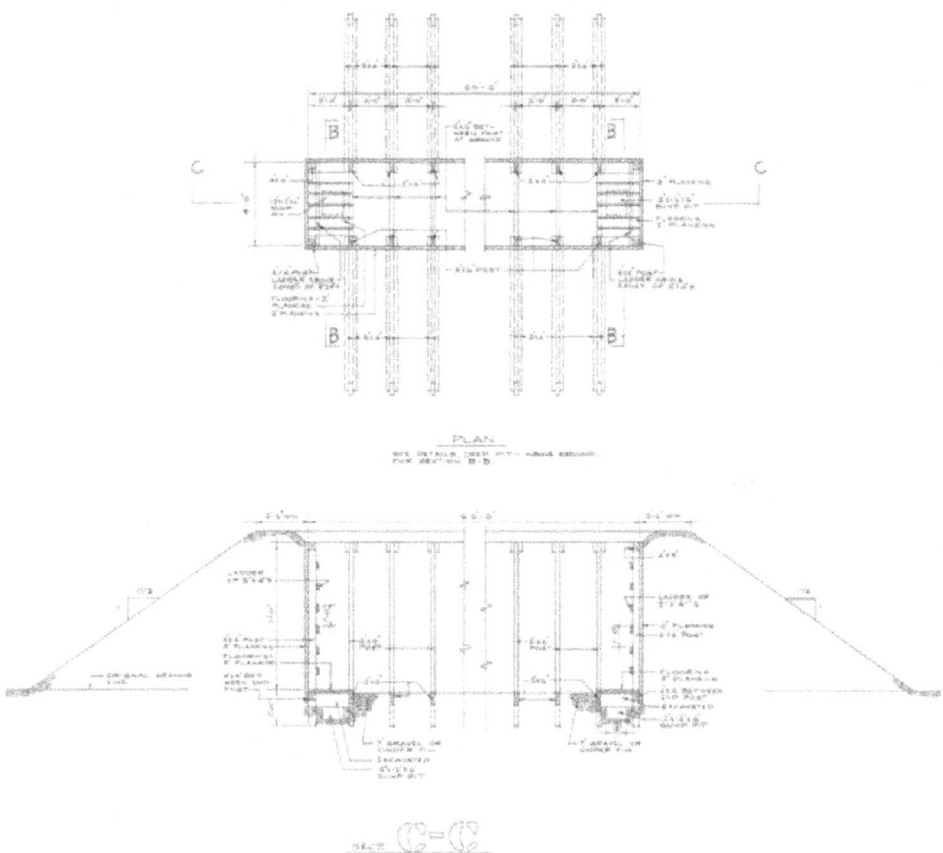

Figure 79. Above ground moving target pit, section and plan, Fort Bragg, NC, 1955 (Standard drawing 28-13-05 sheet 1 of 1, Transition ranges, M1 rifle and MG, above ground pits, construction details, 6 June 1955).

Embankments at firing lines

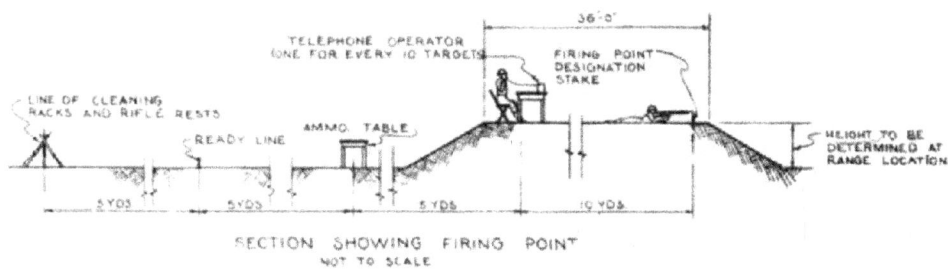

Figure 80. Typical rifle range section showing firing point, Fort Bragg, NC, 1952 (Standard drawing No. 28-13-09 drawing 1 of 7, Range, rifle, known distance, plans and details, 5 January 1952).

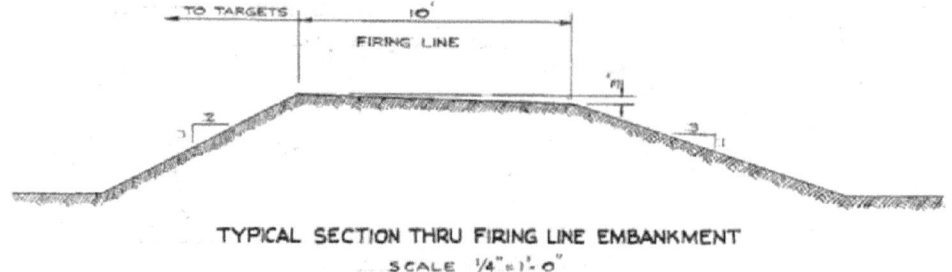

Figure 81. Superimposed transition range, firing line, Fort Bragg, NC, 1954 (Standard drawing 28-13-107 sheet 1 of 1, Superimposed transition range, table VII, 28 March 1954).

Figure 82. Marines on firing line at unknown location, undated (NARA College Park, RG 127-GC, box 5, photo 227310).

Figure 83. Training of marines on the firing line at NAS Jacksonville, 20 May 1942 (NARA College Park, RG 80-G, box 283, photo 64650).

Figure 84. Machine gun crews in action on range at MCAS Cherry Point, NC, 21 August 1943 (NARA College Park, RG 80-G, box 1360, photo 358873).

Embankments between ranges

Figure 85. Castner Range #2, a .30 caliber known distance range at Fort Bliss, TX, 31 August 1953 (NARA College Park, RG 111-SC WWII, box 279, photo SC460727).

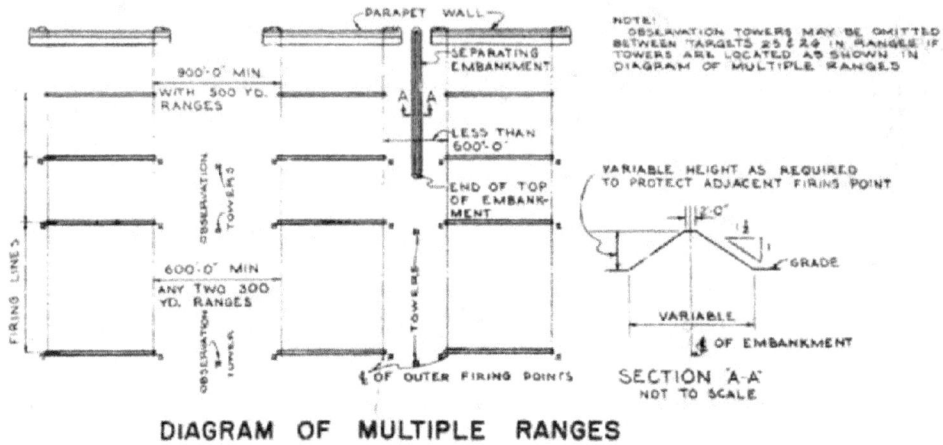

Figure 86. Typical rifle range diagram of multiple ranges, Fort Bragg, NC, 1952 (Standard drawing No. 28-13-09 drawing 1 of 7, Range, rifle, known distance, plans and details, 5 January 1952).

Trenches or foxholes at firing lines

Figure 87. Trench firing line at Camp Wheeler, GA, 1918 (New York Public Library, digital No. 117146).

Figure 88. Throwing live fragmentation hand grenades at the dummy targets at Fort Jackson, SC, 1 November 1943 (NARA College Park, RG 111-SC WWII, box 681, photo SC324452).

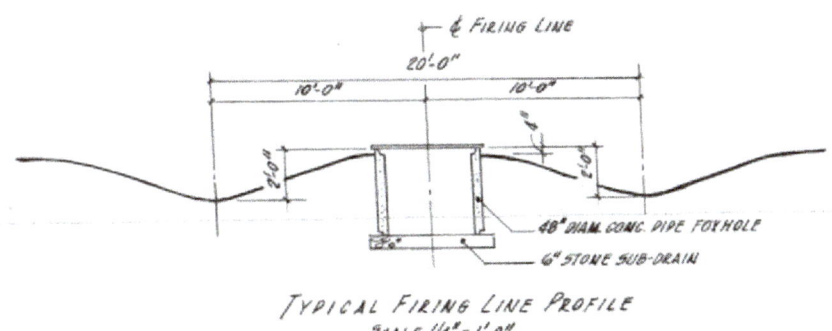

Figure 89. ATC facilities, night firing range "N", firing line, Fort Bragg, NC, 1966 (Standard drawing 28-13-117 sheet 9, Construction of ranges, phase 1, U.S. ATC facilities, Fort Bragg, NC, range "N", night firing range, 28 June 1966).

Trenches or foxholes at target lines

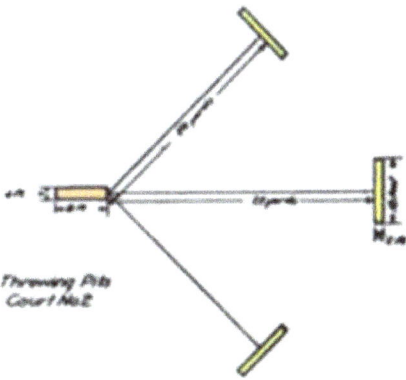

Figure 90. Throwing pits course, circa 1942 (FM 23-30, Grenades, 15 June 1942, p 23).

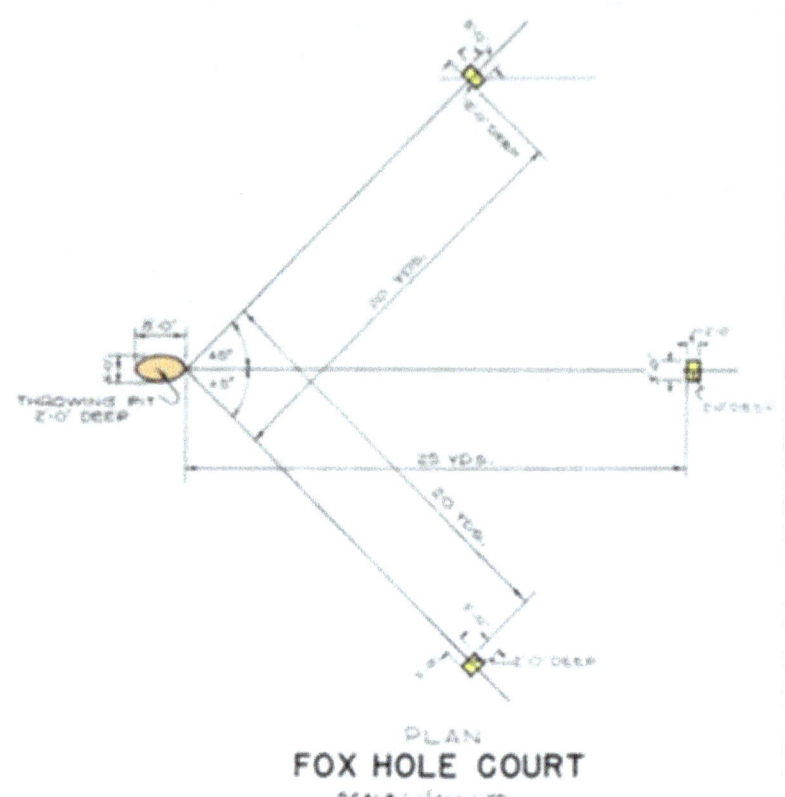

Figure 91. Practice grenade course, Foxhole Court, Fort Bragg, NC, 1951 (Standard drawing 28-13-43 sheet 1 of 1, Practice grenade course layout and details, 21 November 1951).

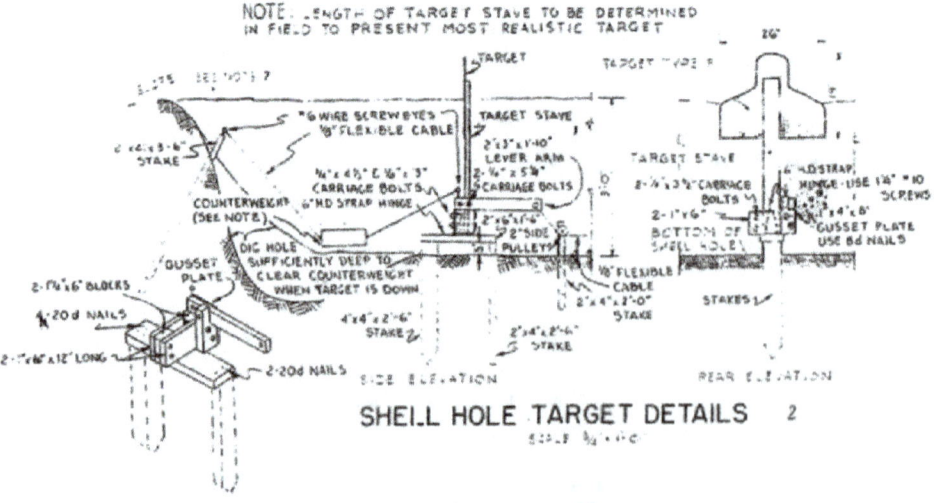

Figure 92. Shell hole target details, Fort Bragg, NC, 1957 (Standard drawing 28-13-05, Close combat course, plan and details, 8 September 1957).

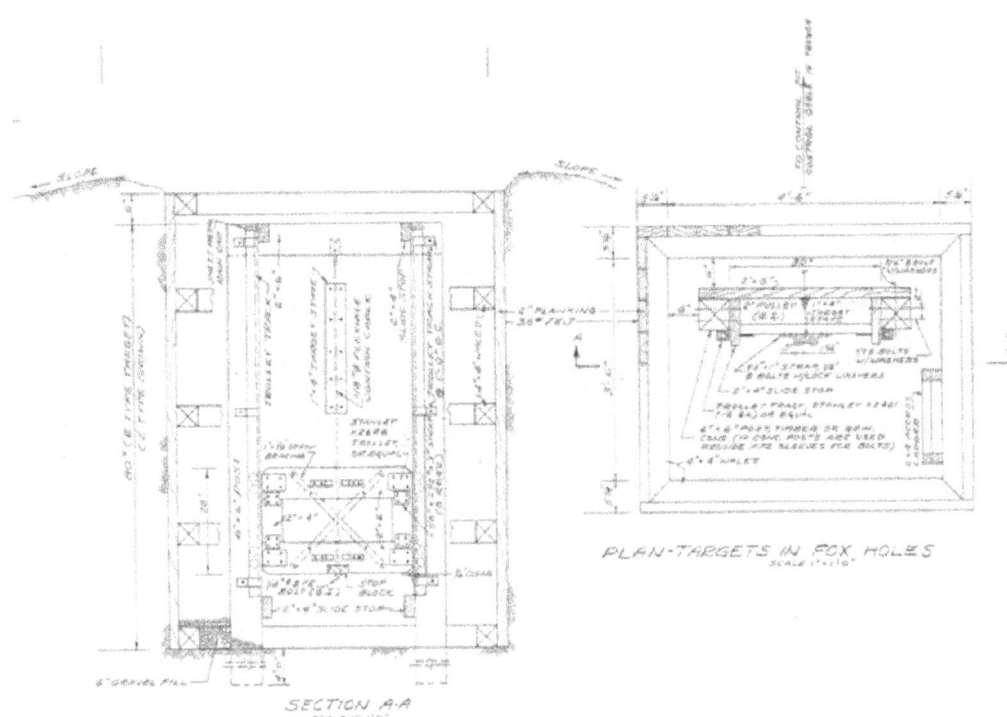

Figure 93. Superimposed transition range, plan of targets in foxholes and section, Fort Bragg, NC, 1954 (Standard drawing 28-13-108 sheet 1 of 1, Superimposed transition range, table VIII, 22 March 1954).

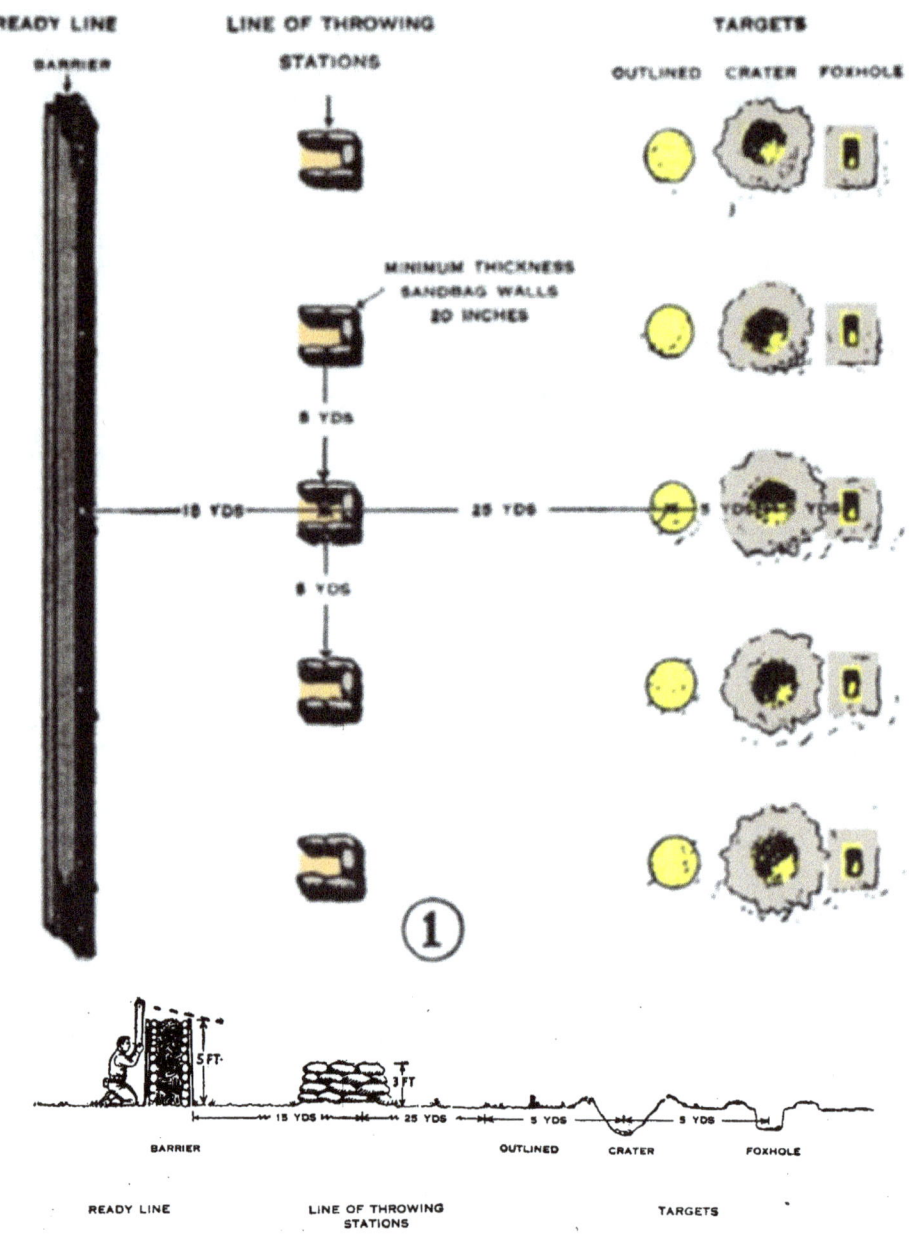

Figure 94. Live grenade practice course, plan and profile, circa 1944-1949 (FM 23-30, Hand and rifle grenades, rocket, AT, HE, 2.36-in., 14 February 1944, pp 35, 36, FM 23-30, Hand and rifle grenades, 14 April 1949, pp 40, 41).

Target control pit

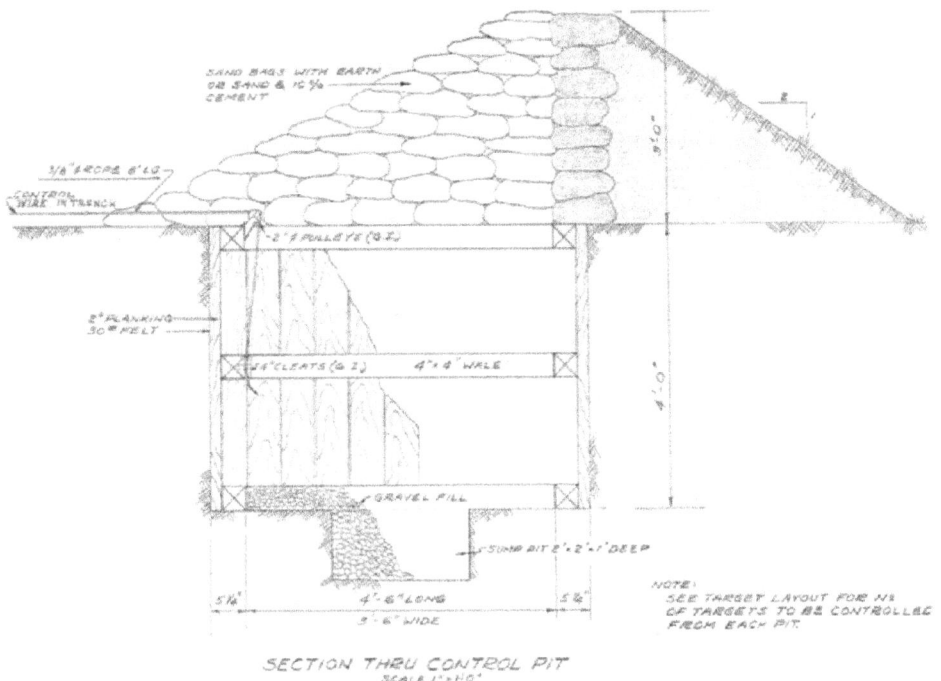

Figure 95. Superimposed transition range, section thru control pit, Fort Bragg, NC, 1954 (Standard drawing 28-13-108 sheet 1 of 1, Superimposed transition range, table VIII, 22 March 1954).

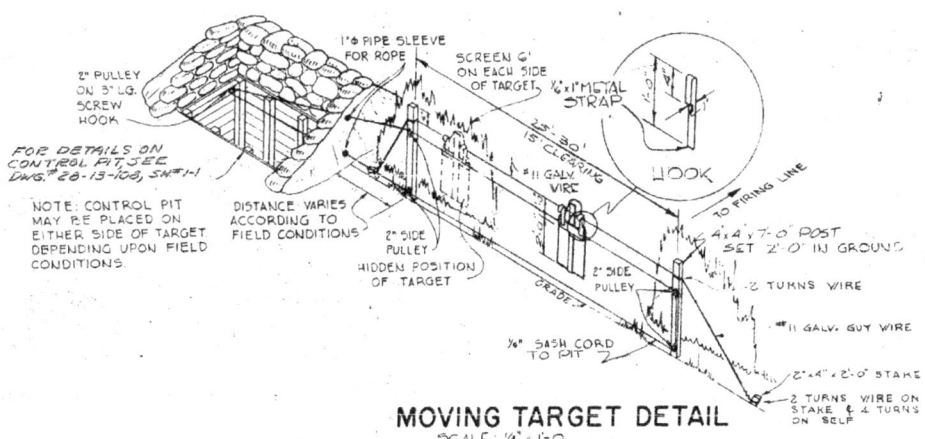

Figure 96. Transition range moving target detail, Fort Bragg, NC, 1955 (Standard drawing 28-13-04 sheet 1 of 1, Range, transition, details, 6 June 1955).

Demolition pits

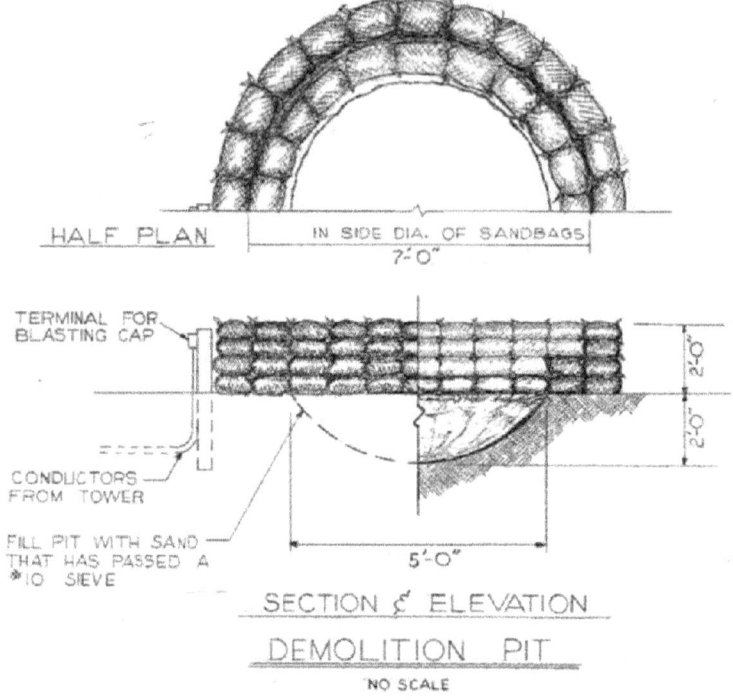

Figure 97. ATC Facilities, range details, demolition pit, Fort Bragg, NC, 1966 (Standard drawing 28-13-117 sheet 14, Construction of ranges, phase 1, U.S. ATC Facilities, Fort Bragg, NC, range details, 28 June 1966).

Buildings

A small arms range may have had a control tower, latrine, target storage building, ammunition storage building, bleachers with storage, other storage sheds, and administrative/maintenance buildings supporting general range functions. The range may also have been part of a larger installation range complex that contained these buildings. Drawings and photographs of several range and support buildings are shown below.

Figure 98. Range at NAS Deland, FL, 23 December 1944 (NARA College Park, RG 80-G, box 1382, photo 363872).

Figure 99. Aerial of gunnery range at NAS Beaufort, SC, 24 January 1945 (NARA College Park, RG 80-G, box 1376, photo 362569).

Observation towers

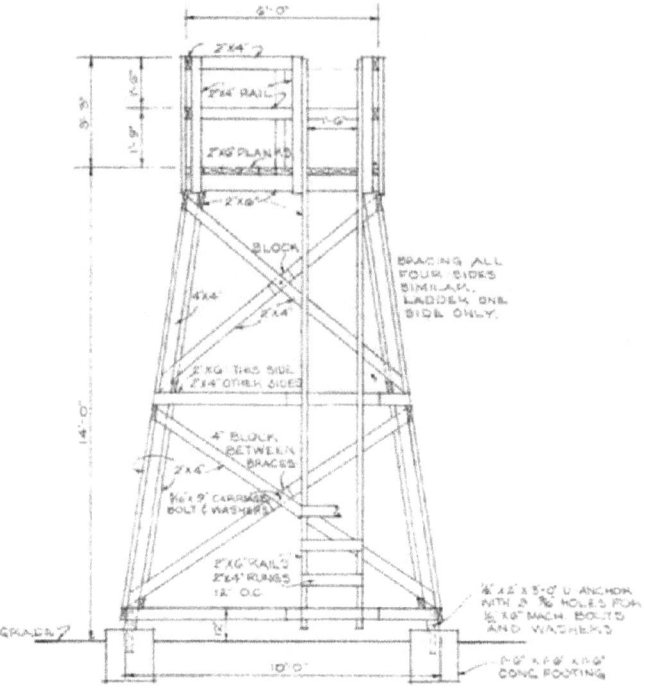

Figure 100. Rifle range observation tower elevation, Fort Bragg, NC, 1952 (Standard drawing No. 28-13-09A drawing 4 of 4, Range, rifle, known distance, observation tower and latrine, 5 January 1952).

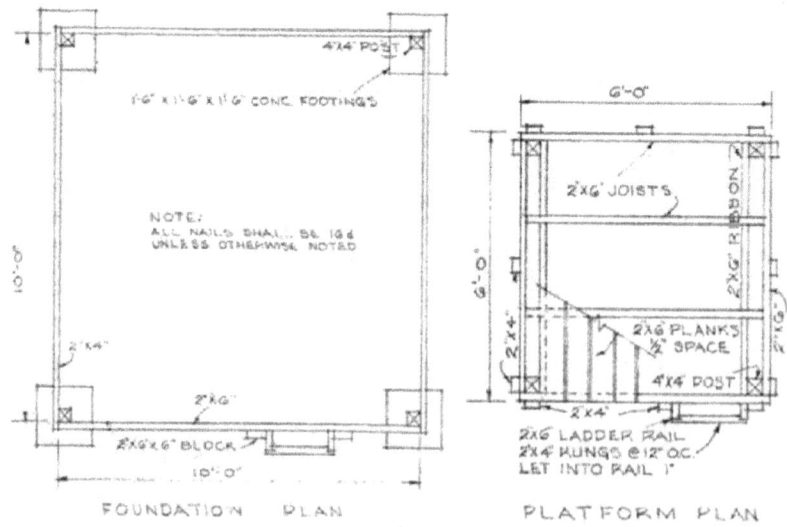

Figure 101. Rifle range observation tower foundation and platform plans, Fort Bragg, NC, 1952 (Standard drawing No. 28-13-09A drawing 4 of 4, Range, rifle, known distance, observation tower and latrine, 5 January 1952).

Control towers

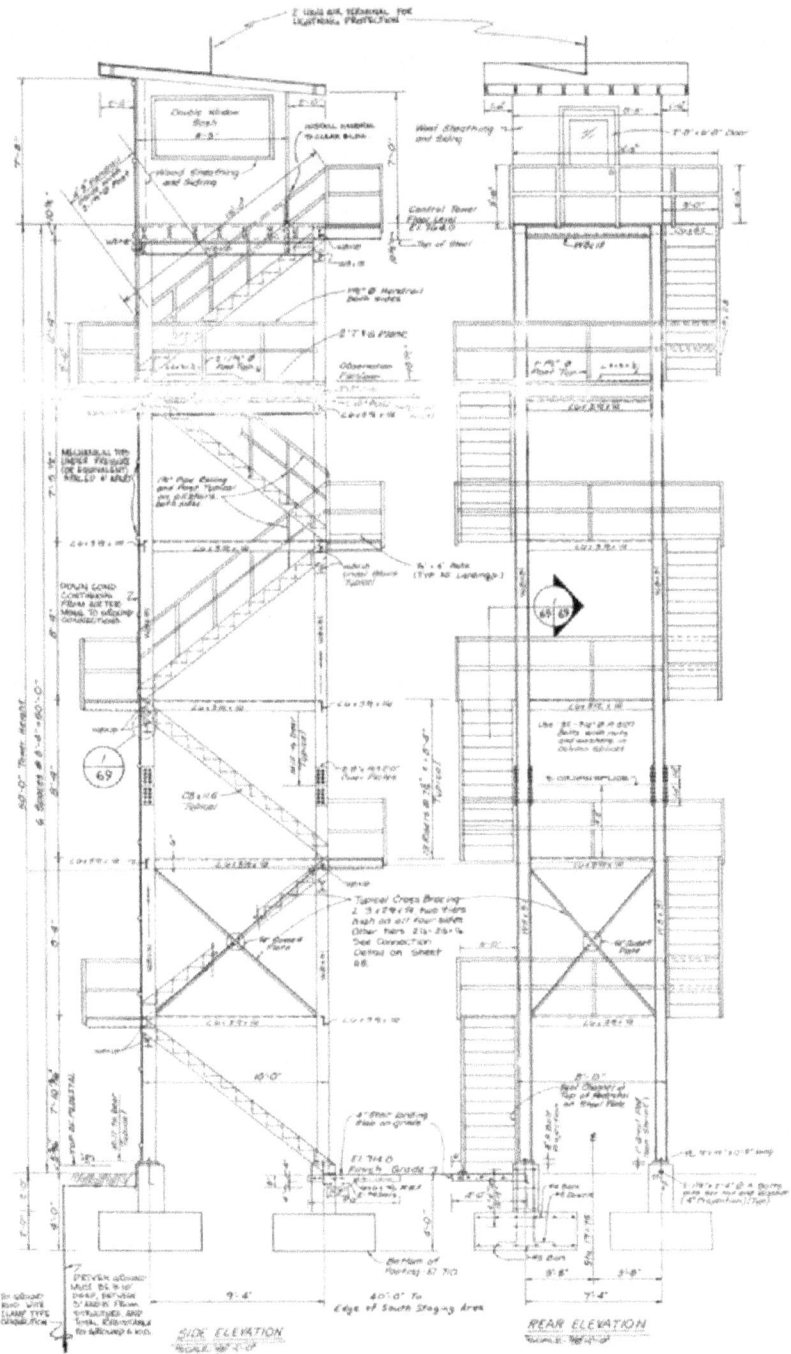

Figure 102. Mendick tollgate range control tower side and rear elevations, Fort Knox, KY, 1981 (Standard drawing 28-13-13, Control tower details, mendick-tollgate range, table V, September 1981).

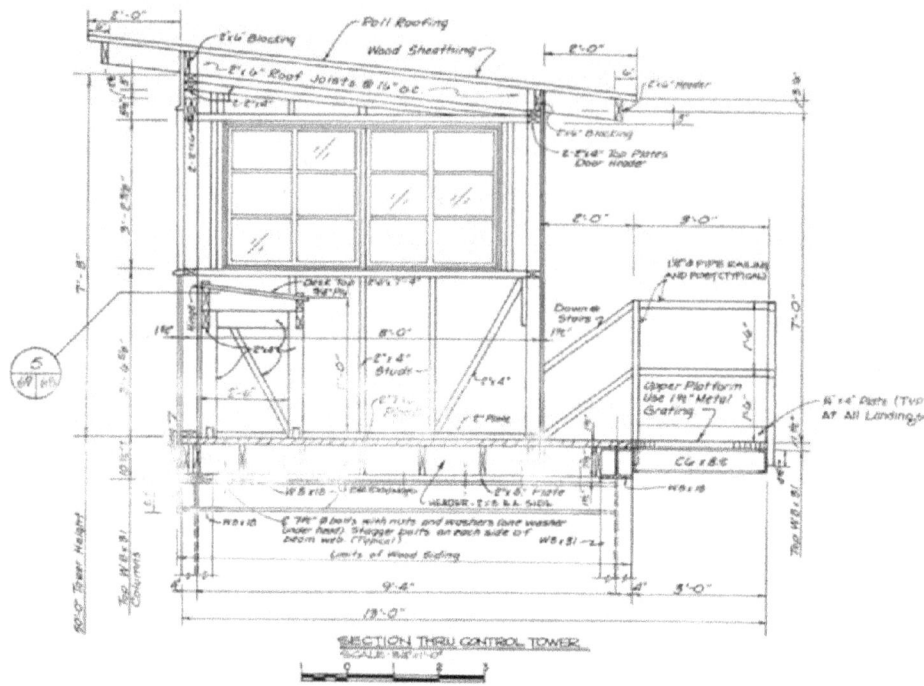

Figure 103. Mendick tollgate range section through control tower, Fort Knox, KY, 1981 (Standard drawing 28-13-13, Control tower details, Mendick-Tollgate Range, table V, September 1981).

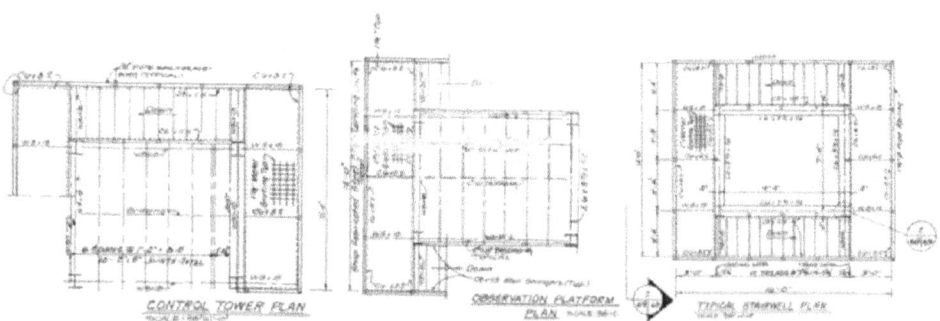

Figure 104. Mendick-Tollgate Range control tower plans, Fort Knox, KY, 1981 (Standard drawing 28-13-13, Control tower details, Mendick-Tollgate Range, table V, September 1981).

Figure 105. The combat rifleman environmental range tower #1 including the "C" portion and part of "B" portion as seen from Tower #2 at Range 208 at MCB Camp Pendleton, CA, 24 July 1962 (NARA College Park, RG 127-GG-2058, box 42, photo A353462).

Figure 106. Controls that operate pop-up targets at Range 208 at MCB Camp Pendleton, CA, 24 July 1962 (NARA College Park, RG 127-GG-2021, box 40, photo A353453).

Figure 107. The combat rifleman environmental range tower #1 and bleachers at Range 208 at MCB Camp Pendleton, CA, 24 July 1962 (NARA College Park, RG 127-GG-2058, box 42, photo A353466).

Figure 108. New rifle range control tower that is an essential component of the rifle ranges at Fort Jackson, SC, 14 January 1964 (NARA College Park, RG 111-SC post-1955, box 384, photo SC607734).

Portable control tower

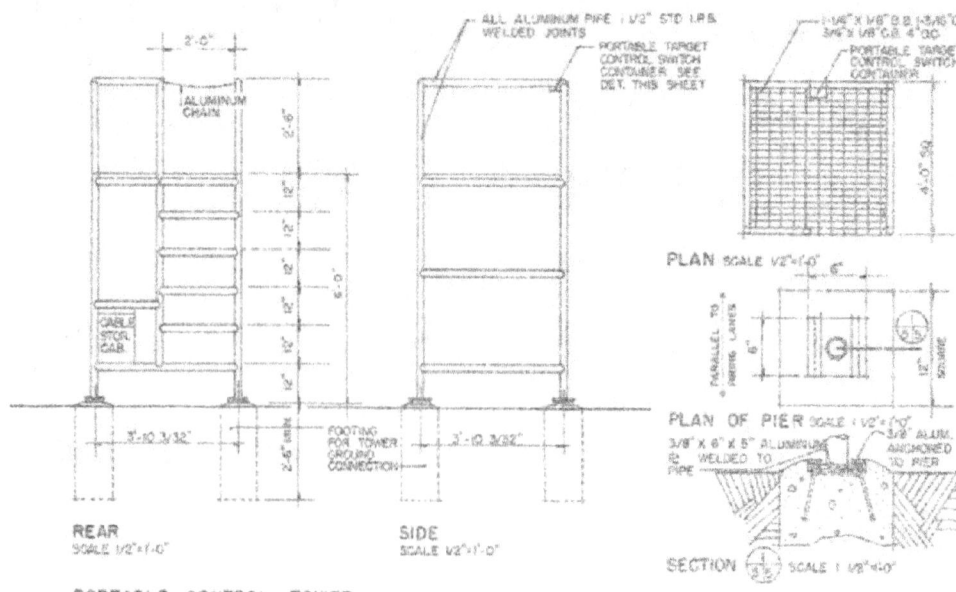

Figure 109. Typical rifle range portable control tower, Fort Bragg, NC, 1952 (Standard drawing No. 28-13-09 drawing 5 of 7, Range, rifle, known distance, details, 5 January 1952).

Latrine

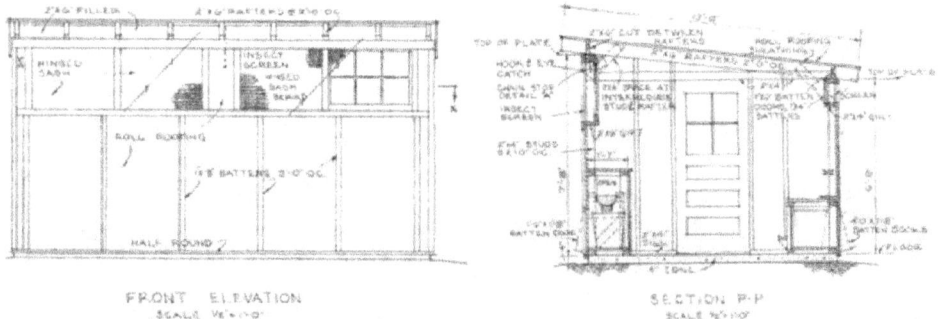

Figure 110. Rifle range latrine front elevation and section, Fort Bragg, NC, 1952 (Standard drawing No. 28-13-09A drawing 4 of 4, Range, rifle, known distance, observation tower and latrine, 5 January 1952).

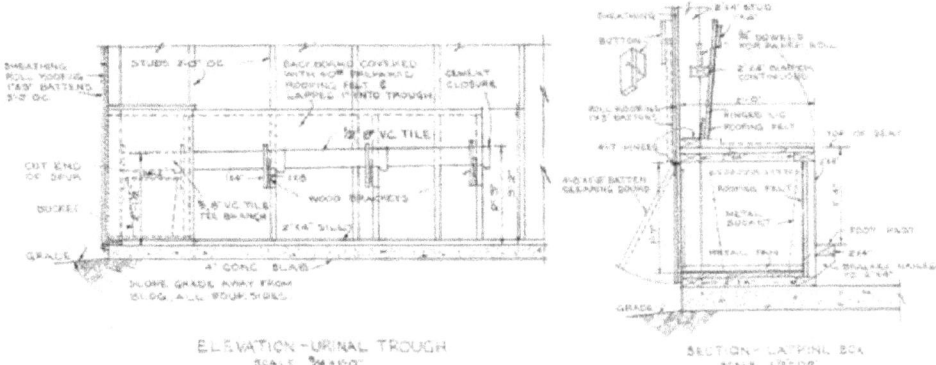

Figure 111. Rifle range latrine elevation of urinal trough and section of latrine box, Fort Bragg, NC, 1952 (Standard drawing No. 28-13-09A drawing 4 of 4, Range, rifle, known distance, observation tower and latrine, 5 January 1952).

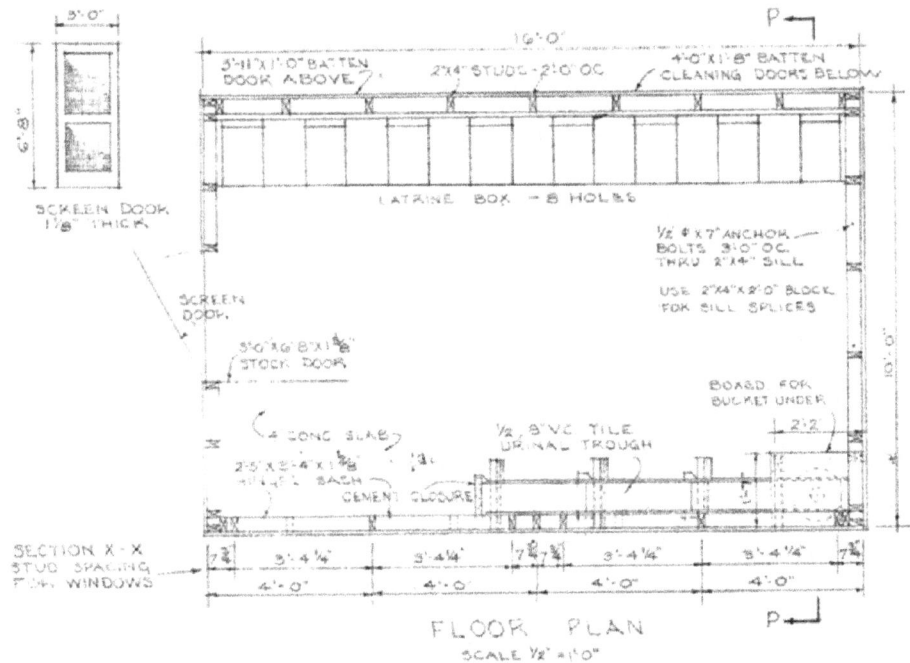

Figure 112. Rifle range latrine floor plan, Fort Bragg, NC, 1952 (Standard drawing No. 28-13-09A drawing 4 of 4, "Range, rifle, known distance, observation tower and latrine, 5 January1952").

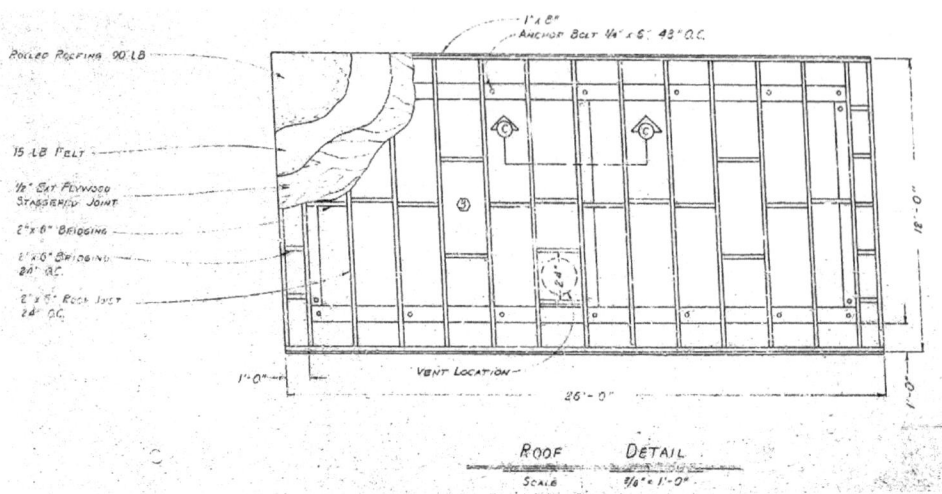

Figure 113. Range latrine, roof detail, Fort Bragg, NC, 1983 (DEH 4122 sheet 2 of 5, Range latrine, 21 December 1983).

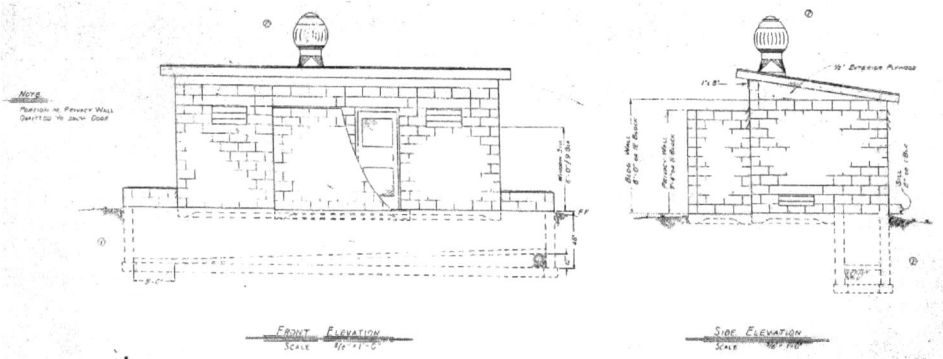

Figure 114. Range latrine, front and side elevations, Fort Bragg, NC, 1983 (DEH 4122 sheet 2 of 5, Range latrine, 21 December 1983).

Target storage and latrine building

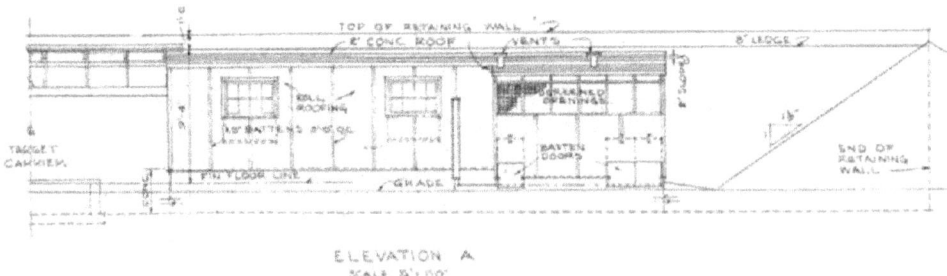

Figure 115. Typical rifle range target storage and latrine building, elevation A, Fort Bragg, NC, 1952 (Standard drawing No. 28-13-09A drawing 3 of 4, Range, rifle, known distance, target storage and latrine Bldg, 5 January 1952).

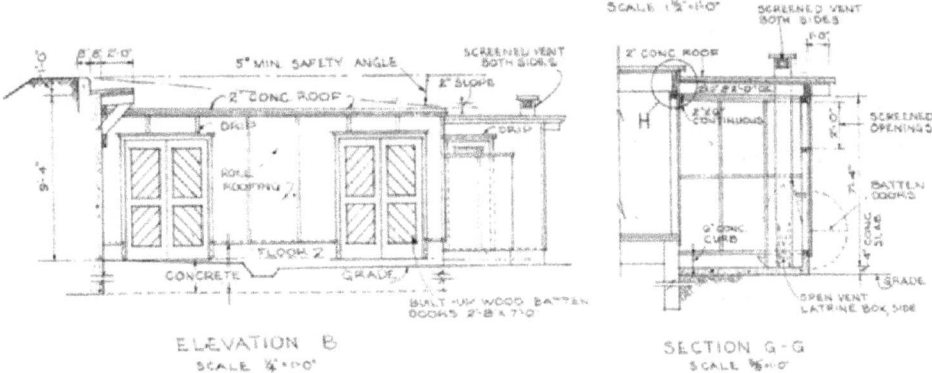

Figure 116. Typical rifle range target and storage building, elevation B and Section G-G, Fort Bragg, NC, 1952 (Standard drawing No. 28-13-09A drawing 3 of 4, Range, rifle, known distance, target storage and latrine Bldg, 5 January 1952).

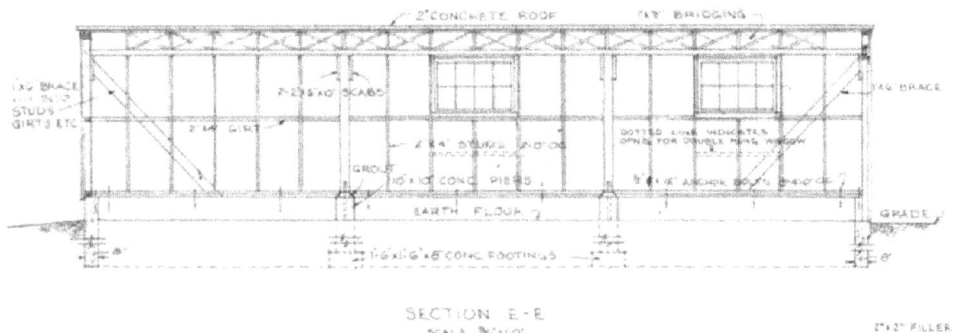

Figure 117. Typical rifle range target storage and latrine building, Section EE, Fort Bragg, NC, 1952 (Standard drawing No. 28-13-09A drawing 3 of 4, Range, rifle, known distance, target storage and latrine Bldg, 5 January 1952).

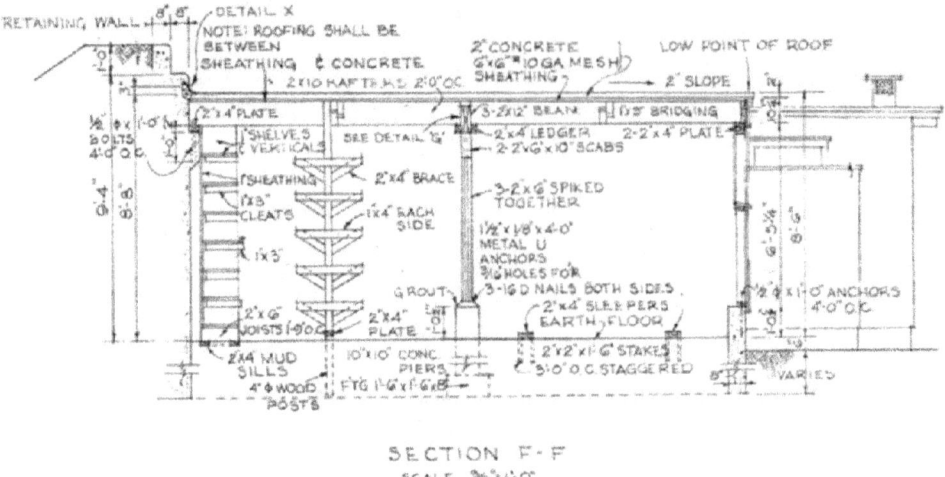

Figure 118. Typical rifle range target storage and latrine building, section FF, Fort Bragg, NC, 1952 (Standard drawing No. 28-13-09A drawing 3 of 4, Range, rifle, known distance, target storage and latrine Bldg, 5 January 1952).

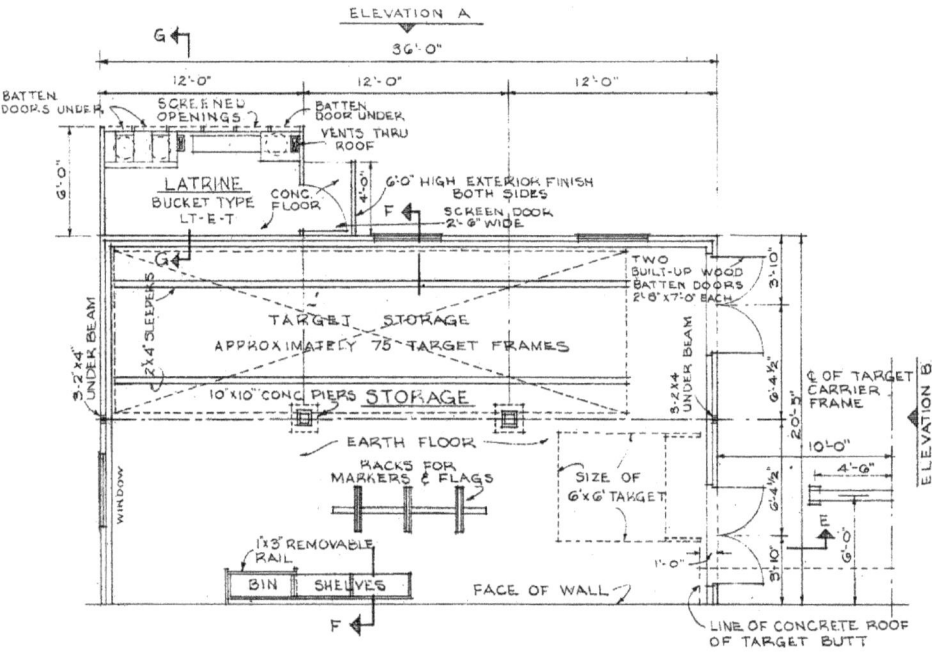

Figure 119. Typical rifle range target and storage building floor plan, Fort Bragg, NC, 1952 (Standard drawing No. 28-13-09A drawing 3 of 4, Range, rifle, known distance, target storage and latrine Bldg, 5 January 1952).

Target storage buildings

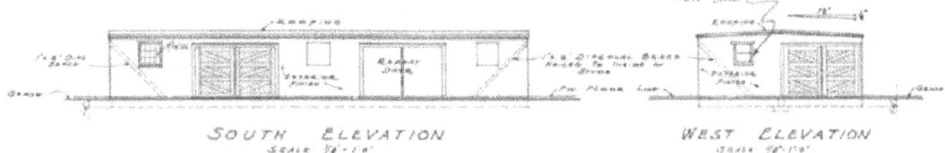

Figure 120. Range target storage and repair shed, south and west elevations, Fort Knox, KY, 1946 (FK-9-2 (Adapted From Standard drawing No. T.O. 700-6013), Range target storage and repair shed, Type S-A-T (modified), plans, elevations, and details, 13 August 1946).

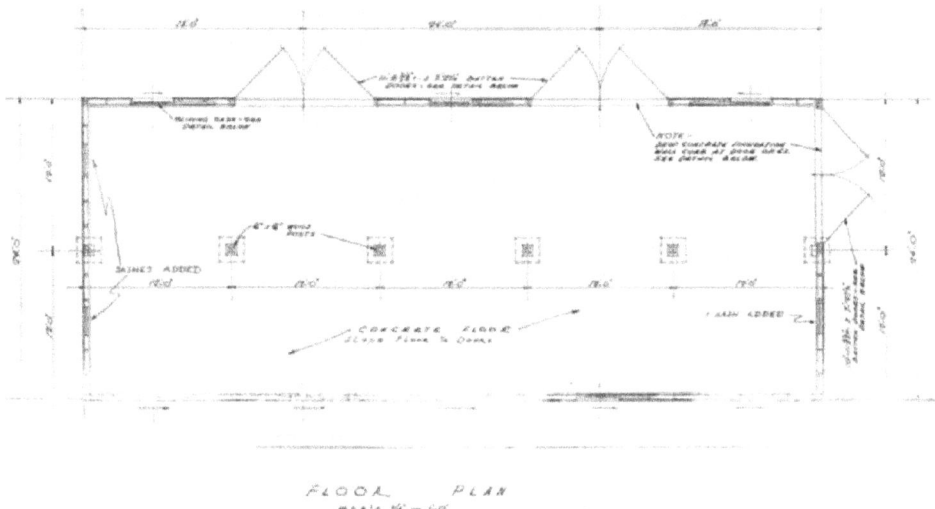

Figure 121. Range target storage and repair shed floor plan, Fort Knox, KY, 1946 (FK-9-2 (adapted from Standard drawing No. T.O. 700-6013), Range target storage and repair shed, type S-A-T (modified), plans, elevations, and details, 13 August 1946).

Figure 122. Target shelter, front and side elevations, Fort Bragg, NC, undated (DEH-4154 sheet o5101, Range 51, target shelter, undated).

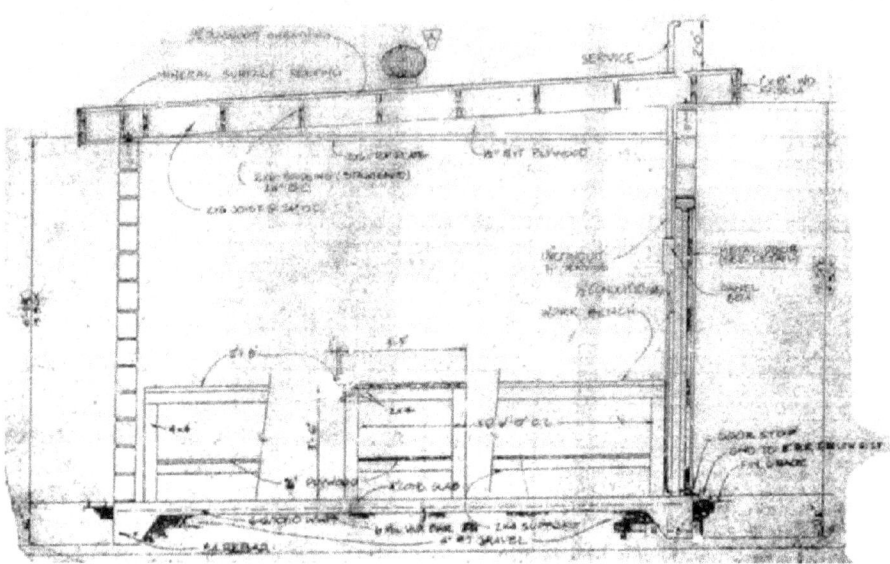

Figure 123. Target shelter, section, Fort Bragg, NC, undated (DEH-4154 sheet o5101, Range 51, target shelter, undated).

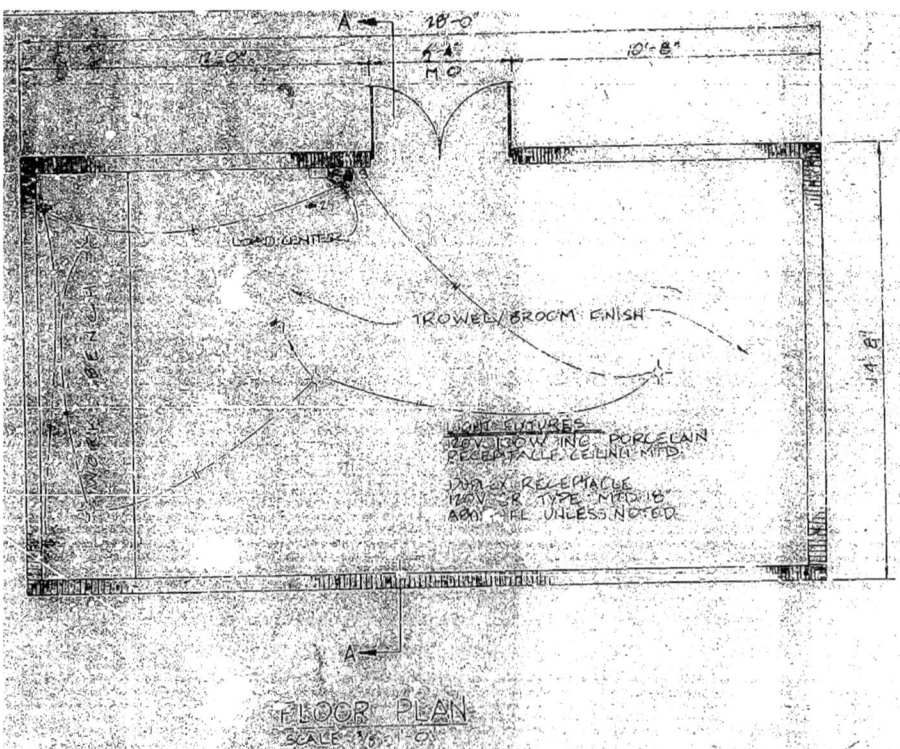

Figure 124. Target shelter, floor plan, Fort Bragg, NC, undated (DEH-4154 sheet o5101, Range 51, target shelter, undated).

Target repair and range house

Masonry construction

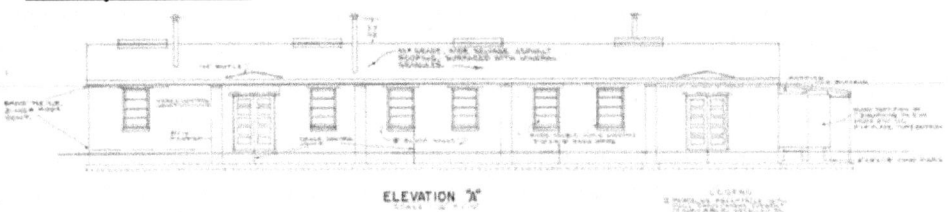

Figure 125. Masonry target repair and range house, elevation A, Fort Bragg, NC, 1952 (Standard drawing 28-13-96 sheet 1 of 3, Target repair and range house, masonry, plan, elevations, and section, 11 April 1952).

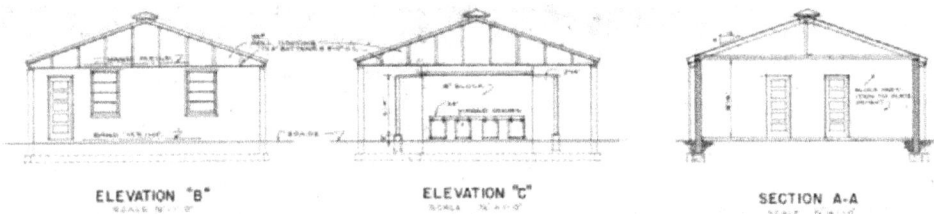

Figure 126. Masonry target repair and range house, elevation A & C, and Section A, Fort Bragg, NC, 1952 (Standard drawing 28-13-96 sheet 1 of 3, Target repair and range house, masonry, plan, elevations, and section, 11 April 1952).

Figure 127. Masonry target repair and range house plan, Fort Bragg, NC, 1952 (Standard drawing 28-13-96 sheet 1 of 3, Target repair and range house, masonry, plan, elevations, and section, 11 April 1952).

Frame construction

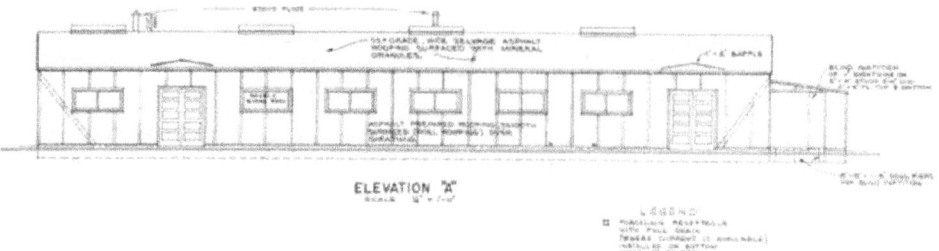

Figure 128. Target repair and range house, elevation, Fort Bragg, NC, 1952 (Standard drawing No. 28-13-29 sheet 1 of 3, Target repair and range house, frame construction, plan, elevations, and section, 11 April 1952).

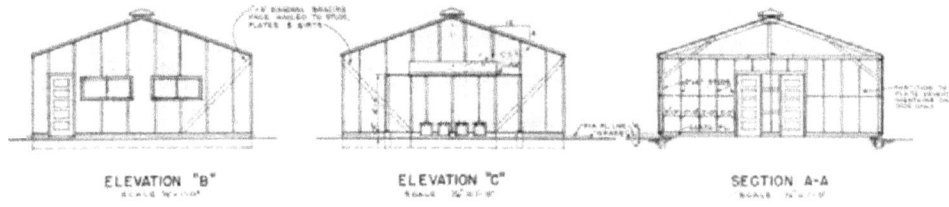

Figure 129. Target repair and range house, elevations, Fort Bragg, NC, 1952 (Standard drawing No. 28-13-29 sheet 1 of 3, Target repair and range house, frame construction, plan, elevations, and section, 11 April 1952).

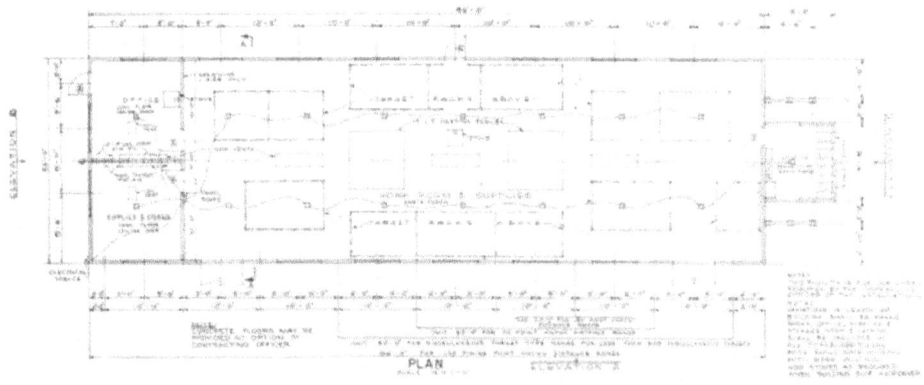

Figure 130. Target repair and range house, plan, Fort Bragg, NC, 1952 (Standard drawing No. 28-13-29 sheet 1 of 3, Target repair and range house, frame construction, plan, elevations, and section, 11 April 1952).

Bleachers

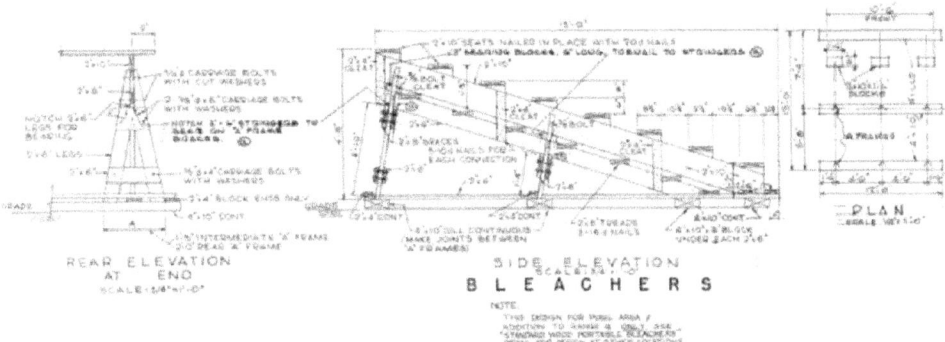

Figure 131. Bleachers, plan, rear elevation, and side elevation, Fort Bragg, NC, 1966 (Standard drawing 28-13-117 sheet 22, Construction of ranges, phase I, U.S. ATC facilities, Fort Bragg, NC, bleachers with shed, 28 June 1966).

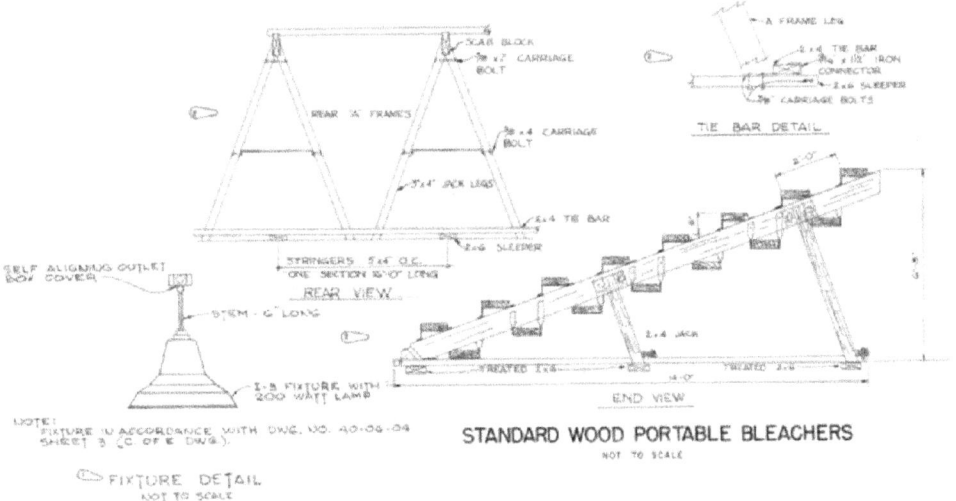

Figure 132. Bleacher end view and details, Fort Bragg, NC, 1966 (Standard drawing 28-13-117 sheet 22, Construction of ranges, phase I, U.S. ATC Facilities, Fort Bragg, NC, Bleachers with shed, 28 June 1966).

Bleacher shed

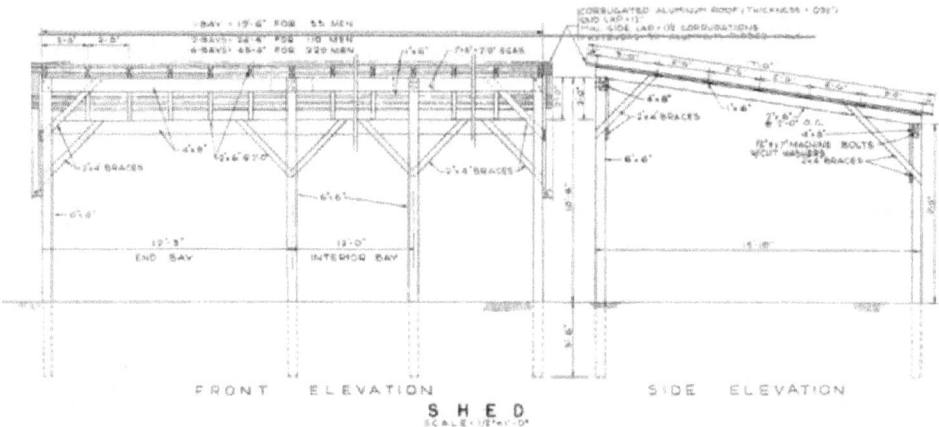

Figure 133. Bleacher shed front and side elevations, Fort Bragg, NC, 1966 (Standard drawing 28-13-117 sheet 22, Construction of ranges, phase I, U.S. ATC Facilities, Fort Bragg, NC, bleachers with shed, 28 June 1966).

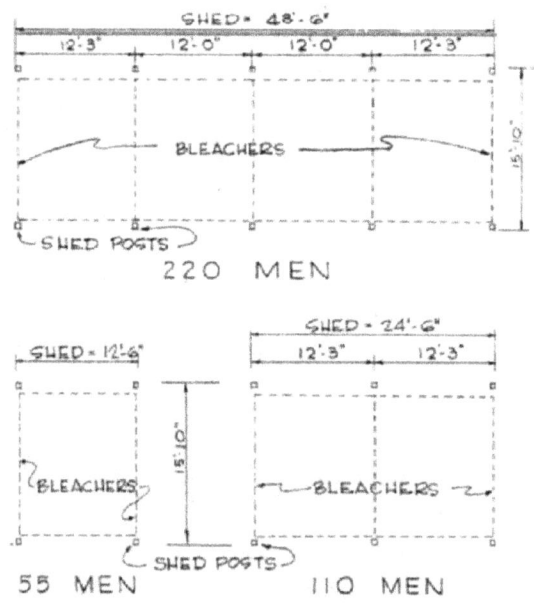

Figure 134. Bleachers layout plan, Fort Bragg, NC, 1966 (Standard drawing 28-13-117 sheet 22, Construction of ranges, phase I, U.S. ATC Facilities, Fort Bragg, NC, bleachers with shed, 28 June 1966).

Ammo storage

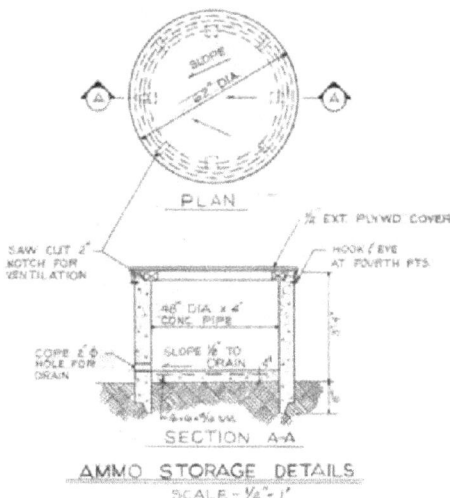

Figure 135. ATC facilities, range details, ammo storage details, Fort Bragg, NC, 1966 (Standard drawing 28-13-117 sheet 14, Construction of ranges, phase 1, U.S. ATC facilities, Fort Bragg, NC, range details, 28 June 1966).

Winch house and counterweight tower

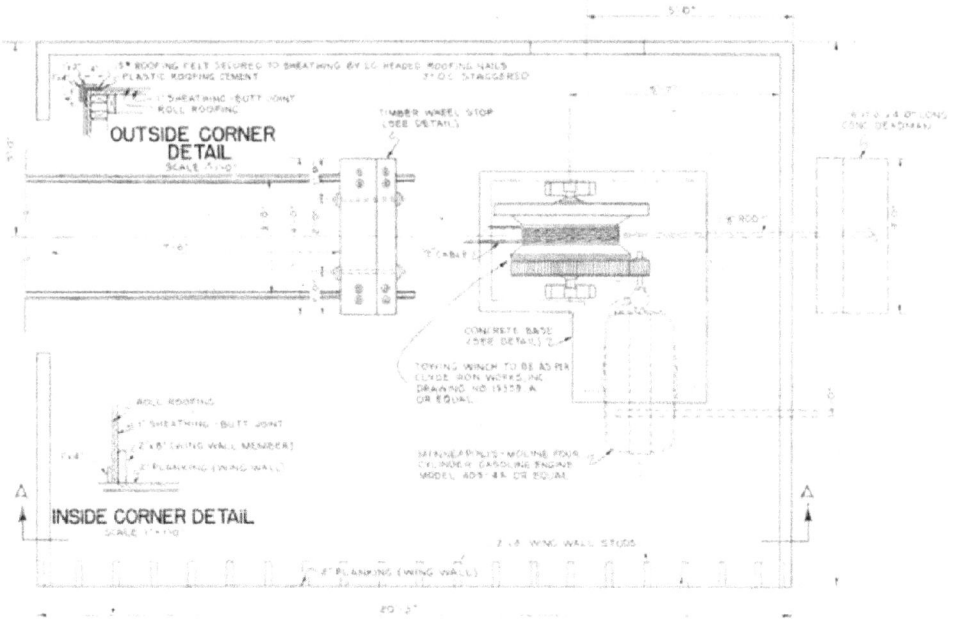

Figure 136. Moving target submachine gun range, winch house plan, Fort Bragg, NC, 1952 (Standard drawing 28-13-14 sheet 3 of 3, Range, moving target, submachine gun, winch house, 20 June 1952).

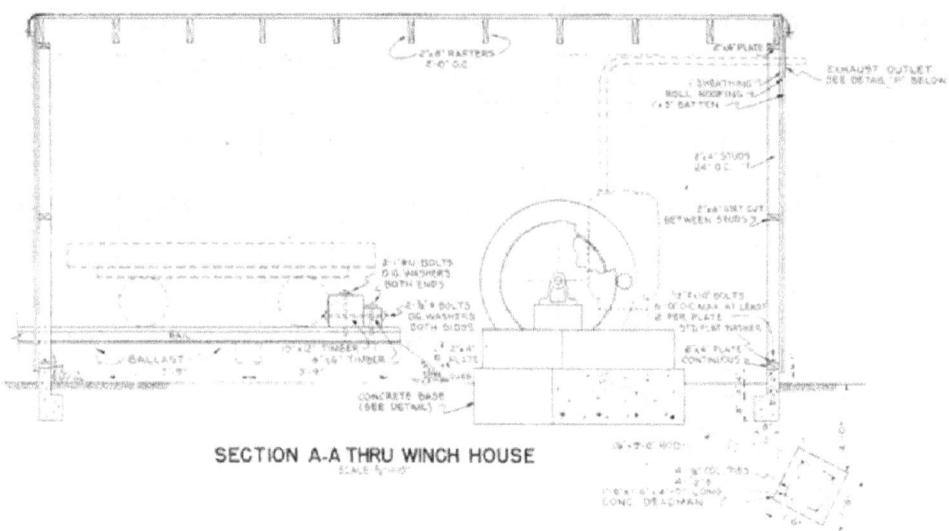

Figure 137. Moving target submachine gun range, winch house section, Fort Bragg, NC, 1952 (Standard drawing 28-13-14 sheet 3 of 3, Range, moving target, submachine gun, winch house, 20 June 1952).

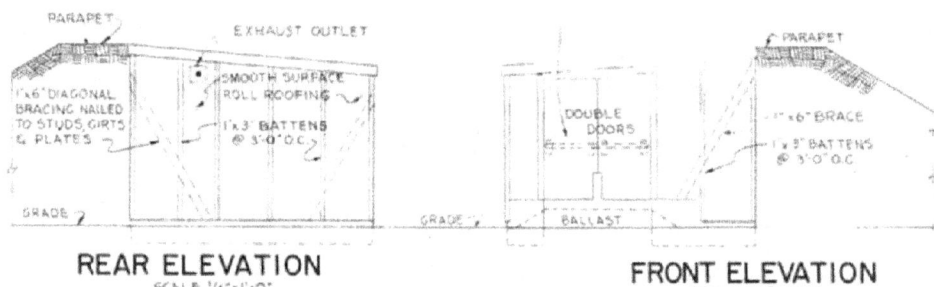

Figure 138. Moving target submachine gun range, winch house rear and front elevations, Fort Bragg, NC, 1952 (Standard drawing 28-13-14 sheet 3 of 3, Range, moving target, submachine gun, winch house, 20 June 1952).

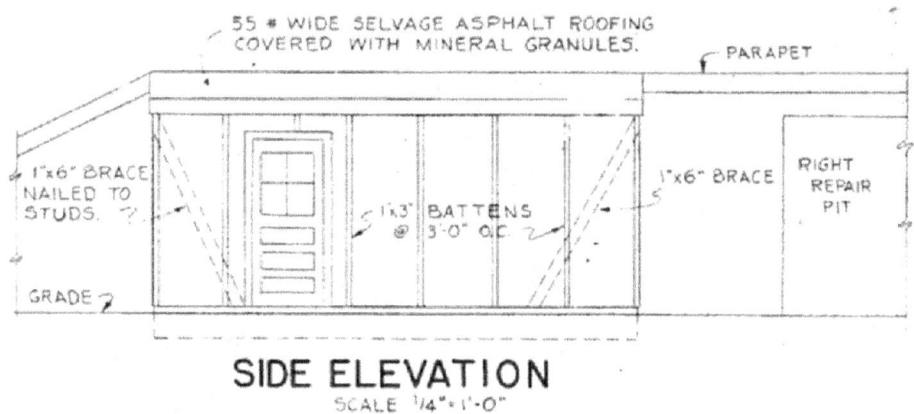

Figure 139. Moving target submachine gun range, winch house side elevation, Fort Bragg, NC, 1952 (Standard drawing 28-13-14 sheet 3 of 3, Range, moving target, submachine gun, winch house, 20 June 1952).

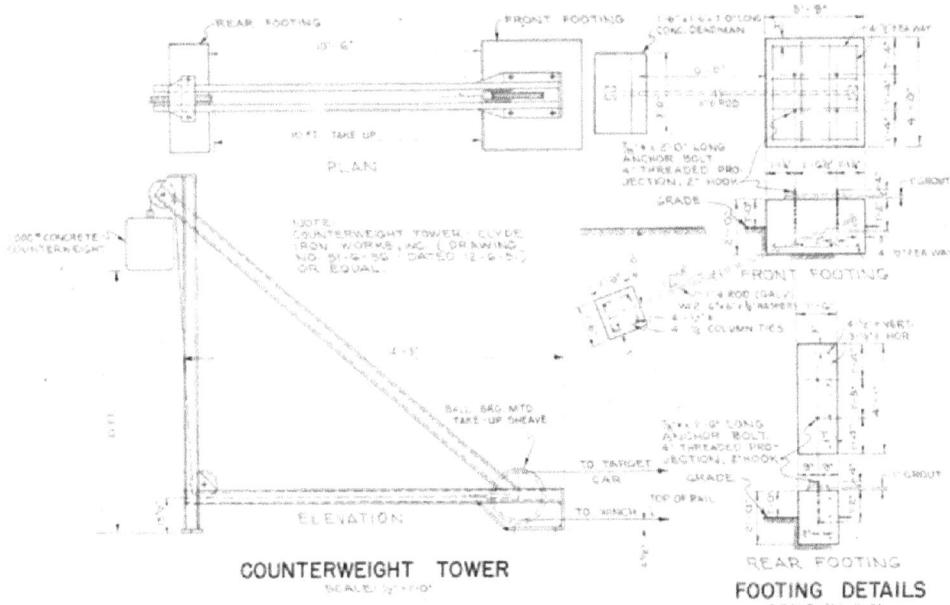

Figure 140. Moving target submachine gun range, counterweight tower, Fort Bragg, NC, 1952 (Standard drawing 28-13-14, sheet 1 of 3, Range, moving target, submachine gun, plans and details, 20 June 1952).

Range buildings

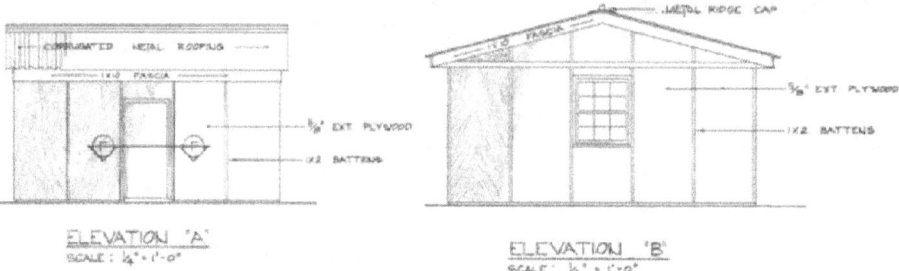

Figure 141. ATC Facilities, range building, elevations, Fort Bragg, NC, 1966 (Standard drawing 28-13-117 sheet 24, Construction of ranges, phase 1, U.S. ATC Facilities, Fort Bragg, NC, range building, 28 June 1966).

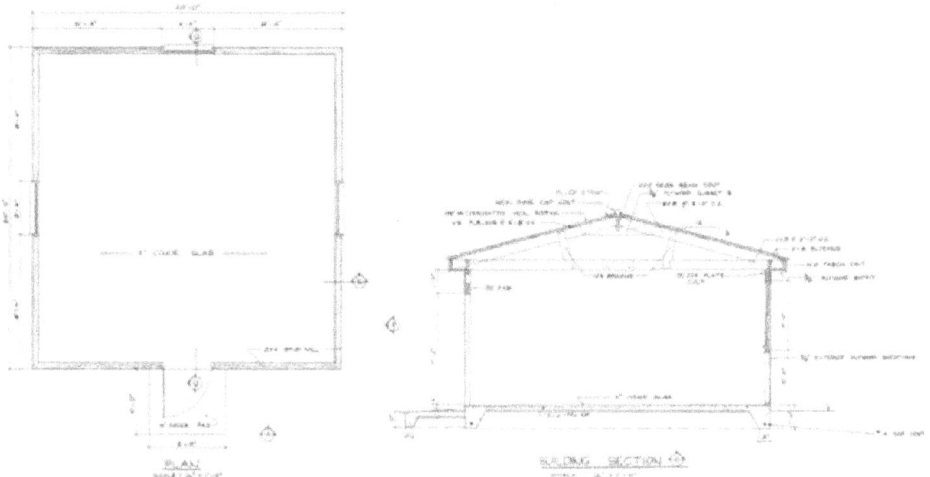

Figure 142. ATC facilities, range building, plan and section, Fort Bragg, NC, 1966 (Standard drawing 28-13-117, sheet 24, Construction of ranges, phase 1, U.S. ATC Facilities, Fort Bragg, NC, range building, 28 June 1966).

Prefabricated metal range building

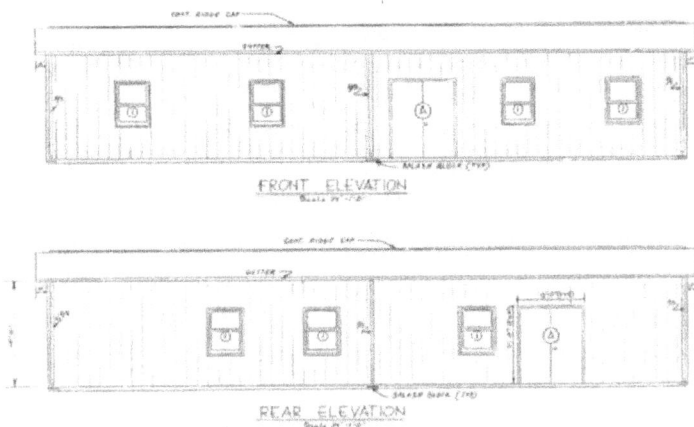

Figure 143. Prefabricated metal range building (range 9), front and rear elevations, Fort Bragg, NC, 1981 (DFE 3629 sheet 3 of 4, Construct metal prefab building [range 9], plan, elevations, and details, 7 May 1981).

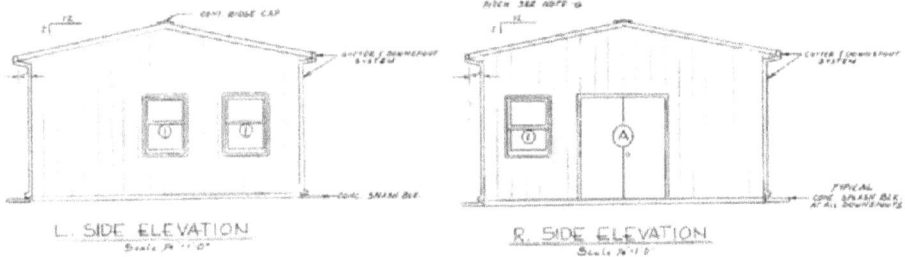

Figure 144. Prefabricated metal range building (range 9), elevations and details, Fort Bragg, NC, 1981 (DFE 3629 sheet 3 of 4, Construct metal prefab building [range 9], plan, elevations, and details, 7 May 1981).

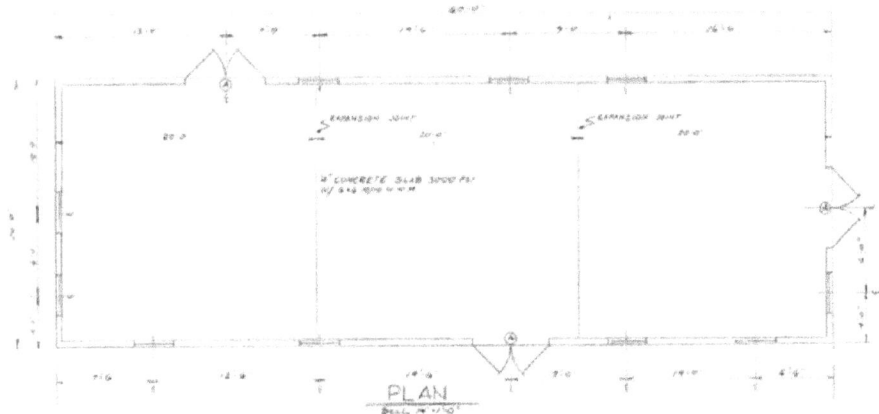

Figure 145. Prefabricated metal range building (range 9), plan, Fort Bragg, NC, 1981 (DFE 3629 sheet 3 of 4, Construct metal prefab building [range 9], plan, elevations, and details, 7 May 1981).

Fixed firing points and fixed targets

1,000-in. fixed target type range

Soldiers were trained on these ranges in basic marksmanship skills including zeroing rifles or machine guns. The range consisted of a firing line with firing points spaced 10 ft apart, a target line 1,000 in. from the firing line with five by 3-ft targets placed 10 ft apart, and an embankment with a sand salvage apron and sod backstop behind the target line (see Figure 147). "When space was limited at an installation, this type of range was used with miniature targets to simulate a known distance range. The range was also used as a landscape range with landscape targets" ("RO-1" 21). This range could also be adjusted for machine gun firing by either moving the target line 500 in. closer to the firing line, or moving the firing line 500 in. closer to the target line. Training was completed with .30 caliber rifles or machine guns.

Layouts

Danger area

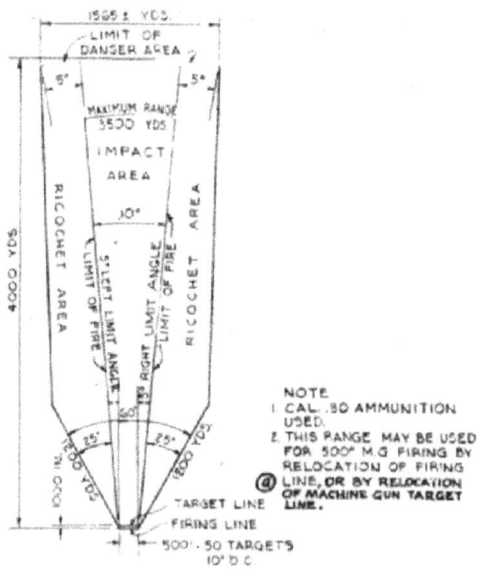

Figure 146. A 1000-in. fixed target type range, danger area, 1952 (Standard drawing No 28-13-07, sheet 1 of 1, Range, 1000-in. fixed target type, 26 February 1952).

Typical layout

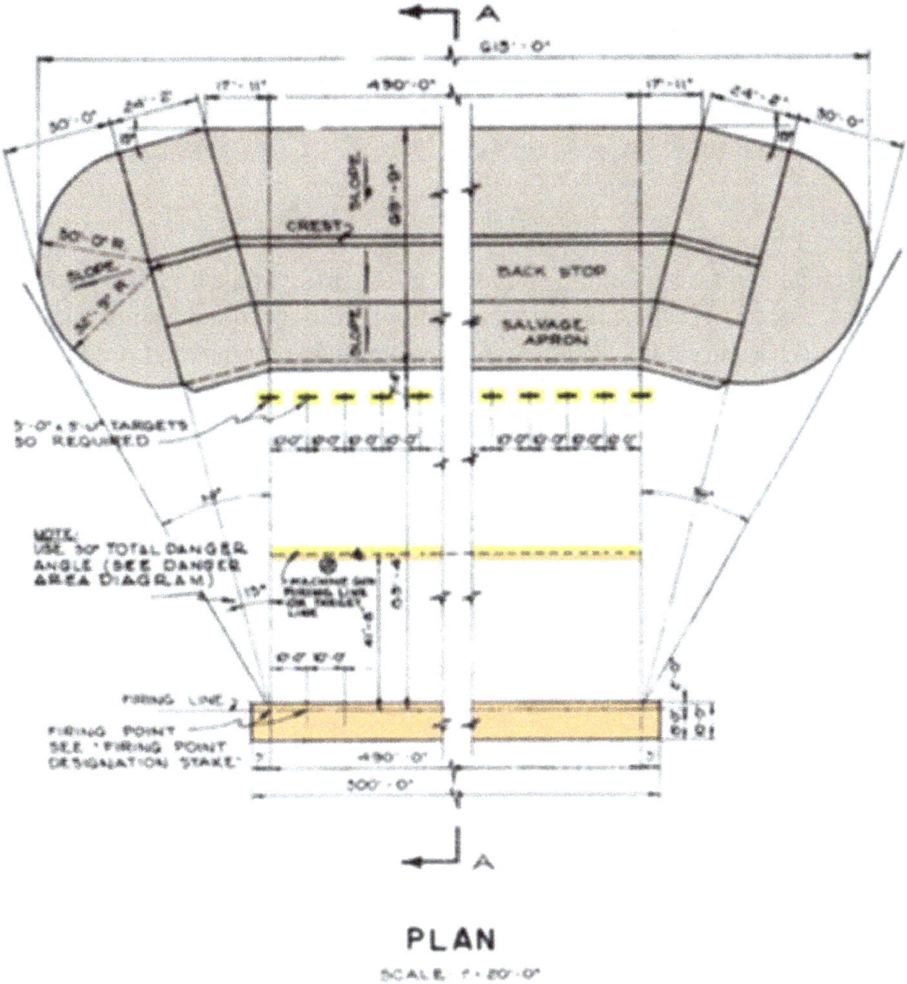

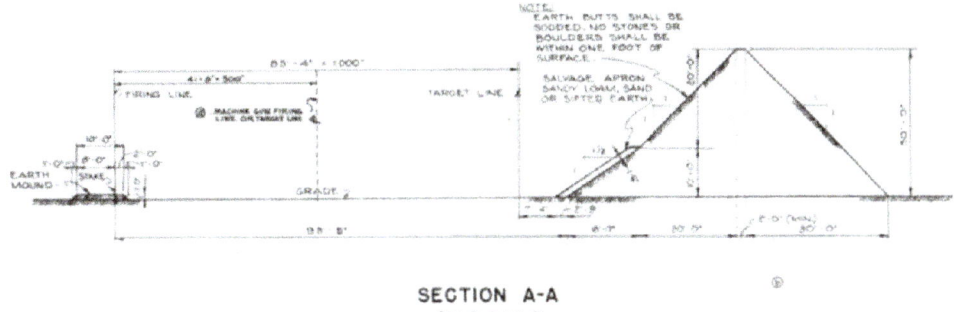

Figure 147. A 1000-in. fixed target type range, plan, 1952 (Standard drawing No 28-13-07 sheet 1 of 1, Range, 1000-in. fixed target type, 26 February 1952).

Figure 148. 1000-in. fixed target type range, section, 1952 (Standard drawing No 28-13-07 sheet 1 of 1, Range, 1000-in. fixed target type, 26 February 1952).

Firing lines

This range consisted of a mounded firing line with firing points spaced 10 ft apart. The range could also be adjusted for machine gun firing by either moving the target line 500 in. closer to the firing line, or moving the firing line 500 in. closer to the target line.

Targets

This range had a target line 1,000 in. from the firing line with five by 3-ft targets placed 10 ft apart. Panel targets mounted on wooden frames were typical (see Figure 245). "When space was limited at an installation, this type of range was used with miniature targets to simulate a known distance range. The range was also used as a landscape range with landscape targets" ("RO-1" 21).

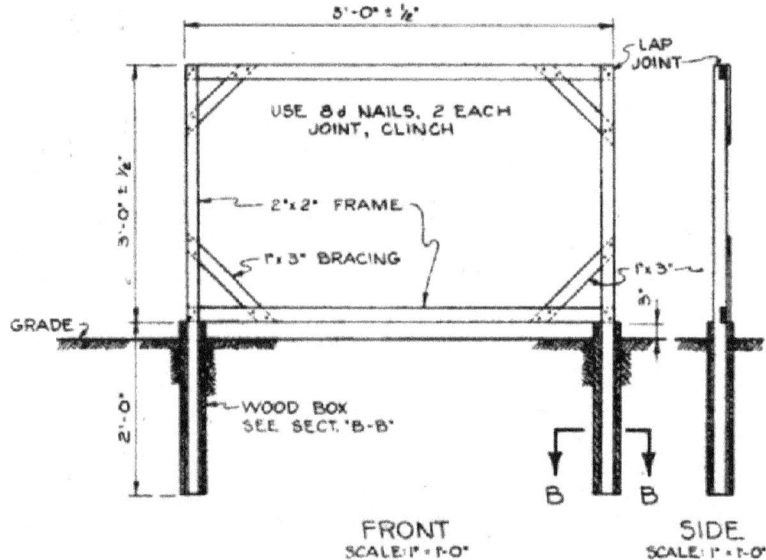

Figure 149. A 1000-in. fixed target type range, target frame, 1952 (Standard drawing No 28-13-07 sheet 1 of 1, Range, 1000-in. fixed target type, 26 February 1952).

Embankments/trenches/etc.

A 30-ft tall and at least 68-ft wide embankment with a sandy apron and sod backstop was constructed behind the target line (see Figure 150).

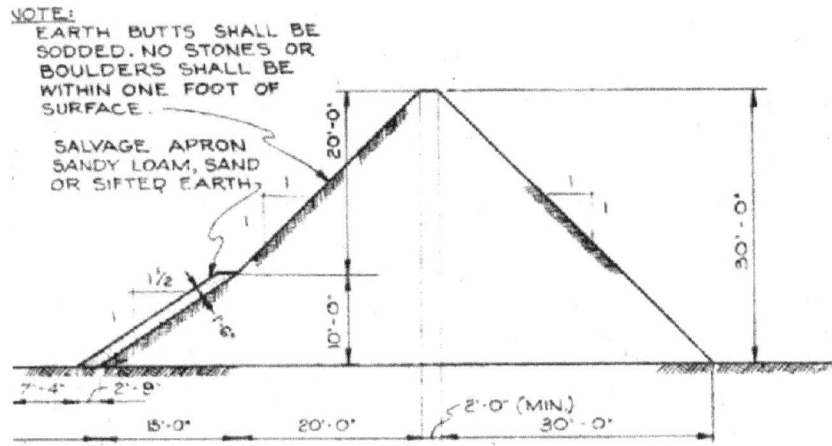

Figure 150. A 1000-in. fixed target type range, section, 1952 (Standard drawing No 28-13-07 sheet 1 of 1, Range, 1000-in. fixed target type, 26 February 1952).

Buildings

No buildings are mentioned in the standard plans for these ranges. However, a range may have had a control tower, latrine, target storage building, ammunition storage building, other storage sheds, and administrative/maintenance buildings supporting general range functions. The range may also have been part of a larger installation range complex that contained these buildings.

Known distance ranges

Soldiers were trained in basic marksmanship on these ranges. "The standard known distance range had firing lines at various distances from a fixed target line, and possibly a continuous embankment behind targets (see Figure 151). A rifleman could become fairly effective against individual objects at moderate distances, probably up to about 600 yds. A small proportion of men could extend their proficiency up to 1000 or 1200 yds. The distance on known distance ranges was reduced over time. At some newer ranges, distances were only 300 meters long" ("RO-1" 19). Training was completed with rifles.

Layouts

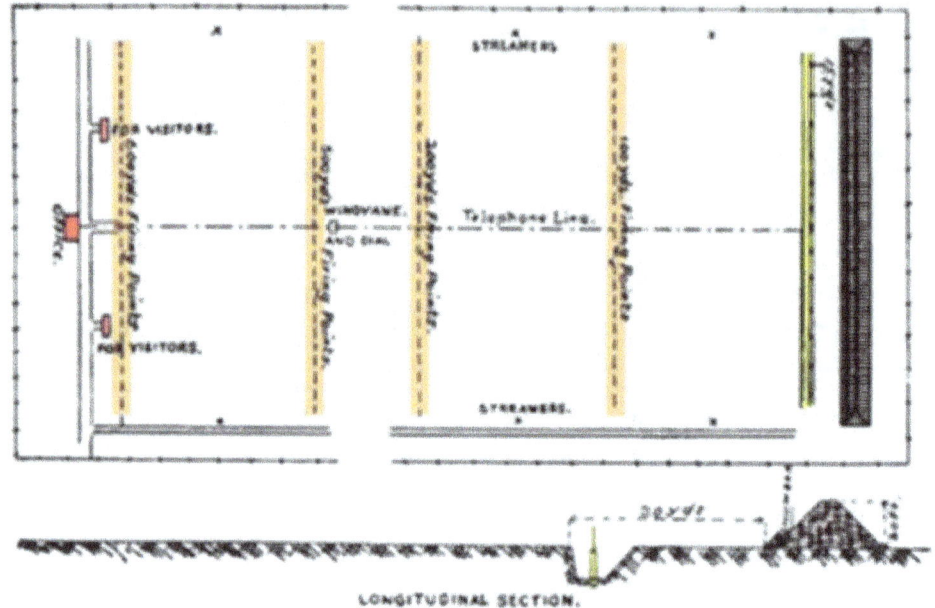

Figure 151. Known distance range plan, circa 1906 (War Department Document No. 261, Small arms firing regulations – 1906, 1 January 1906, p 142).

Figure 152. Castner Range #2, A .30 caliber known distance range at Fort Bliss, TX, 31 August 1953 (NARA College Park, RG 111-SC WWII, box 279, photo SC460727).

Firing lines

The standard known distance range had firing lines at various distances from a fixed target line. The distance of the firing lines on known distance ranges decreased over time.

Targets

Targets on known distance likely included stationary panel and E- and F-type silhouette targets made of wooden frames and cloth, metal, plastic, or other materials.

Embankments/trenches/etc.

An embankment was typically constructed behind the target line in order to stop fired projectiles. This was often a dirt mound reinforced with wooden beams or concrete, and may have included some area for storage (see rifle ranges). Embankments may also have been built along the left and rights sides of ranges (see Figure 152).

Buildings

A range office building and bleachers for visitors are mentioned in the standard plan for these ranges. However, a range may have had a control tower, latrine, target storage building, ammunition storage building, other storage sheds, and administrative or maintenance buildings supporting general range functions. The range may also have been part of a larger installation range complex that contained these buildings.

Machine gun squares

Soldiers were trained on these ranges in the basics of machine gun use. The range consisted of an instructor's platform in the center of a large square. Target lines extended out from the instructors platform, dividing the square into four grass quadrants. Firing positions were placed around the perimeter of the square, and the entire square was surrounded by a 10-ft wide path made of sand (see Figure 153). After receiving instruction, soldiers practiced dry firing or "snapping in" on targets. Training was completed with unloaded machine guns.

Layouts

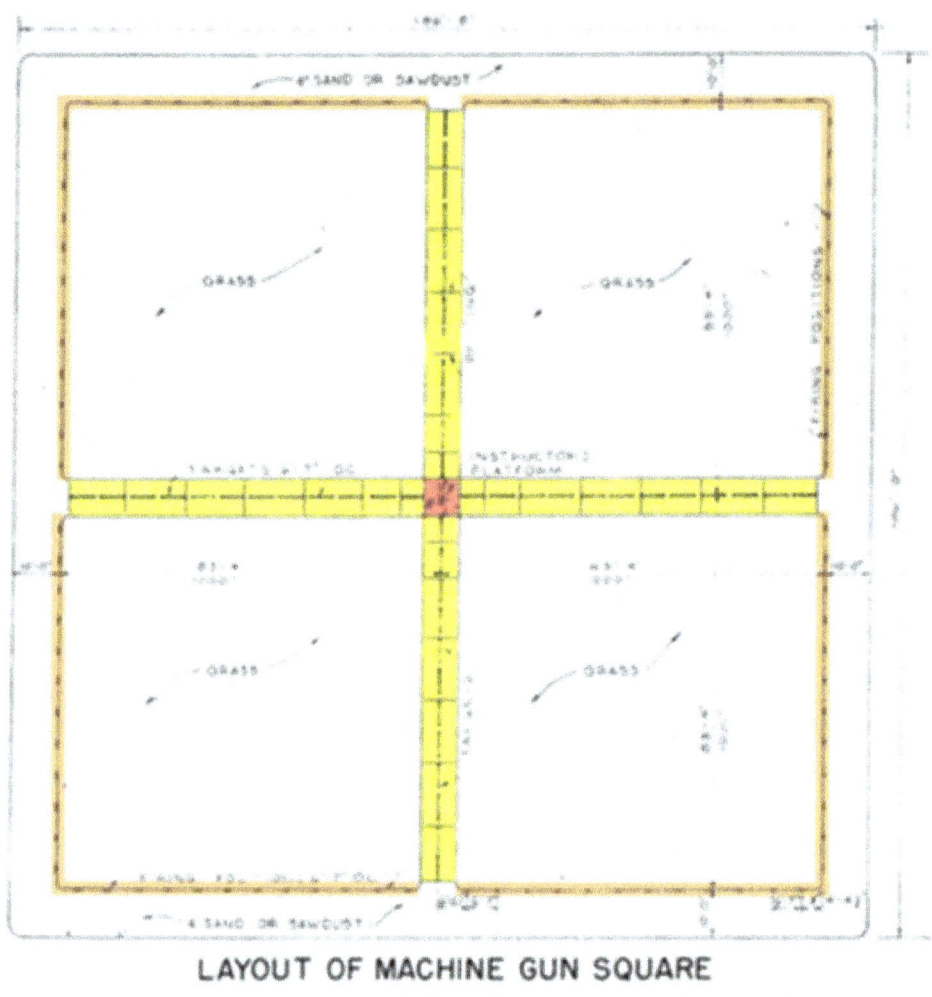

Figure 153. Machine gun squares, layout, Fort Bragg, NC, 1951 (Standard drawing No. 28-13-27 sheet 1 of 1, Machine gun squares, plan and details, 21 November 1951).

Firing lines

Firing points were placed every 6 ft, 7 in. on center along the outside perimeters of the four grass quadrants of the square.

Targets

Target lines extended out in four directions from the instructor's platform, and included rectangular panel targets mounted in a continuous wooden frame (see Figures 154 and 155).

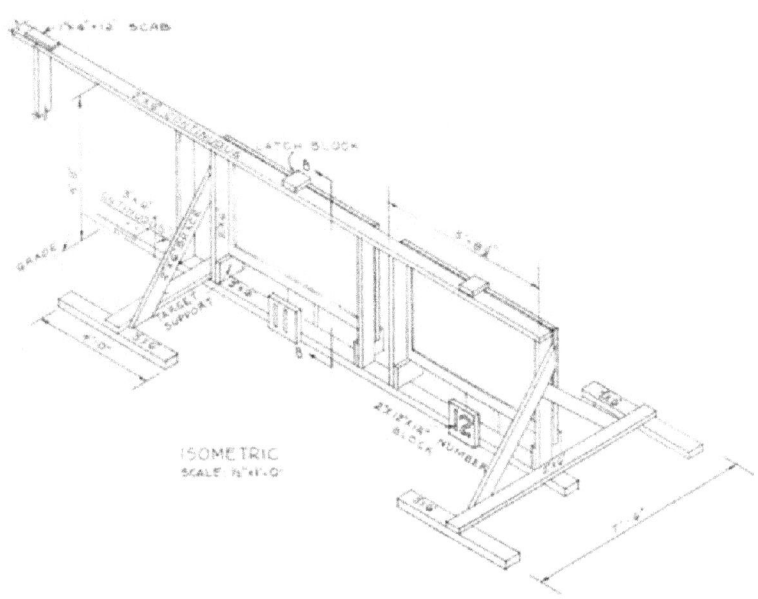

Figure 154. Machine gun squares, target frame, Fort Bragg, NC, 1951 (Standard drawing No. 28-13-27 sheet 1 of 1, Machine gun squares, plan and details, 21 November 1951).

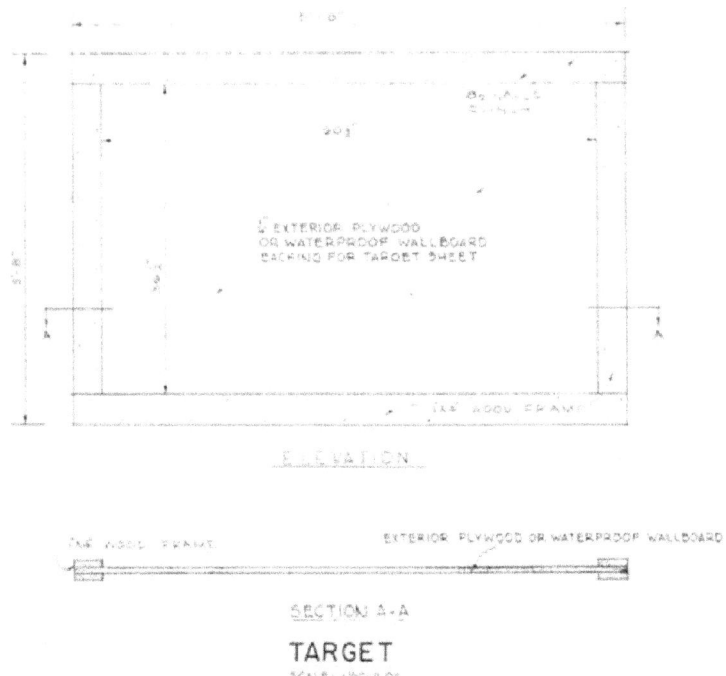

Figure 155. Machine gun squares, target, Fort Bragg, NC, 1951 (Standard drawing No. 28-13-27 sheet 1 of 1, Machine gun squares, plan and details, 21 November 1951).

Embankments/trenches/etc.

None.

Buildings

No buildings other than an instructor's platform (Figure 156) are mentioned in the standard plans for these ranges. However, a range may have had a latrine, target storage building, other storage sheds, and administrative/maintenance buildings supporting general range functions. The range may also have been part of a larger installation range complex that contained these buildings.

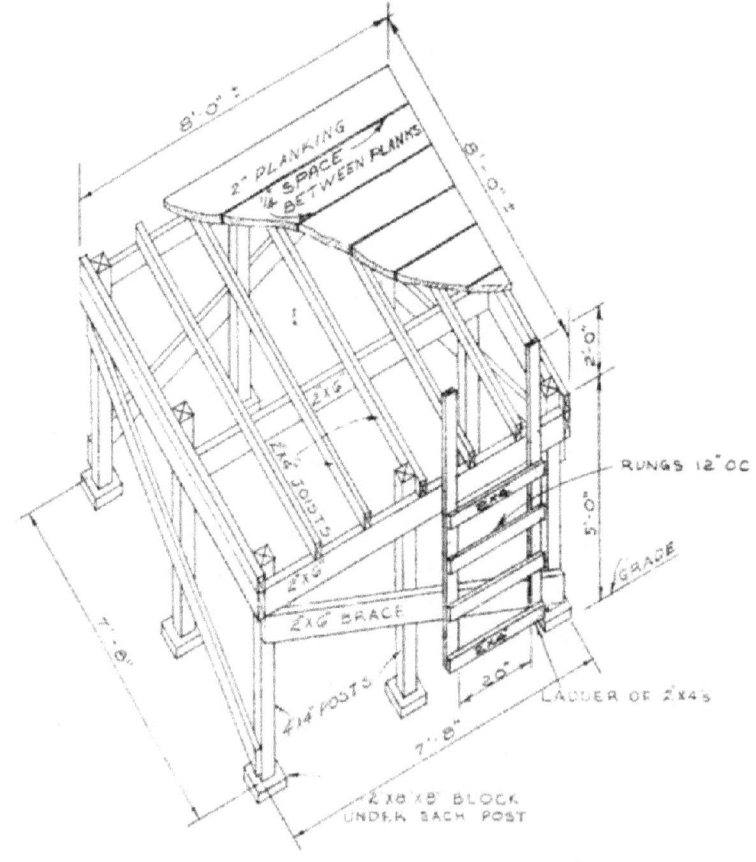

Figure 156. Machine gun squares, instructors platform, Fort Bragg, NC, 1951 (Standard drawing No. 28-13-27 sheet 1 of 1, Machine gun squares, plan and details, 21 November 1951).

Pistol ranges

"The pistol range was a variation of the known distance range modified for pistols" ("RO-1" 20). Soldiers trained on these ranges to fire pistols from various distances at stationary targets. The range consisted of firing lines set at 5, 10, 15, and 25 ft from a target line, stationary and bobbing targets, and embankments built behind target lines (see Figure 158). Training was completed with .22 and .45 caliber pistols.

Layouts

<u>Danger area</u>

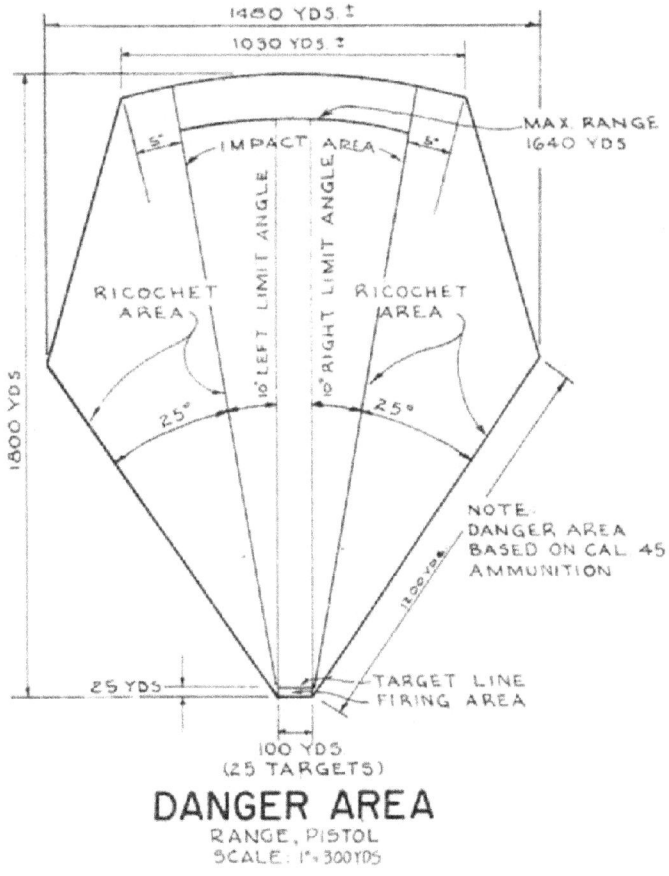

Figure 157. Pistol range danger area safety fan, Fort Bragg, NC, 1951 (Standard drawing No. 28-13-12 drawing 1 of 1, Range, pistol, landscape targets, plans and details, 21 November 1951).

Typical layout

Figure 158. Pistol range plan, Fort Bragg, NC, 1951 (Standard drawing No. 28-13-12 drawing 1 of 1, Range, pistol, landscape targets, plans and details, 21 November 1951).

Figure 159. CMTC students on the pistol range at Camp Vail, NJ, 1923 (NARA College Park, RG 111-SC WWI, box 700, photo 94912).

Figure 160. Perhaps the most important day in the recruit's boot training is qualification day, he will not only fire the rifle for an official score to be entered in his record book, he will also assist in operating some part of the firing line like phoning to the butts or keeping score at Camp Matthews, MCRD San Diego, CA, 6 June 1955 (NARA College Park, RG 127-GC, box 5, photo A227310).

Figure 161. High angle view of the electric pistol range at MCRD Parris Island, SC, 10 February 1958 (NARA College Park, RG 127-GC, box 34, photo A600627).

Firing lines

Firing lines were set at 5, 10, 15 and 25 yds away from target lines. Firing points were placed every 12 ft on center along the firing line (Figure 158). Firing lines often included a small stand or bench at the firing position (Figures 164, 168, 169), and some firing lines were covered.

Figure 162. CMTC students during pistol practice at Fort Sam Houston, TX, undated (NARA College Park, RG 111-SC WWI, box 700, photo 94878).

Figure 163. A .45 caliber pistol firing line at Fort Jackson, SC, 23 April 1943 (NARA College Park, RG 111-SC WWII, box 155, photo SC173966).

Figure 164. Firing position on pistol range during training at Amphibious Training Base Fort Pierce, FL, 23 November 1943 (NARA College Park, RG 80-G, box 861, photo 264381).

Figure 165. Pistol firing range at NATC Corpus Christi, TX, November 1942 (NARA College Park, RG 80-G, box 1694, photo 417646).

Figure 166. Pistol firing line on the 50-yd range at Finger Bay, Adak, AK, 23 June 1945 (NARA College Park, RG 80-G, box 1285, photo 342166).

Figure 167. Pistol Firing Range at San Antonio Aviation Cadet Center, TX, 9 April 1943 (NARA College Park, RG 342-FH, box 2207, photo 4A-18406).

Figure 168. Platoon 23 firing the .22 caliber pistol on L Range at Camp Matthews, MCRD San Diego, CA, 11 February 1952 (NARA College Park, RG 127-GC, box 5, photo A219142).

Figure 169. Platoon 381 on the pistol range at Camp Matthews, MCRD San Diego, CA, 29 November 1951 (NARA College Park, RG 127-GC, box 5, photo A77625).

Targets

Pistol ranges typically used stationary E- and F-type silhouette targets, L-type bobbing targets (Figure 171), and panel targets mounted on rectangular frames (Figure 173).

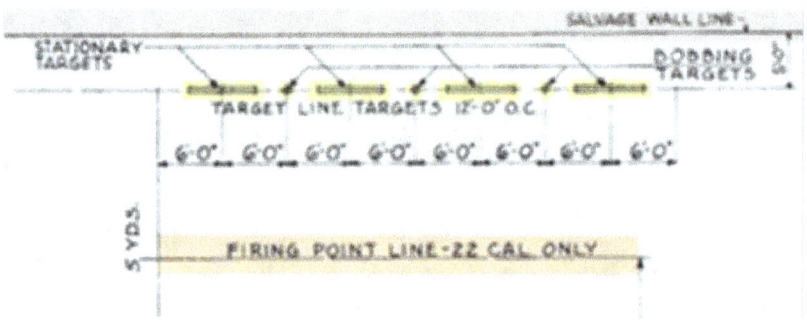

Figure 170. Pistol range plan, Fort Bragg, NC, 1951 (Standard drawing No. 28-13-12 drawing 1 of 1, Range, pistol, landscape targets, plans and details, 21 November 1951).

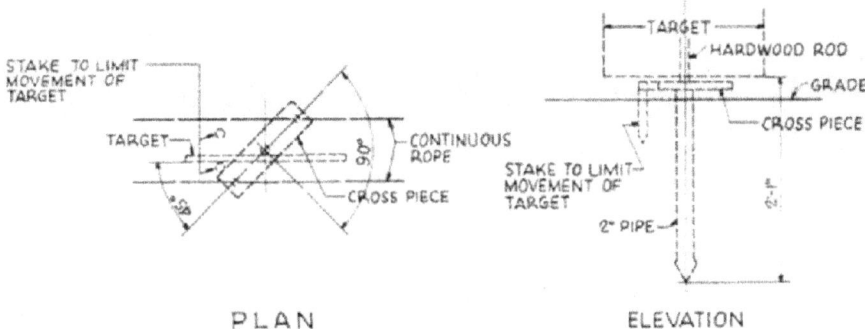

Figure 171. Pistol range details for bobbing targets, Fort Bragg, NC, 1951 (Standard drawing No. 28-13-12 drawing 1 of 1, Range, pistol, landscape targets, plans and details, 21 November 1951).

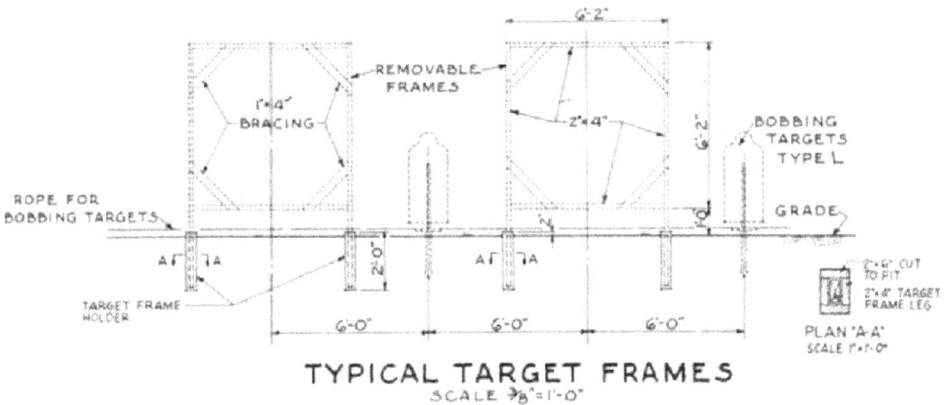

Figure 172. Pistol range typical target frames, Fort Bragg, NC, 1951 (Standard drawing No. 28-13-12 drawing 1 of 1, Range, pistol, landscape targets, plans and details, 21 November 1951).

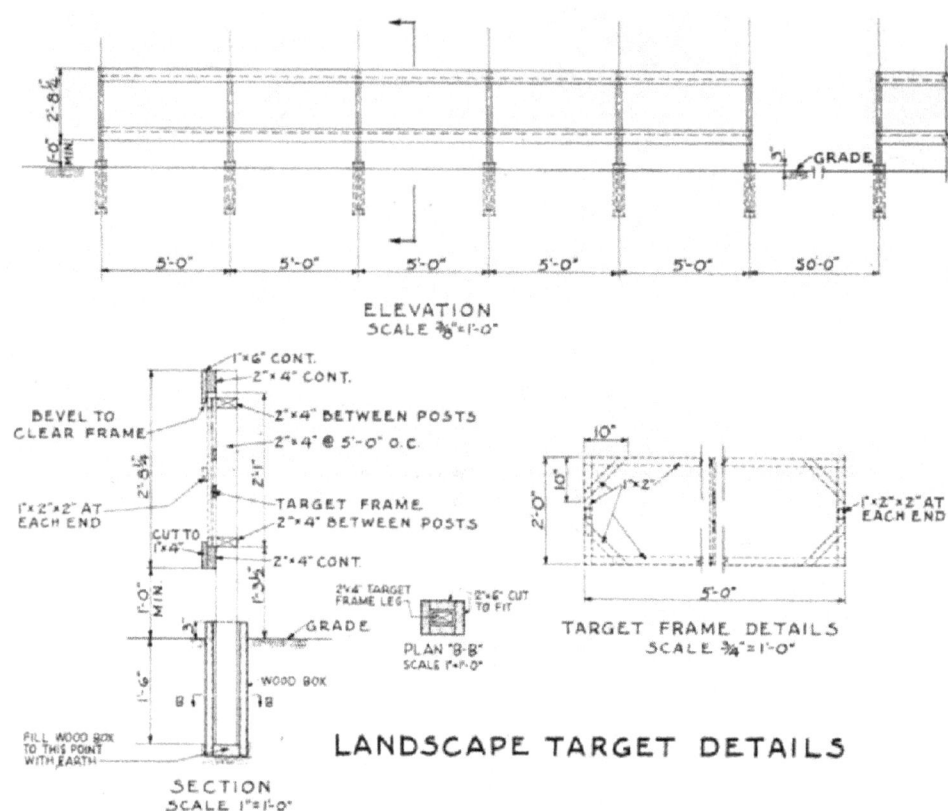

Figure 173. Pistol range landscape target details, Fort Bragg, NC, 1951 (Standard drawing No. 28-13-12 drawing 1 of 1, Range, pistol, landscape targets, plans and details, 21 November 1951).

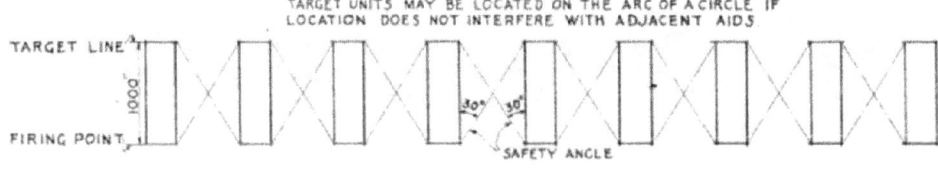

Figure 174. Pistol range landscape targets plan, Fort Bragg, NC, 1951 (Standard drawing No. 28-13-12 drawing 1 of 1, Range, pistol, landscape targets, plans and details, 21 November 1951).

Figure 175. Official scores and markers inspecting the targets on the pistol range at Camp Matthews, MCRD San Diego, CA, March 1949 (NARA College Park, RG 127-GC, box 5, photo A224845).

Embankments/trenches/etc.

Embankments were often built behind the target line (Figure 176), and the backside of the embankments often included space for target storage (Figure 177).

Figure 176. Small arms target practice for cadets at NAS Corpus Christi, TX, 23 July 1941 (NARA College Park, RG 80-G, box 1978, photo 463689).

Figure 177. Range at NAS Deland, FL, 23 December 1944 (NARA College Park, RG 80-G, box 1382, photo 363872).

Buildings

No buildings are mentioned in the standard plan for these ranges. However, a range may have had a control tower, latrine, target storage building, ammunition storage building, other storage sheds, and administrative/maintenance buildings supporting general range functions. The range may also have been part of a larger installation range complex that contained these buildings.

Figure 178. Each recruit must be proficient with basic infantry weapons, here familiarity with the .22 caliber pistol at Camp Matthews, MCRD San Diego, CA, 6 June 1955 (NARA College Park, RG 127-GC, box 5, photo A227316).

Preliminary rifle instruction circle

New recruits were trained on these ranges in the fundamentals of rifle marksmanship. The range consisted of a central circle of sustained fire two men targets with an instructor's platform and bleachers inside, a second circle of numbered firing positions facing inward, a third circle of numbered firing positions facing outward, and an outer circle lined with sustained fire four men targets. The surface of the range was often grass and the circles may have been lined with sandbags (see Figure 179). A simpler version included a field of grass with a target in the center of the circle around which soldiers sat in concentric circles (see Figure 180). Soldiers received instruction as they practiced firing positions and "snapping in" or dry firing at the targets. Training was completed with unloaded rifles.

Layouts

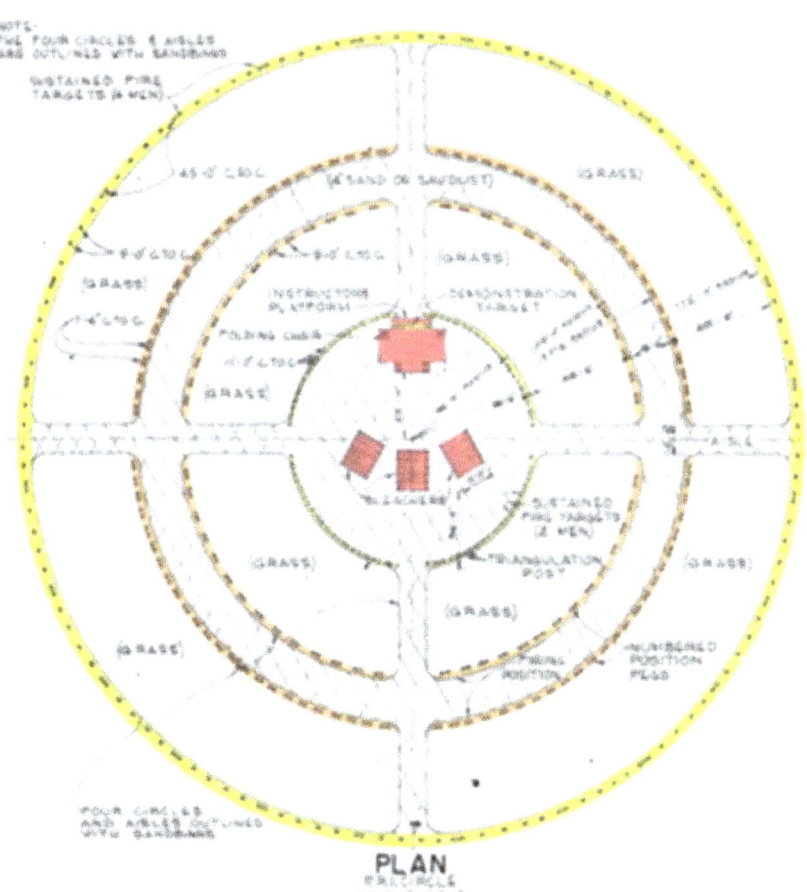

Figure 179. Preliminary rifle instruction circle, plan, Fort Bragg, NC, 1951 (Standard drawing 28-13-28 sheet 1 of 1, Preliminary rifle instruction circle, plan and details, 7 December 1951).

Figure 180. A platoon of recruits "snap in" during their first week at the rifle range at MCRD Parris Island, SC, 1967 (NARA College Park, RG 127-GG-921, box 33, photo A601744).

Firing lines

Two interior concentric circles were lined with numbered firing positions (see Figure 181).

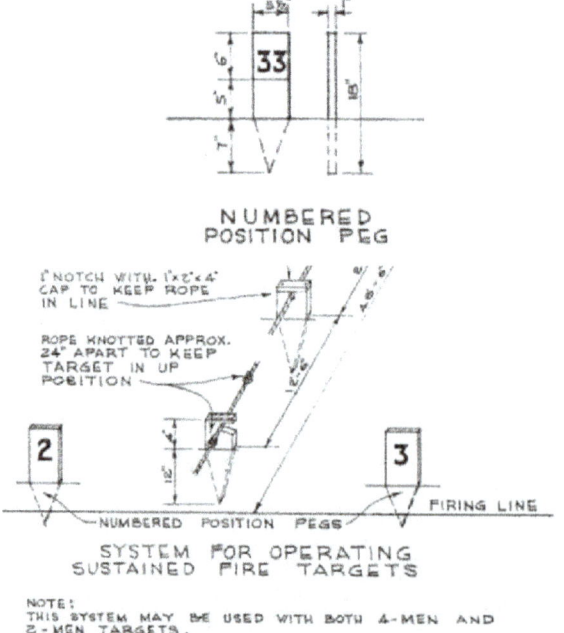

Figure 181. Preliminary rifle instruction circle, numbered firing position pegs, Fort Bragg, NC, 1951 (Standard drawing 28-13-28 sheet 1 of 1, Preliminary rifle instruction circle, plan and details, 7 December 1951).

Targets

Sustained fire two and four men targets were used on these ranges (see Figures 182 and 183).

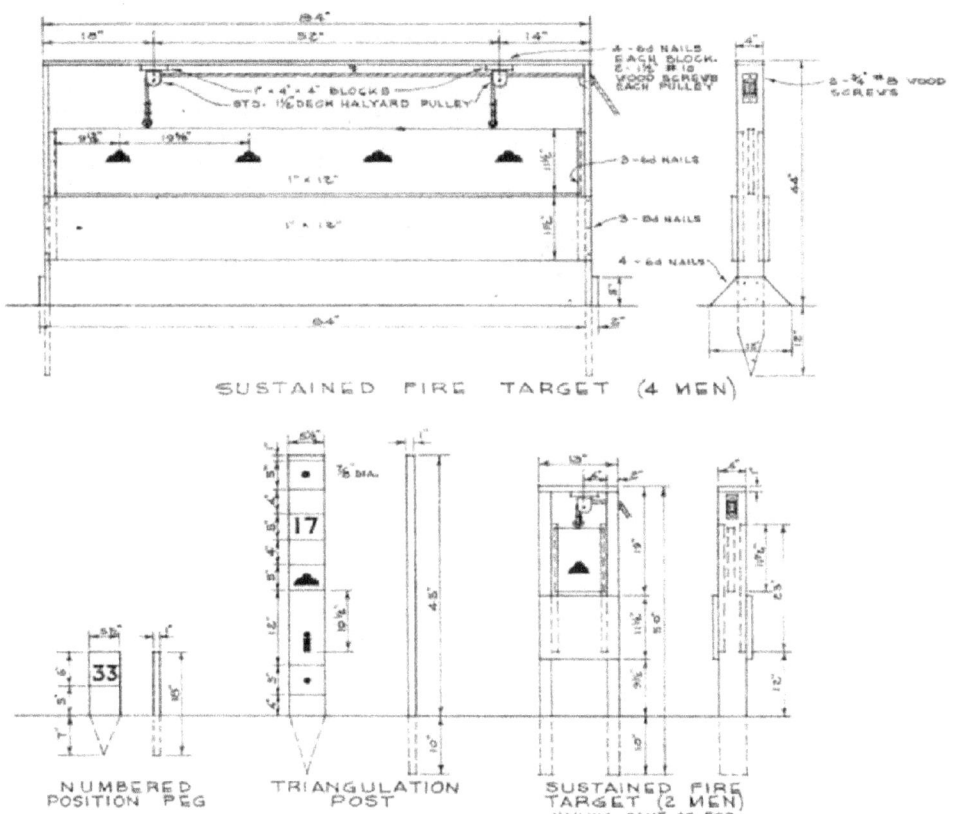

Figure 182. Preliminary rifle instruction circle, target details, Fort Bragg, NC, 1951 (Standard drawing 28-13-28 sheet 1 of 1, Preliminary rifle instruction circle, plan and details, 7 December 1951).

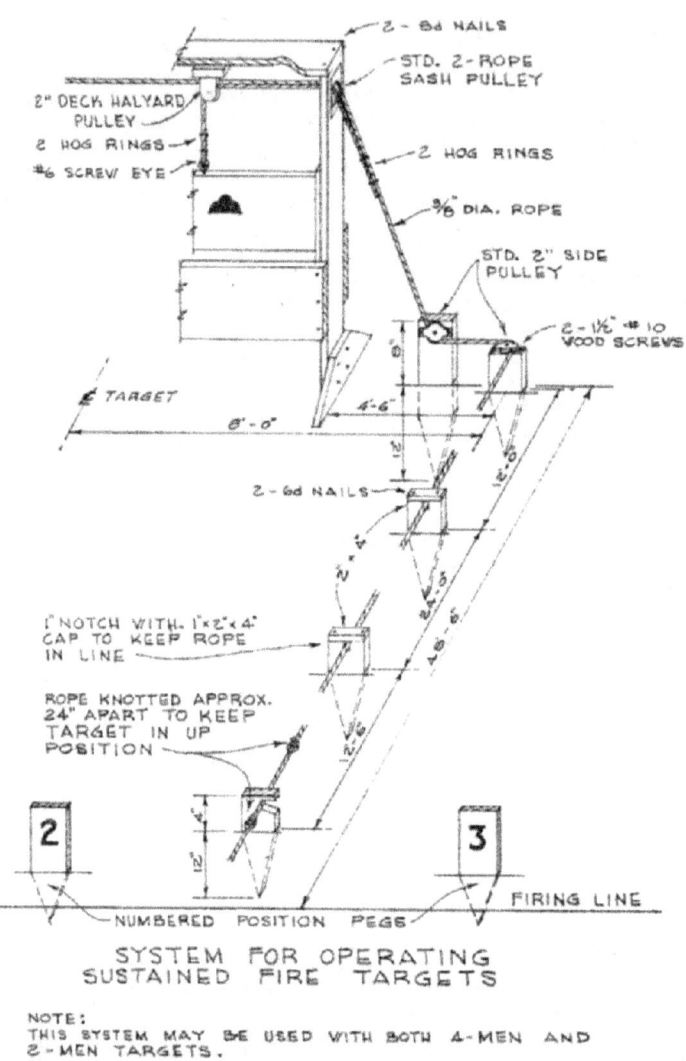

Figure 183. Preliminary rifle instruction circle, system for operating sustained fire targets, Fort Bragg, NC, 1951 (Standard drawing 28-13-28 sheet 1 of 1, Preliminary rifle instruction circle, plan and details, 7 December 1951).

Embankments/trenches/etc.

None.

Buildings

An instructor's platform (Figure 184) and bleachers (Figures 185 and 186) are mentioned in the standard plans for these ranges. However, a range may also have had a latrine, target storage building, other storage sheds, and administrative/maintenance buildings supporting general range functions. The range may also have been part of a larger installation range complex that contained these buildings.

Instructor's platform

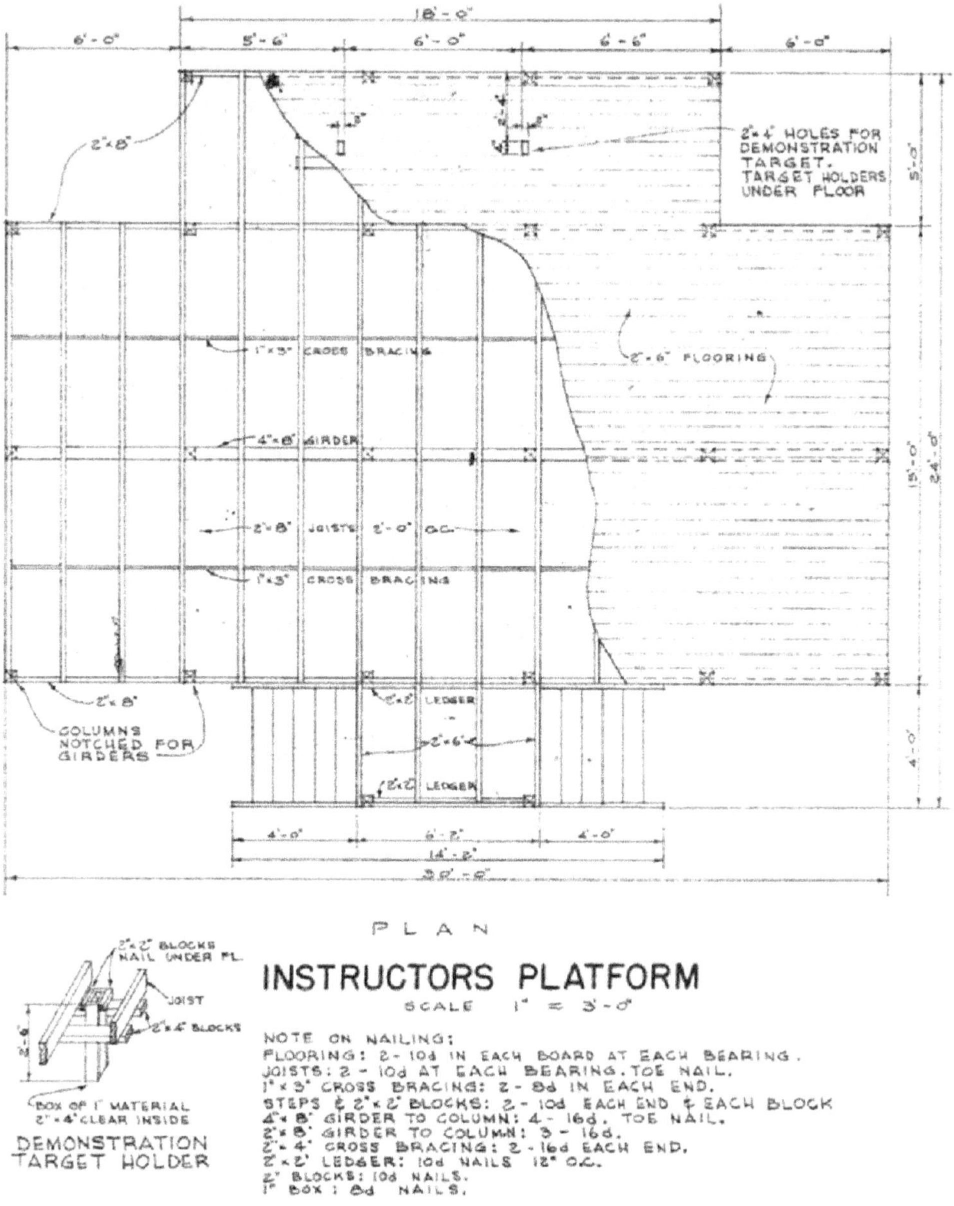

Figure 184. Preliminary rifle instruction circle, instructors platform plan, Fort Bragg, NC, 1951 (Standard drawing 28-13-28 sheet 1 of 1, Preliminary rifle instruction circle, plan and details, 7 December 1951).

Bleachers

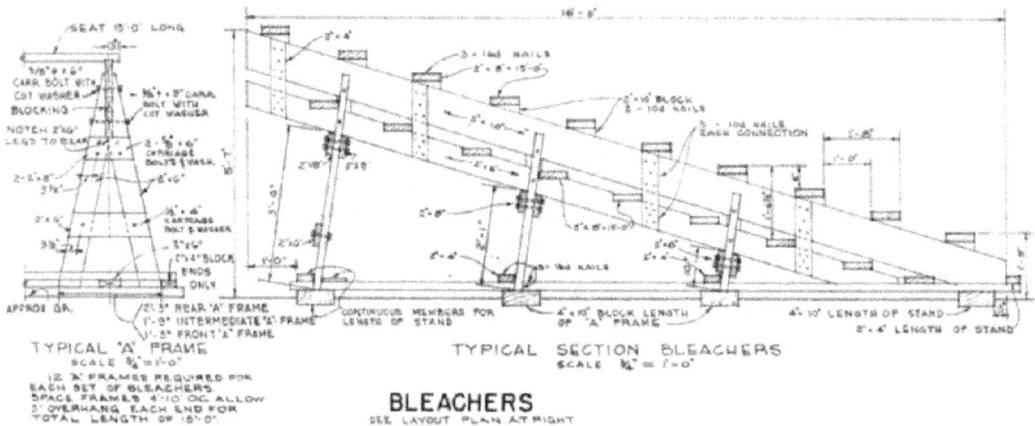

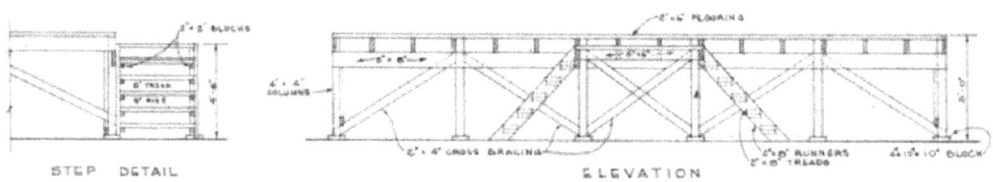

Figure 185. Preliminary rifle instruction circle, bleachers, Fort Bragg, NC, 1951 (Standard drawing 28-13-28 sheet 1 of 1, Preliminary rifle instruction circle, plan and details, 7 December 1951).

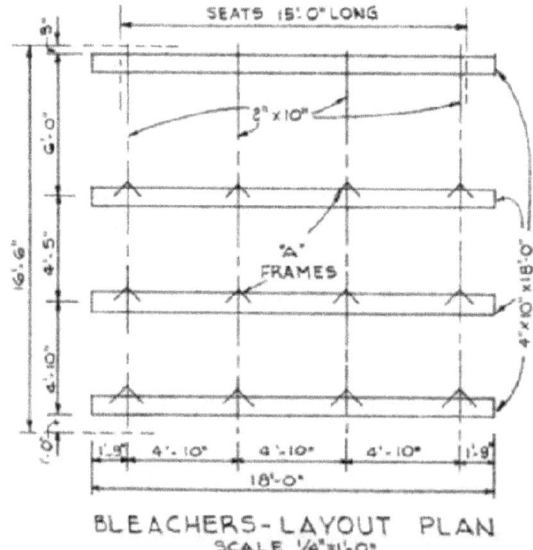

Figure 186. Preliminary rifle instruction circle, bleachers layout plan, Fort Bragg, NC, 1951 (Standard drawing 28-13-28 sheet 1 of 1, Preliminary rifle instruction circle, plan and details, 7 December 1951).

Rifle ranges

Soldiers were trained on these ranges in basic marksmanship to fire rifles at fixed targets. The range consisted of firing lines set at 300, 600, 900, and 1500 yds from target lines (sometimes covered or built up with an embankment), stationary panel targets, pull-up panel targets with target pits, or pop-up silhouette targets, and embankments built behind targets, as part of target butts or in between ranges (see Figure 188). Training was completed with .30 caliber rifles.

Layouts

<u>Danger area</u>

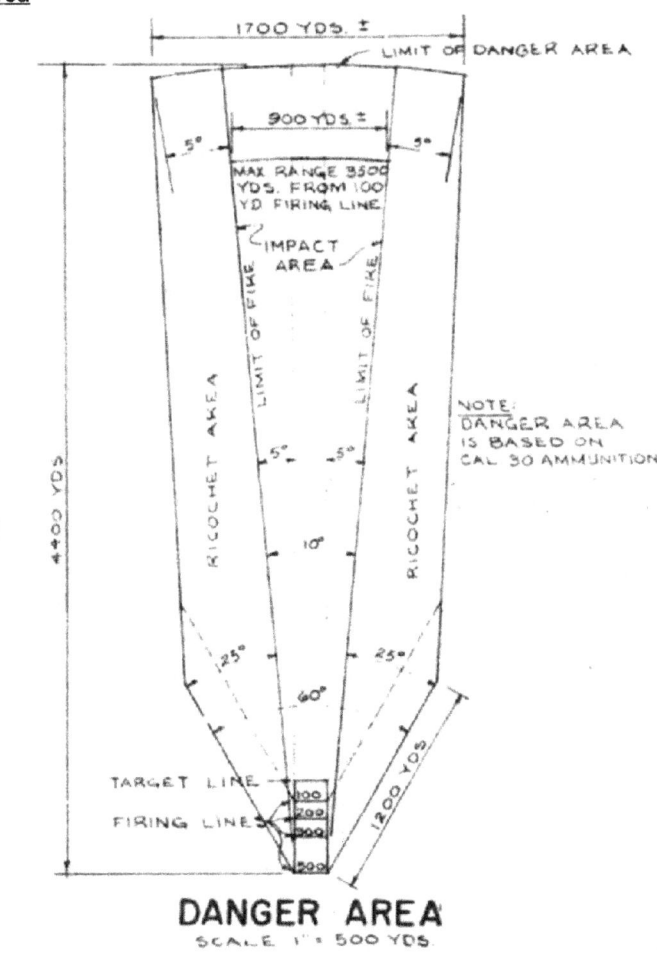

Figure 187. Typical rifle range danger area safety fan, Fort Bragg, NC, 1952 (Standard drawing No. 28-13-09 drawing 1 of 7, Range, rifle, known distance, plans and details, 5 January 1952).

Typical layout

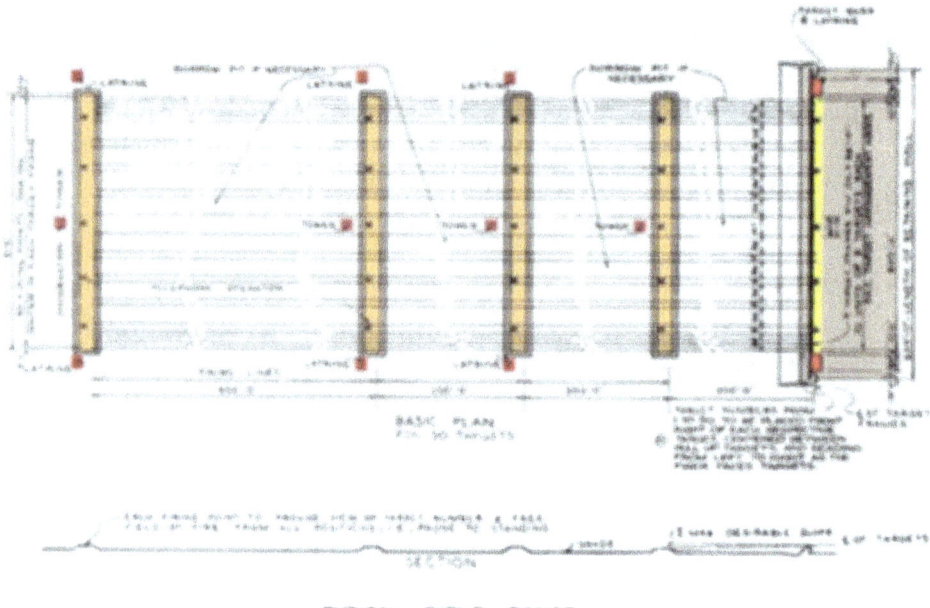

Figure 188. Typical rifle range basic plan and section, Fort Bragg, NC, 1952 (Standard drawing No. 28-13-09 drawing 1 of 7, Range, rifle, known distance, plans and details, 5 January 1952).

Figure 189. Winners of the Marine Corps rifle match at USMC Winthrop Rifle Range, Indian Head, MD, 1913 (NARA College Park, RG 127-G-100b, box 23, photo 516689).

"A prototype for a modern Marine Corps installation was the rifle range at Winthrop, MD. Although the range was located at the Navy's Indian Head Proving Ground across from Mattawoman Creek, the Marines referred to it as Winthrop. They stationed a small number of Marines at the site, with appropriate facilities. It was used to train Marines stationed on the East Coast from Philadelphia to Charleston. This rifle range (Figure 190) was one more indication that the Marines were acquiring their own real estate as training areas" (Goodwin 69).

Figure 190. View of Winthrop Rifle Range, Indian Head, MD, 1915 (NARA College Park, RG 127-G-100b, box 23, photo 521532).

Figure 191. Rifle matches at Camp Curtis Guild, Wakefield, MA, August 1927 (NARA College Park, RG 127-G-100c, box 23, photo 524720).

Figure 192. Rifle team at Camp Curtis Guild, Wakefield, MA, 1929 (NARA College Park, RG 127-G-100c, box 23, photo 527257).

Figure 193. Rifle range and pistol range at Finger Bay, Adak, AK, 23 June 1945 (NARA College Park, RG 80-G, box 1285, photo 342161).

Figure 194. Typical "boot" camp scene at MCRD San Diego, CA, June 1946 (NARA College Park, RG 127-GC, box 5, photo 400734).

Figure 195. Rifle range "A" at Range Area looking toward the butts and New River with personnel firing on the 200-yd line at Camp Lejeune, NC, July 1957 (NARA College Park, RG 127-GC, box 22, photo 340630).

Figure 196. Rifle range "B" at Range Area looking toward the butts and New River with personnel firing on the 200-yd line at Camp Lejeune, NC, July 1957 (NARA College Park, RG 127-GC, box 22, photo 340620).

Figure 197. One of the newer techniques in recruit rifle marksmanship training is the 900-in. firing line that is designed to familiarize the recruit with the M-14 rifle and to give him the proper windage and elevation for the 200-yd line at MCRD Parris Island, SC, 1967 (NARA College Park, RG 127-GG-921, box 33, photo A601726).

Firing lines

Firing lines were set at various distances from target lines including 300, 600, 900, and 1500 yds. Each firing point was to provide a view of target numbers and free field of fire from all positions prone to standing (see Figure 188). Behind firing lines may have been a desk with a telephone operator (one for every ten targets), an ammunition table, and rifle resting racks (Figure 198). Some firing lines were covered (Figure 204). Embankments may have been built to stabilize firing lines (Figure 205), to give soldiers a clear view of the entire range (see Figure 189), or to support individual firing positions (Figure 206). Several devices were used to train soldiers how to properly site and aim their rifles (see Figures 212 through 217).

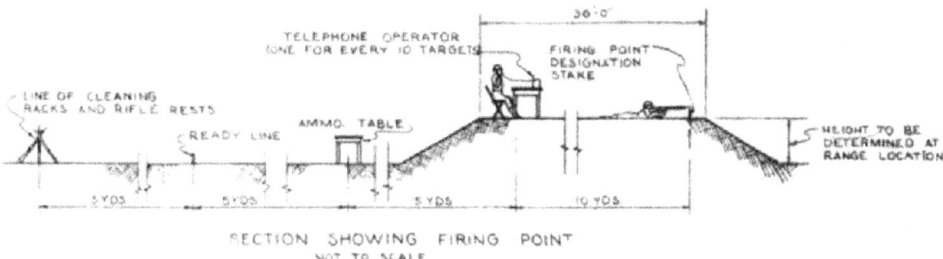

Figure 198. Typical rifle range section showing firing point, Fort Bragg, NC, 1952 (Standard drawing No. 28-13-09 drawing 1 of 7, Range, rifle, known distance, plans and details, 5 January 1952).

Figure 199. Firing line at Camp Meade, MD, 1918 (New York Public Library, digital No. 117098).

Figure 200. Dr. Samuel Scott served as private-gunnery sergeant at the USMC Winthrop Rifle Range, Indian Head, MD, 1915 (NARA College Park, RG 127-G-100b, box 23, photo 523539).

Figure 201. Rifle range at Camp Curtis Guild, Wakefield, MA, 1917, http://is.noblenet.org/images/wak/camp_curtis_wakefield_rifle_range.jpg.

Figure 202. Major J.C. Smith, USMC rifle team captain at Camp Curtis Guild, Wakefield, MA, 1929 (NARA College Park, RG 127-G, box 32, photo 527261).

Figure 203. Firing a Springfield 03 rifle at Camp Perry, Port Clinton, OH, 1939 (NARA College Park, RG 127-G-126N, box 32, photo 521798).

Figure 204. Rifle range firing line at Finger Bay, Adak, AK, 23 June 1945 (NARA College Park, RG 80-G, box 1285, photo 342163).

Figure 205. Marines on firing line at unknown location, undated (NARA College Park, RG 127-GC, box 5, photo 227310).

Figure 206. Training of Marines on the firing line at NAS Jacksonville, 20 May 1942 (NARA College Park, RG 80-G, box 283, photo 64650).

Figure 207. Looking down the firing line at Camp Lejeune, NC, 18 May 1949 (NARA College Park, RG 127-GC, box 22, photo 506751).

Figure 208. Recruits in the sitting position fire from the 200-yd line at MCRD Parris Island, SC, 9 February 1961 (NARA College Park, RG 127-GG-925, photo 601432).

Figure 209. The control tower NCO tells the shooters how much time they have to finish their string of slow fire at the 300-yd line at MCRD Parris Island, SC, 1967 (NARA College Park, RG 127-GG-921, box 33, photo A601829).

The loudspeaker system can be heard above the noise of firing, but coaches and PMI's make sure their shooters know how much time is left. The shooters have 12 minutes to fire their 10 rounds of slow fire at the 300-yd line, and 50 seconds for their 10 rounds of rapid fire. The tower NCO gives two and three minute warnings (see Figure 209).

Figure 210. A rifle range coach assists a recruit in marking a sighting change on the 500-yd range at MCRD Parris Island, SC, 1967 (NARA College Park, RG 127-GG-921, box 33, photo A601729).

Figure 211. Trainees participating in the M-16 rifle qualification on Range #12 at Fort Ord, CA, 26 September 1969 (NARA College Park, RG 111-CRB box 86, photo SC67552).

Aiming support devices

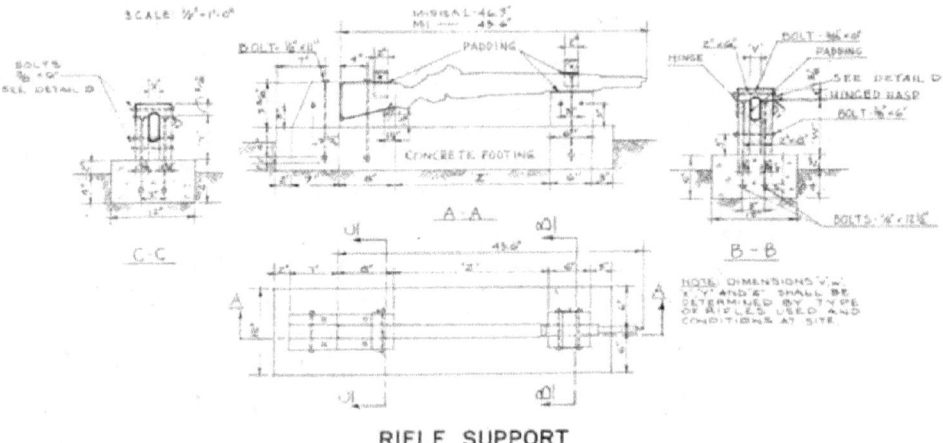

Figure 212. Small arms fire course, rifle support, Fort Bragg, NC, 1952 (Standard drawing 28-13-37 sheet 2 of 2, Small arms fire course, details, 20 June 1952).

Figure 213. Platoon 23 at an exercise on the sighting bars, a mirror is at the end of the bar and the recruits obtain a sight picture by moving the target with his right hand while sighting at the reflection in the mirror at MCRD San Diego, CA, 6 February 1952 (NARA College Park, RG 127-GC, box 5, photo A219137).

Figure 214. Every person who fires the rifle in the Marine Corps gets practice at the sighting and aiming bar which determines if a man is getting the correct sight picture at MCRD Parris Island, SC, 10 February 1958 (NARA College Park, RG 127-GC, box 34, photo A600628).

Figure 215. Recruits are given instruction in the use of aiming devices at MCRD Parris Island, SC, 9 February 1961 (NARA College Park, RG 127-GG-925, photo 610427).

Figure 216. Recruit using aiming device at MCRD Parris Island, SC, 1970 (NARA College Park, RG 127-GG-919, box 32, photo A601880).

Figure 217. A Marine recruit utilizes the M-16 rifle equipped with laser marksmanship rifle trainer device at MCRD Parris Island, SC, 1 March 1977 (NARA College Park, RG 127-GG-913, box 32, photo A602764).

Targets

Rifle ranges typically had panel targets that were stationary or pulled up from a trench underneath the target. Ranges may have contained ten units of five targets each, with target frames 9 ft on center for 36 ft (see Figure 218). Target numbers from one to fifty were to be placed from the right of each perspective target, centered between pull-up targets, and reading from left to right as the firer faced the targets (Figure 219). Ranges may have also used pop-up silhouette targets.

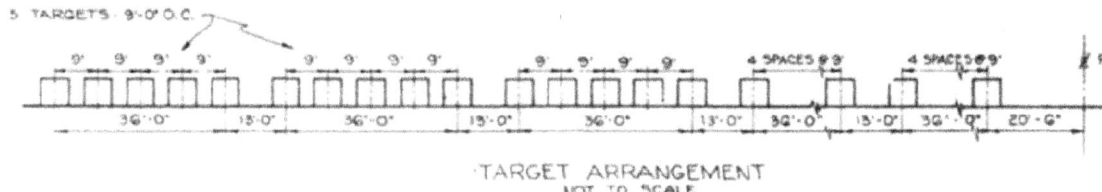

Figure 218. Typical rifle range target arrangement, Fort Bragg, NC, 1952 (Standard drawing No. 28-13-09 drawing 1 of 7, Range, rifle, known distance, plans and details, 5 January 1952).

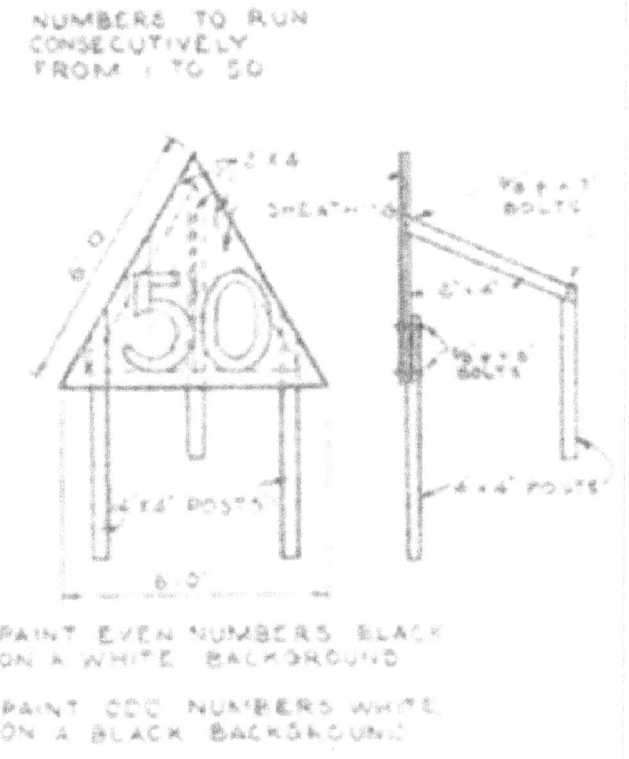

Figure 219. Typical rifle range target number, Fort Bragg, NC, 1952 (Standard drawing No. 28-13-09 drawing 2 of 7, Range, rifle, known distance, details, 5 January 1952).

Figure 220. Rifle range with butts and target shed at Finger Bay, Adak, AK, 23 June 1945 (NARA College Park, RG 80-G, box 1285, photo 342162).

Figure 221. Target line at Sioux Falls Army Air Field, SD, 1942 (NARA College Park, RG 342-FH, box 2202, photo 4A-17256).

Figure 222. Butts at rifle range at Camp Matthews, MCRD San Diego, CA, 6 February 1946 (NARA College Park, RG 127-GC, box 5, photo 401016).

M31A1 pop-up target

If M31A1 pop-up silhouette targets were used, the range contained pop-up target mechanisms built into embankments on the target line (Figures 223 through 226), and portable control towers to control the targets (see Figures 223 and 227).

Figure 223. Typical rifle range, M31A1 target device, basic plan for fifty targets enlarged, Fort Bragg, NC, 1963 (Standard drawing No. 28-13-09 drawing 5 of 7, Range, rifle, known distance, layout and details for installation of M31A1 Target Device, 6 November 1963).

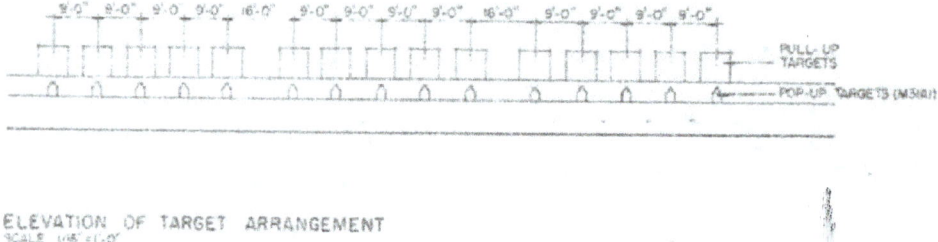

Figure 224. Typical rifle range, M31A1 target device, elevation of target arrangement, Fort Bragg, NC, 1963 (Standard drawing No. 28-13-09 drawing 5 of 7, Range, rifle, known distance, layout and details for installation of M31A1 target device, 6 November 1963).

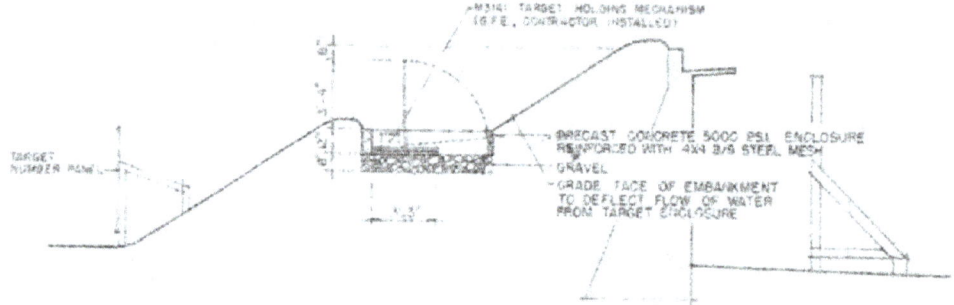

Figure 225. Typical rifle range, M31A1 target device, section through target butt and pop-up target enclosure, Fort Bragg, NC, 1963 (Standard drawing No. 28-13-09 drawing 5 of 7, Range, rifle, known distance, layout and details for installation of M31A1 target device, 6 November 1963).

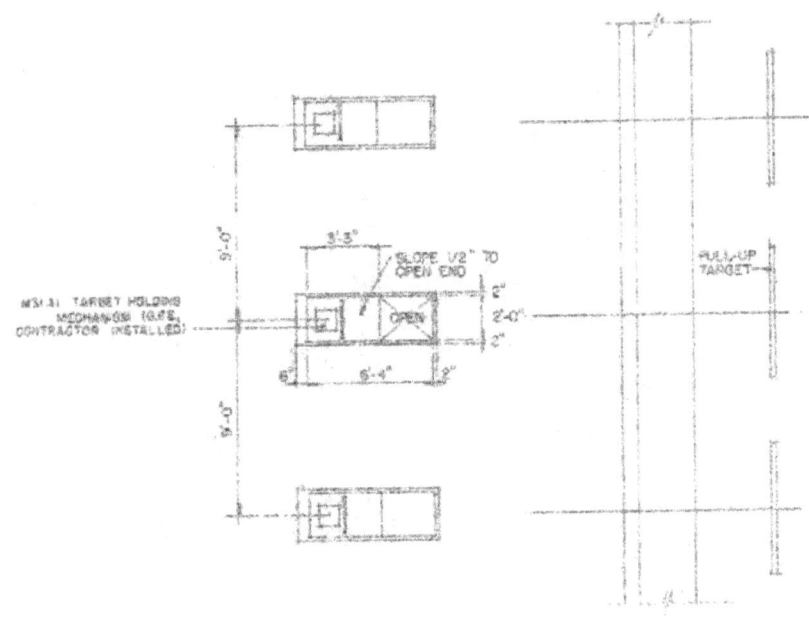

Figure 226. Typical rifle range, M31A1 target device, plan of pop-up target enclosure, Fort Bragg, NC, 1963 (Standard drawing No. 28-13-09 drawing 5 of 7, Range, rifle, known distance, layout and details for installation of M31A1 target device, 6 November 1963).

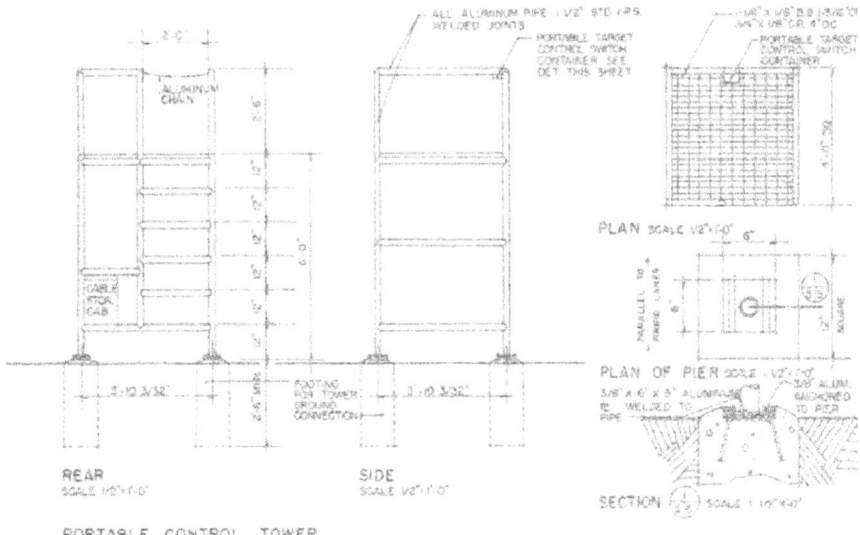

Figure 227. Typical rifle range, M31A1 target device, portable control tower, Fort Bragg, NC, 1963 (Standard drawing No. 28-13-09 drawing 5 of 7, Range, rifle, known distance, layout and details for installation of M31A1 target device, 6 November 1963).

Embankments/trenches/Etc.

Embankments may have been built to stabilize firing lines, to give soldiers a clear view of the entire range, or to support individual firing positions (see "Firing Lines" above). Embankments were also built behind or as part of target lines (target butts, see Figures 228 through 235). Large trenches were often excavated to form a target pit from which pull up targets could be raised. The dirt from the trench went into constructing the embankment behind targets (Figures 236 through 239). Figure 240 below shows how an embankment could also be built between side-by-side rifle ranges to allow them to be safely spaced more closely together.

Timber wall target butts

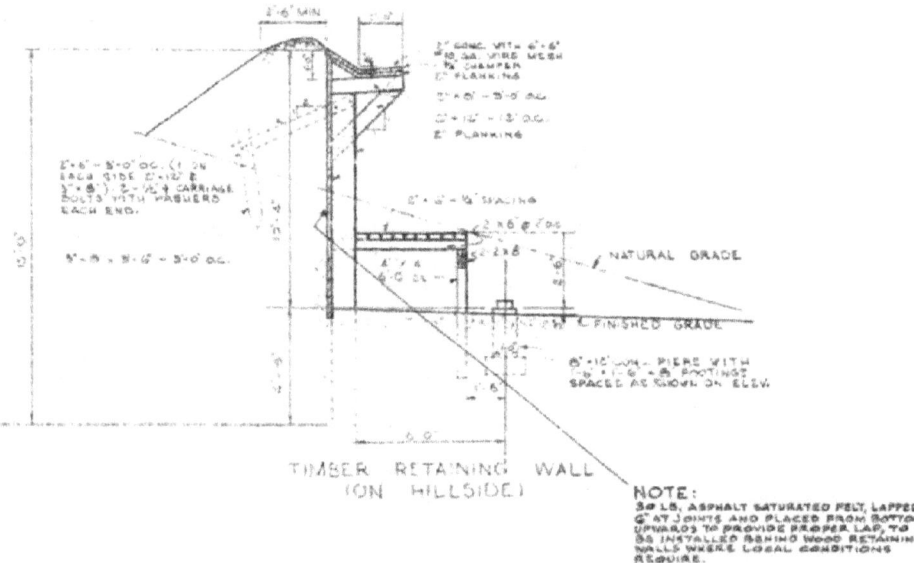

Figure 228. Typical rifle range target butts timber wall (on hillside), Fort Bragg, NC, 1952 (Standard drawing No. 28-13-09 drawing 2 of 7, Range, rifle, known distance, details, 5 January 1952).

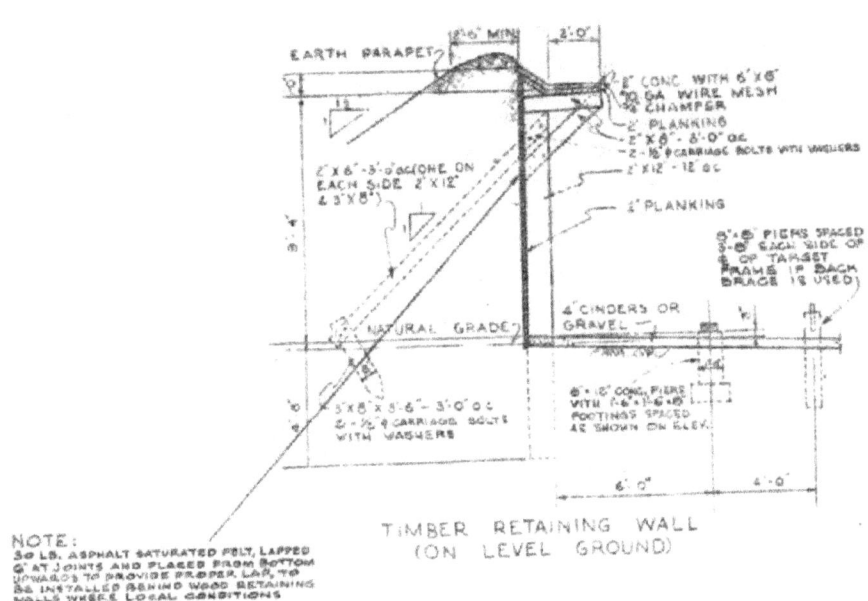

Figure 229. Typical rifle range target butts timber wall (on level ground), Fort Bragg, NC, 1952 (Standard drawing No. 28-13-09 drawing 2 of 7, Range, rifle, known distance, details, 5 January 1952).

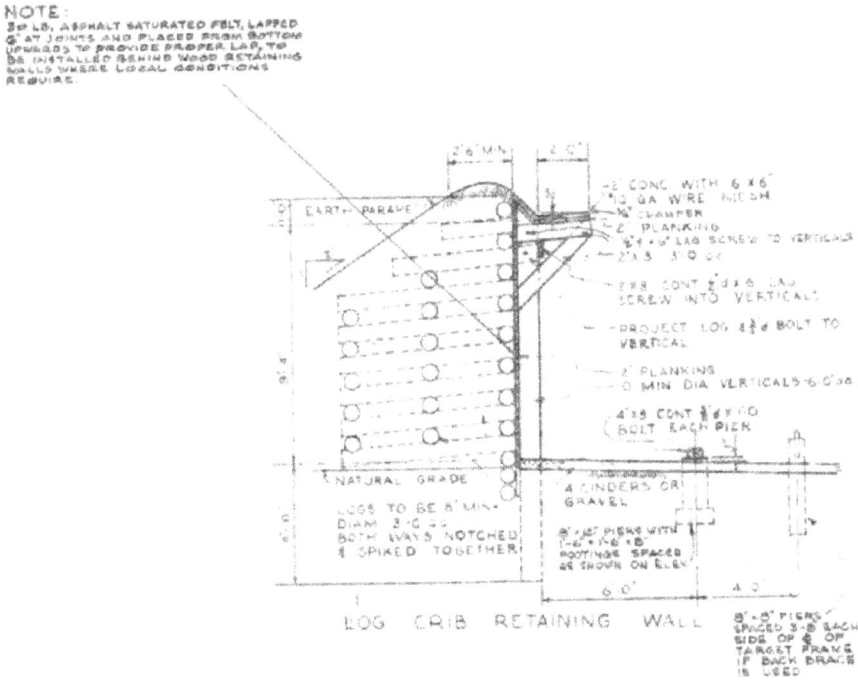

Figure 230. Typical rifle range target butts log crib retaining wall, Fort Bragg, NC, 1952 (Standard drawing No. 28-13-09 drawing 2 of 7, Range, rifle, known distance, details, 5 January 1952).

Concrete wall target butts

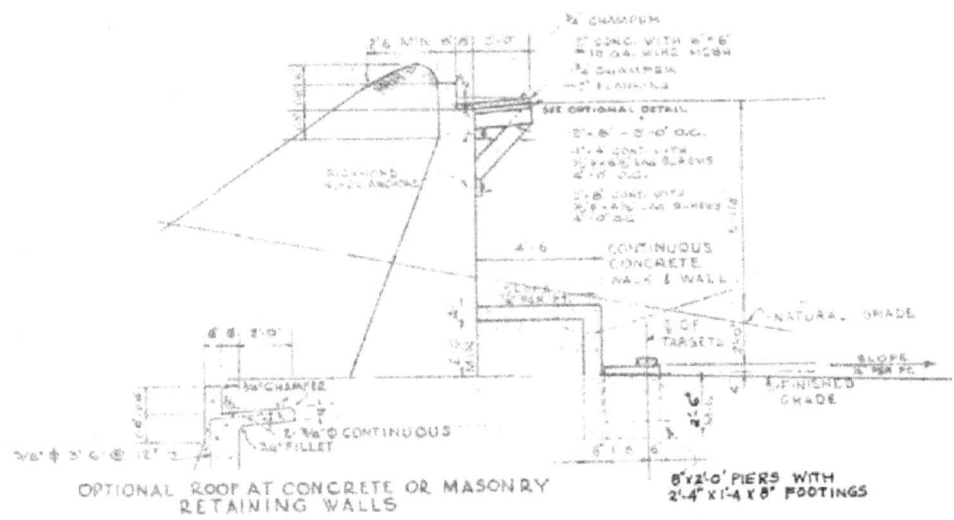

Figure 231. Typical Rifle Range Target Butts Concrete Retaining Wall, Fort Bragg, NC, 1952 (Standard drawing No. 28-13-09 drawing 2 of 7, Range, rifle, known distance, details, 5 January 1952).

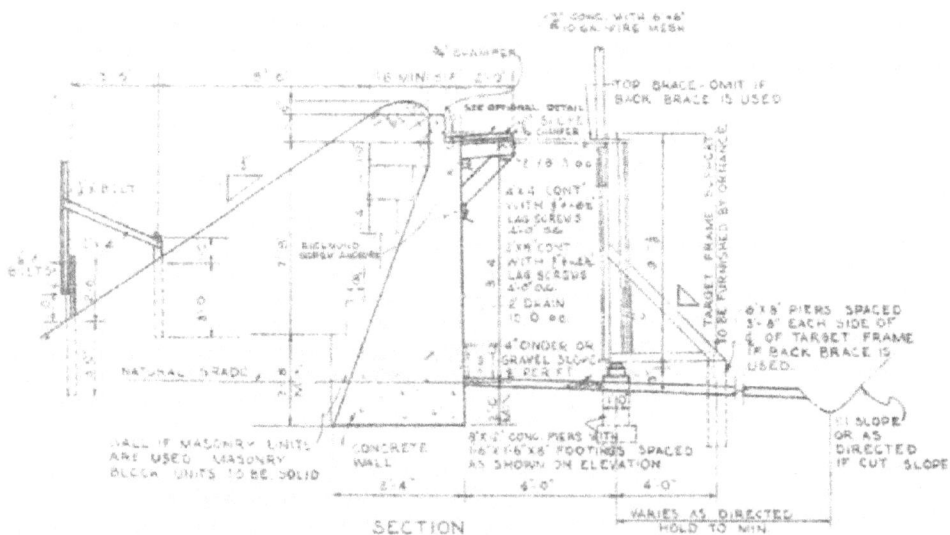

Figure 232. Typical rifle range target butts section detail, Fort Bragg, NC, 1952 (Standard drawing No. 28-13-09 drawing 2 of 7, Range, rifle, known distance, details, 5 January 1952).

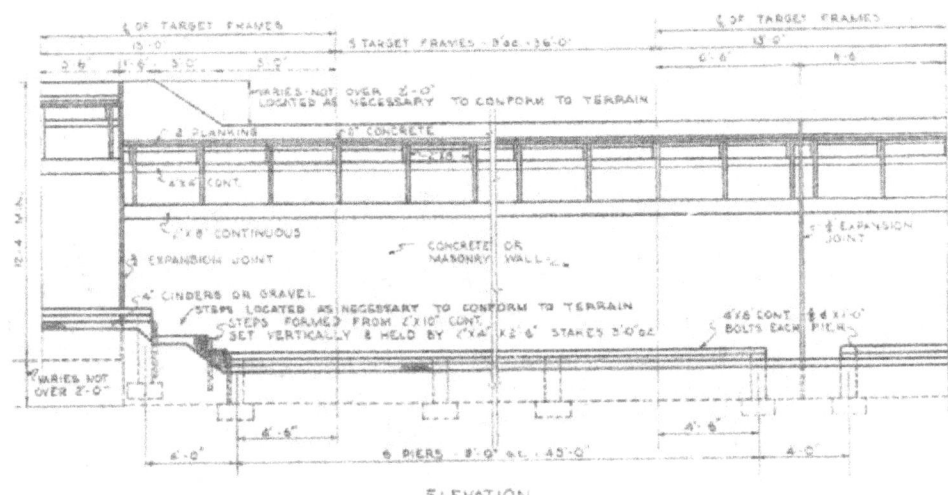

Figure 233. Typical rifle range target butts section elevation, Fort Bragg, NC, 1952 (Standard drawing No. 28-13-09 drawing 2 of 7, Range, rifle, known distance, details, 5 January 1952).

Figure 234. Target line at the USMC Winthrop Rifle Range, Indian Head, MD, 1915 (NARA College Park, RG 127-G-100b, box 23, photo 521528).

Figure 235. Ground rifle practice in standing position at Miami Army Air Field, FL, 2 October 1943 (NARA College Park, RG 342-FH, box 2207, photo 4A-18407).

Figure 236. Horses digging target trench, Fort Knox, KY, undated (scans from Knox, training 5, vol. 5).

Figure 237. Horses digging target trench, Fort Knox, KY, undated (scans from Knox, training 6, vol. 5).

Figure 238. Target complex built on top of trench, Fort Knox, KY, undated (scans from Knox, training 7, vol. 5).

Figure 239. Looking up out of target trench (rifle range pits), Fort Knox, KY, undated (scans from Knox, training 8, vol. 5, #358, Kirkpatrick photo).

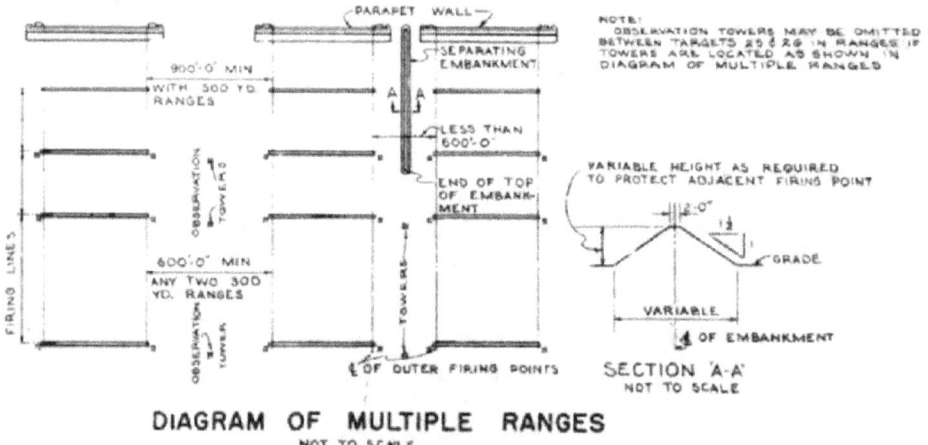

Figure 240. Typical rifle range diagram of multiple ranges, Fort Bragg, NC, 1952 (Standard drawing No. 28-13-09 drawing 1 of 7, Range, rifle, known distance, plans and details, 5 January 1952).

Buildings

"In 1942, the standard plans for a rifle range included a target storage building and latrine. In 1952, an observation tower was included in the plans. Additionally, an installation range complex might have had latrines, storage sheds, and administrative/maintenance buildings supporting general range functions. If present, these buildings and facilities were centrally located beyond the firing range limits and impact areas" ("RO-1" 14-15). The ranges may also have been part of a larger installation range complex that contained these buildings.

Fixed firing points and moving targets

Anti-aircraft towed target range

Soldiers were trained on anti-aircraft towed target ranges to fire small and large arms at aerial targets. "Limited construction was required for this range. The range required only a cleared area for firing positions and markers for the left and right firing limits. Soldiers fired at targets that were either towed behind an aircraft or were miniature remote controlled aircraft (Figure 251)" ("RO-1" 27). Rifles and machine guns were the small arms weapons used in training.

Layouts

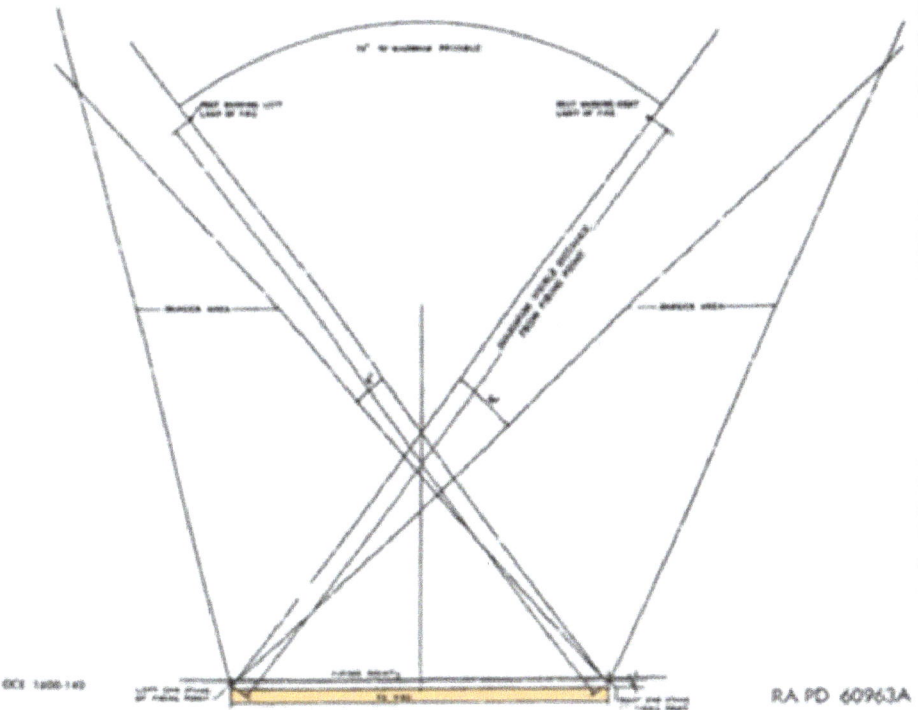

Figure 241. Anti-aircraft towed target range, circa 1942 (Standard drawing No. 1600-140, Field target range A-A towed target range, 17 November 1942).

Figure 242. Firing range at Anti-Aircraft Training and Test Center Dam Neck, VA, December 1943 (NARA College Park, RG 80-G, box 8, photo 2316).

Figure 243. Beach firing range at NAS Fort Lauderdale, FL, 28 February 1944 (NARA College Park, RG 80-G, box 1513, photo 388275).

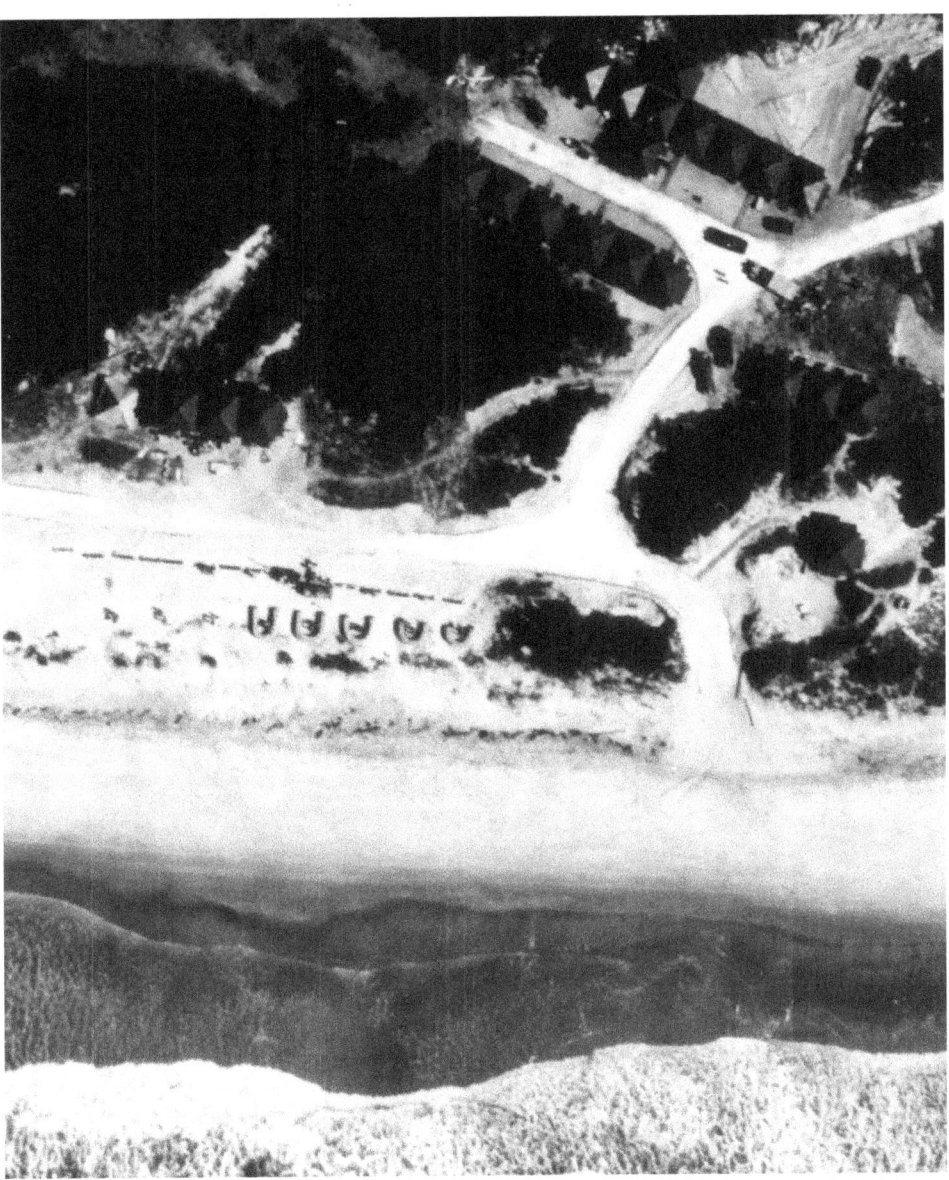

Figure 244. Aerial of machine gun range at MCAS Cherry Point, NC, 9 August 1943 (NARA College Park, RG 80-G, box 1495, photo 384685).

Firing lines

Firing lines and positions may have been stabilized with an embankment or sandbags, particularly when located on a beach.

Figure 245. Machine gun crews in action on range at MCAS Cherry Point, NC, 21 August 1943 (NARA College Park, RG 80-G, box 1360, photo 358873).

Figure 246. Marines operating machine gun at MCAS Cherry Point, NC, 21 August 1943 (NARA College Park, RG 80-G, box 1360, photo 358872).

Figure 247. Sailors at gunnery range at NAS Wildwood, NJ, 30 May 1944 (NARA College Park, RG 80-G, box 1487, photo 383360).

Targets

Targets were either towed behind an aircraft or were miniature remote controlled aircraft.

Embankments/trenches/etc.

Firing lines and positions may have been stabilized with an embankment or sandbags.

Buildings

No buildings are mentioned in the standard plans for these ranges. However, a range may have had a control tower, latrine, target storage building, ammunition storage building, other storage sheds, and administrative/maintenance buildings supporting general range functions. The range may also have been part of a larger installation range complex that contained these buildings.

Machine gun ranges

Soldiers were trained on these ranges to fire machine guns from fixed positions at fixed and moving targets. Ranges consisted of a variety of firing lines and firing positions, targets, and embankments as shown in the examples below. Training was completed with .30 and .50 caliber machine guns.

Layouts

Figure 248. CMTC students on machine gun range at Camp Del Monte, CA, 13 August 1925 (NARA College Park, RG 111-SC WWI, box 700, photo 94896).

<u>Machine gun course</u>

"The machine gun course (Figure 259 below) was standard instruction for using machine guns" ("RO-1" 22).

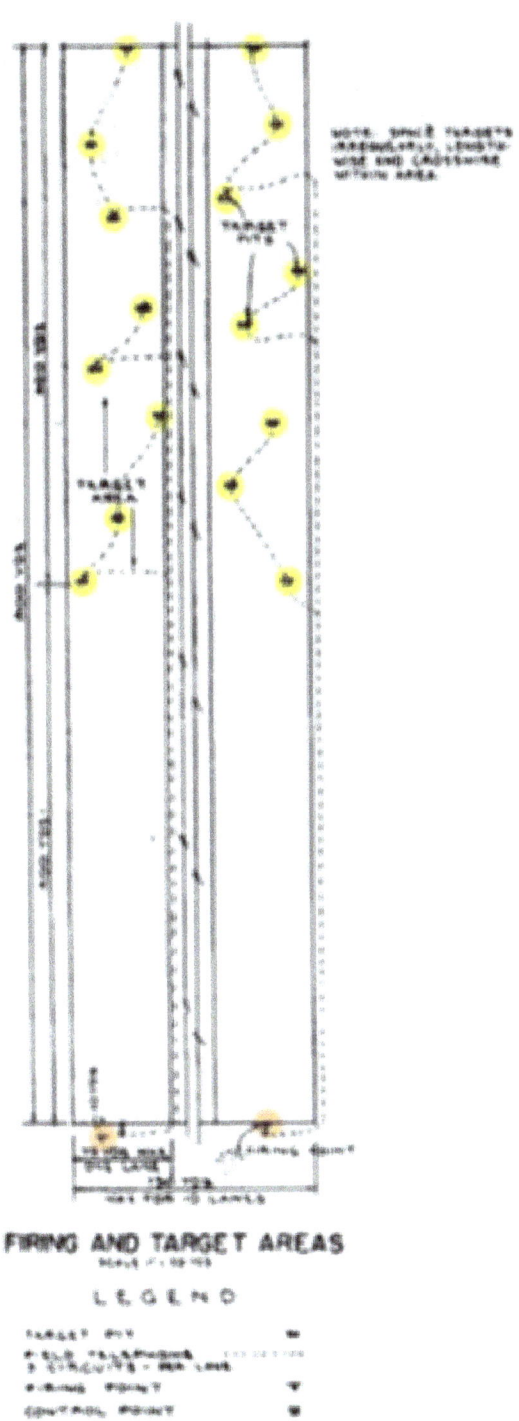

Figure 249. Machine gun course, circa 1952 (Standard drawing No. 28-13-05, Machine gun course, table II, course A, 4 January 1952).

Circular range

Figure 250. Circular machine gun ranges at Naval Air Test Center Corpus Christi, VA, 12 November 1942 (NARA College Park, RG 80-G, box 276, photo 63318).

Harmonization range

Figure 251. .30 caliber machine gun Harmonization Range #3 at Tyndall Army Air Field (Panama City), FL, 11 June 1942 (NARA College Park, RG 342-FH, box 2202, photo 4A-17261).

Coastal range

Figure 252. Machine gun range at Coast Guard Air Station Port Angeles, WA, 8 January 1957 (NARA College Park, RG 80-G, box 39, photo 7163).

Firing lines

Several firing supports were used on machine gun ranges, such as platforms (Figures 253 and 254), swivel mounts, and other stands. Some firing lines were covered.

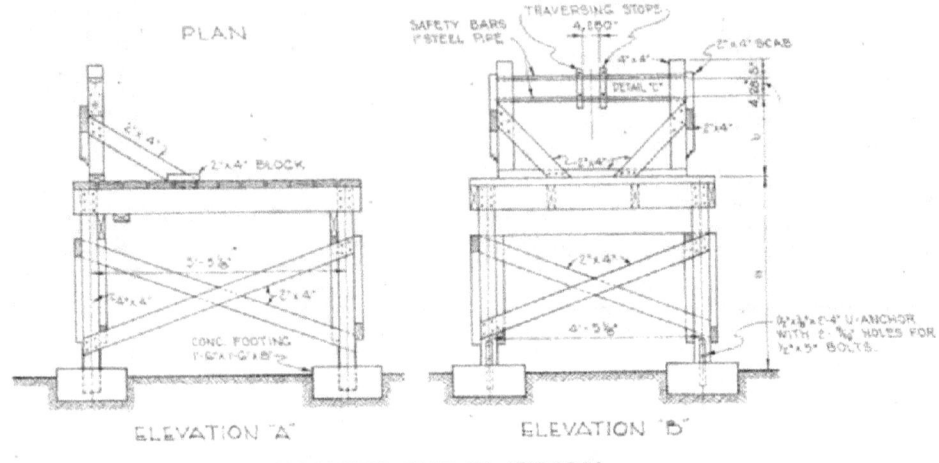

Figure 253. Small arms fire course, machine gun platform, Fort Bragg, NC, 1952 (Standard drawing 28-13-37 sheet 2 of 2, Small arms fire course, details, 20 June 1952).

Figure 254. Training enlisted personnel on the Machine Gun Range at Amphibious Training Base Fort Pierce, FL, 23 November 1943 (NARA College Park, RG 80-G, box 862, photo 264377).

Figure 255. Firing a Lewis machine gun from shoulder at Camp Wheeler, GA, 1917 (New York Public Library, digital No. 117105).

Figure 256. Soldiers operating a Lewis machine gun at unknown location, February 1918 (New York Public Library, digital No. 117103).

Figure 257. Fixed machine gun range at NAS Wildwood, NJ, 29 April 1944 (NARA College Park, RG 80-G, box 1487, photo 383347).

Figure 258. Machine gun practice for cadets at NAS Corpus Christi, TX, 23 July 1941 (NARA College Park, RG 80-G, box 1978, photo 463691).

Figure 259. Trainees fire the 30 caliber machine guns (in background) on swivel at fixed targets 200 yds away first, then trainees fire the 50 caliber machine guns at unknown location, July 1942 (NARA College Park, RG 342-FH, box 2202, photo 4A-17268).

Figure 260. Machine gun firing range at unknown location, undated (NARA College Park, RG 342-FH, box 2230, photo 4A-24093).

Figure 261. A group of trainees fire .30 caliber light machine guns on one of the ranges at Fort Knox, KY, 1947 (NARA College Park, RG 111-SC WWII, box 602, photo SC299052).

Targets

A variety of fixed panel targets, silhouettes, and pop-up targets were used on these ranges.

Embankments/trenches/etc.

Embankments may have been built at firing lines, behind target lines, in front of moving target assemblies, and between ranges.

Buildings

No buildings are mentioned in the standard plans for these ranges. However, a range may have had a control tower, latrine, target storage building, ammunition storage building, other storage sheds, and administrative/maintenance buildings supporting general range functions. The range may also have been part of a larger installation range complex that contained these buildings.

1,000-in. machine gun rolling target range

Soldiers were trained on 1,000-in. machine gun rolling target ranges (also known as Machine Gun 1,000-in. Ranges or 1,000-in. Fixed Target Track Type Ranges) to fire machine guns at approaching targets. The Range consisted of a mounded firing line, a cable system and target track with panel targets mounted on target cars, and an embankment behind the target line (see Figure 263). Soldiers fired at targets as they were pulled toward the firing position ("RO-1" 23). Training was completed with .30 caliber machine guns.

Layouts

Danger area

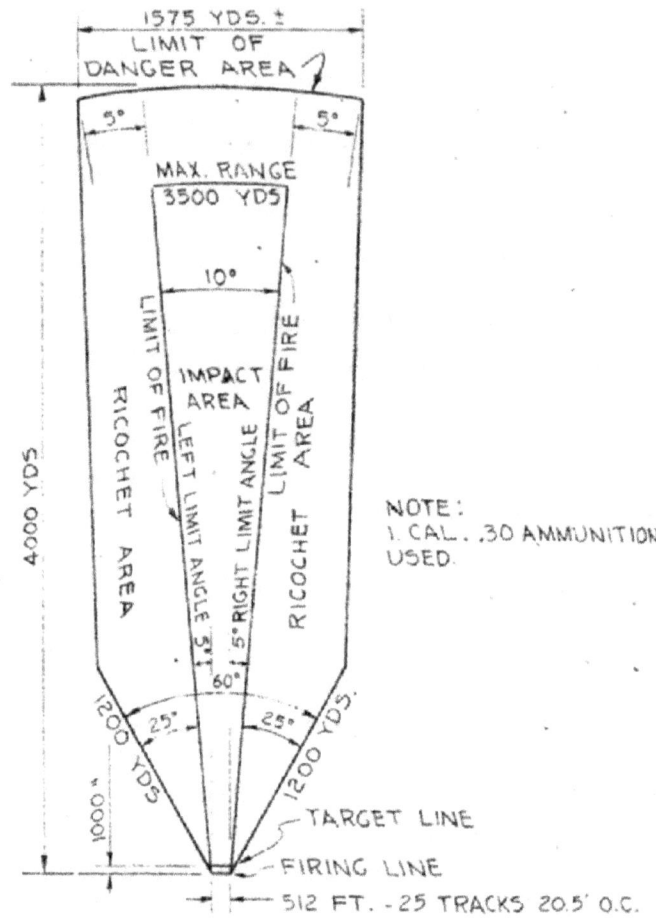

Figure 262. A 1000-in. fixed target track type range, danger area, Fort Bragg, NC, 1952 (Standard drawing 28-13-06 sheet 1 of 2, Range, 1000-in. Fixed target, track type, plans and details, 20 June 1952).

Typical layout

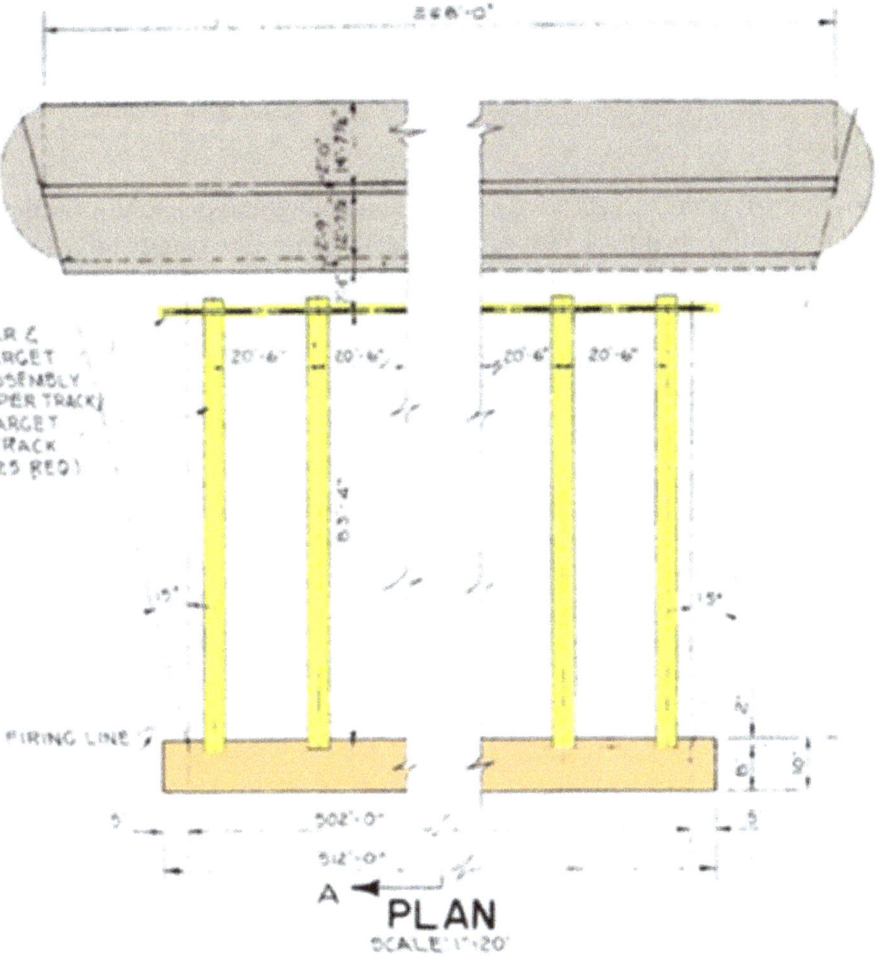

Figure 263. A 1000-in. fixed target track type range, plan, Fort Bragg, NC, 1952 (Standard drawing 28-13-06 sheet 1 of 2, Range, 1000-in. fixed target, track type, plans and details, 20 June 1952).

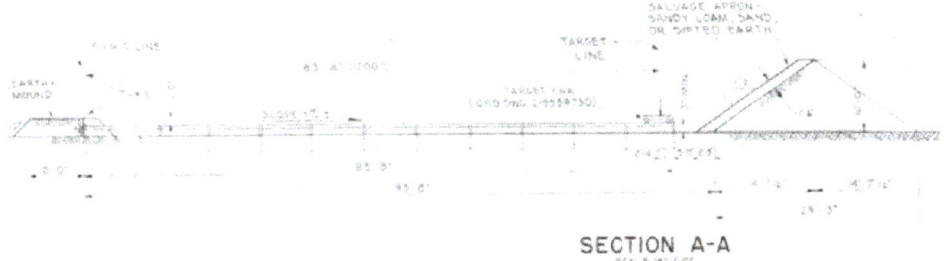

Figure 264. A 1000-in. fixed target track type range, section, Fort Bragg, NC, 1952 (Standard drawing 28-13-06 sheet 1 of 2, Range, 1000-in. Fixed target, track type, plans and details, 20 June 1952).

Firing lines

This range had a mounded firing line 1,000 in. from the target line (see Figure 264).

Targets

Five by 3-ft panel targets were mounted on dollies and pulled toward the firing position on a track by a cable system. There were twenty target tracks required per range, with one car and target assembly per track (Figures 263, 265, 266, and 267).

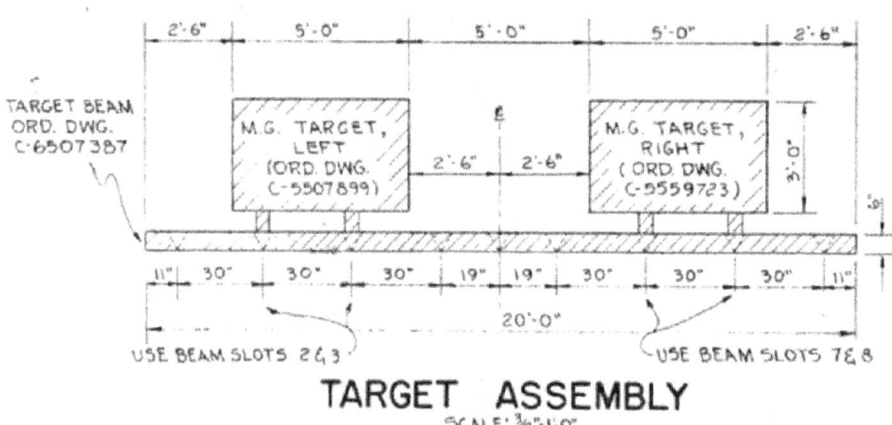

Figure 265. 1000-in. fixed target track type range, target assembly, Fort Bragg, NC, 1952 (Standard drawing 28-13-06 sheet 1 of 2, Range, 1000-in. fixed target, track type, plans and details, 20 June 1952).

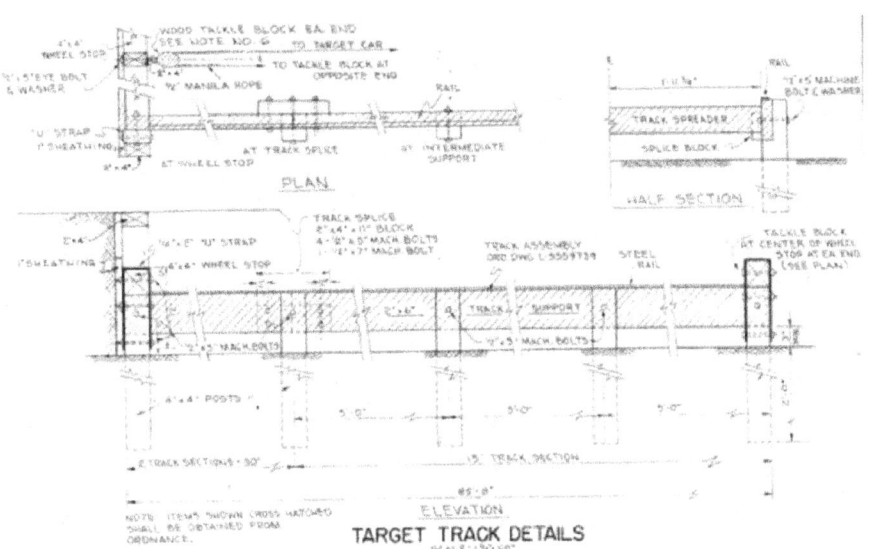

Figure 266. A 1000-in. fixed target track type range, target track details, Fort Bragg, NC, 1952 (Standard drawing 28-13-06 sheet 1 of 2, Range, 1000-in. fixed target, track type, plans and details, 20 June 1952).

Embankments/trenches/etc.

This range had a 10-ft high continuous embankment behind target lines (Figure 267).

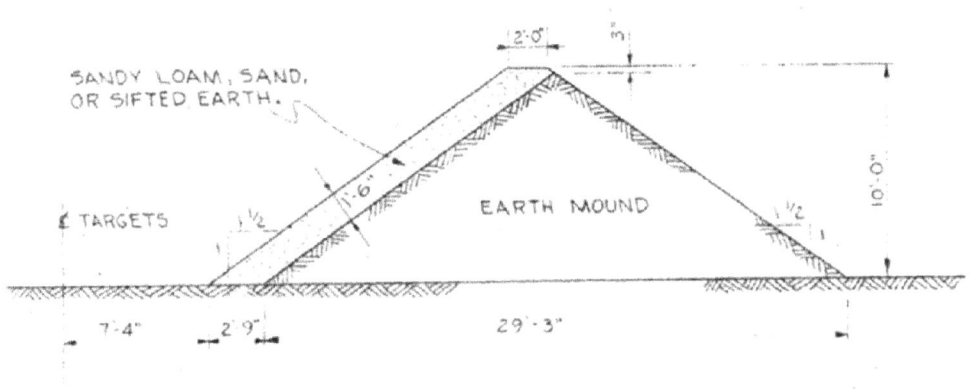

Figure 267. A 1000-in. fixed target track type range, section through salvage wall, Fort Bragg, NC, 1952 (Standard drawing 28-13-06 sheet 1 of 2, Range, 1000-in. fixed target, track type, plans and details, 20 June 1952).

Buildings

No buildings are mentioned in the standard plans for these ranges. However, a range may have had a control tower, latrine, target storage building, ammunition storage building, other storage sheds, and administrative/maintenance buildings supporting general range functions. The range may also have been part of a larger installation range complex that contained these buildings.

1000-in. miniature anti-tank range

Anti-tank and tank gunners were trained on these ranges to fire at moving targets. The range consisted of firing positions, a target track, and moving targets attached to a dolly (Figure 269). Gunners fired at the targets as they were pulled across the track. "Anti-tank and tank guns fired sub-caliber ammunition at this range" ("RO-1" 25).

Layouts

Danger area

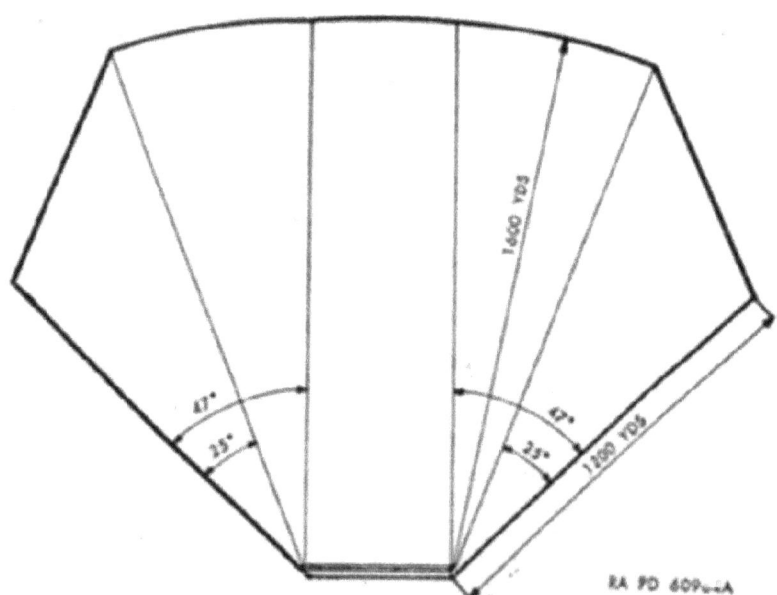

Figure 268. A 1000-in. miniature anti-tank range, circa 1951 (Standard drawing No. 1600-120, 1000-in. Miniature anti-tank range, 28 October 1942, TM 9-855, Targets, target material and training course layouts, 1 November 1951, p 32).

Typical layout

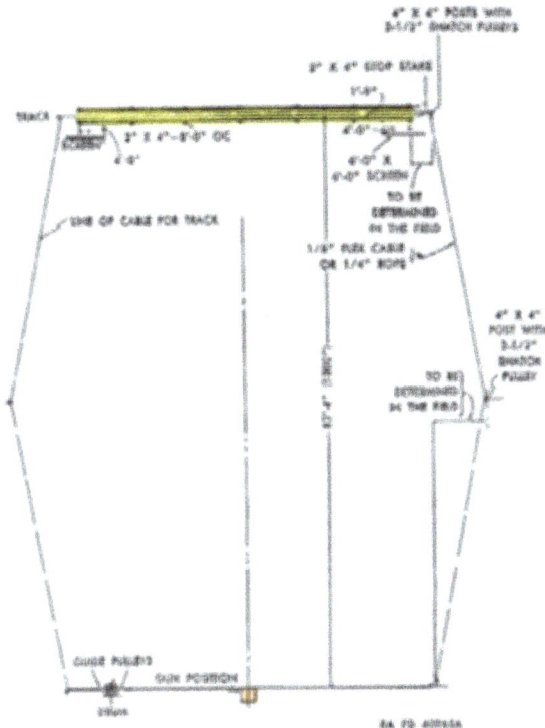

Figure 269. A 1000-in. miniature anti-tank range, circa 1951 (Standard drawing No. 1600-120, 1000-in. miniature anti-tank range, 28 October 1942, TM 9-855, Targets, target material and training course layouts, 1 November 1951, p 32).

Firing lines

Soldiers fired from stationary firing positions 83' 4" perpendicular of the target track.

Targets

A moving target, attached to a dolly, was pulled across the target track.

Embankments/trenches/etc.

Screens were placed at the beginning and end of the target track to block targets.

Buildings

No buildings are mentioned in the standard plans for these ranges. However, a range may have had a control tower, latrine, target storage building, ammunition storage building, other storage sheds, and administrative/maintenance buildings supporting general range functions. The range may also have been part of a larger installation range complex that contained these buildings.

Miniature anti-aircraft range

Soldiers were trained on these ranges to fire at moving aerial targets. The Range consisted of a firing line, and "a set of targets on pulleys designed to simulate aircraft movement (diving, horizontal flight, etc)." Soldiers fired at targets as they moved across the pulley system (see Figure 270). "Firing was limited to using a .22-caliber Long Rifle" ("RO-1" 24).

Layouts

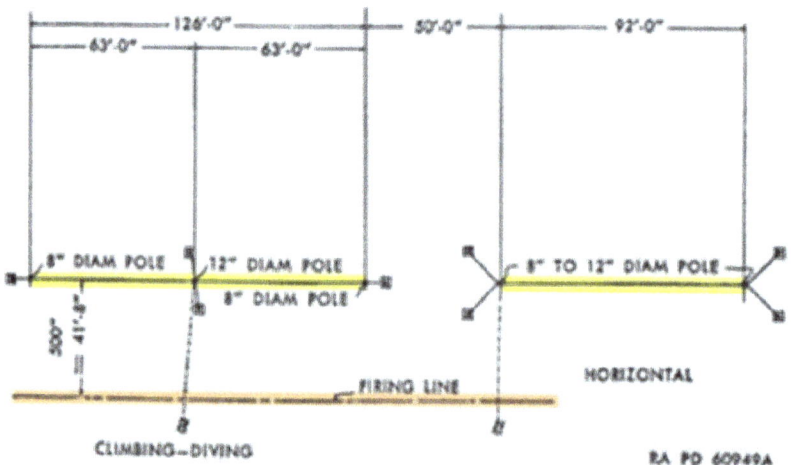

Figure 270. Miniature anti-aircraft range, plan, circa 1942-1951 (Standard drawing No. 1600-130/131, Training aids, A. A. range miniature, 14 October 1942, TM 9-855, Targets, target material and training course layouts, 1 November 1951, pp 28, 29).

Firing lines

Soldiers fired from a single firing line set 41' 8" from the pulley target system (see Figure 270).

Targets

This range had "a set of targets on pulleys designed to simulate aircraft movement (diving, horizontal flight, etc) (see Figures 271 and 272)" ("RO-1" 24).

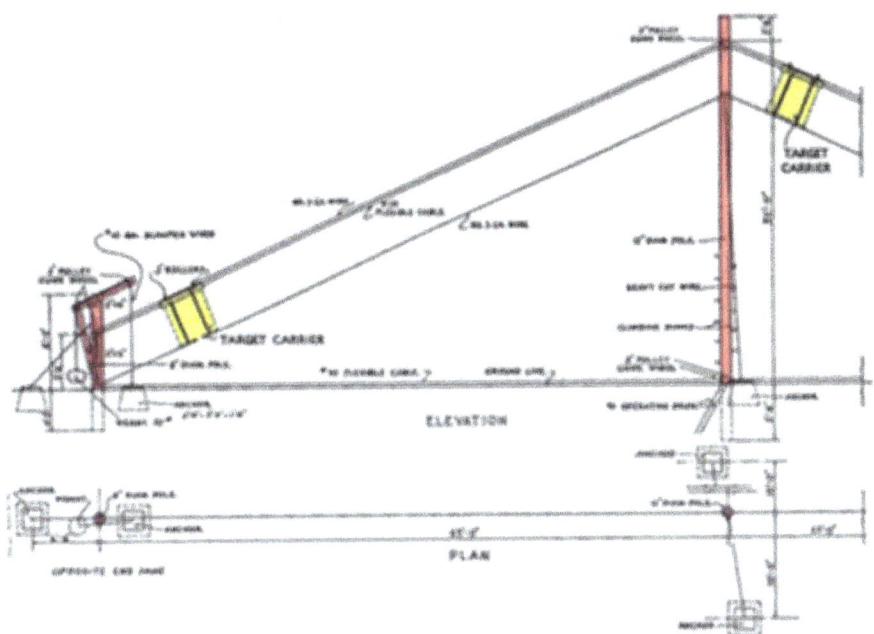

Figure 271. Miniature anti-aircraft range, target pulley system plan and elevation, circa 1942-1951 (Standard drawing No. 1600-130/131, Training aids, A. A. range miniature, 14 October 1942, TM 9-855, Targets, target material and training course layouts, 1 November 1951, pp 28, 29).

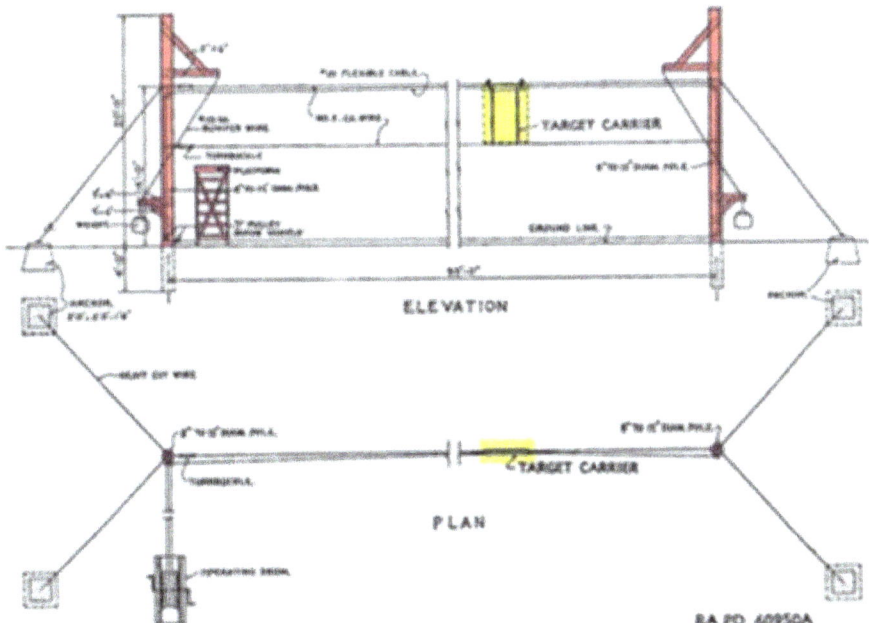

Figure 272. Miniature anti-aircraft range, target pulley system plan and elevation, circa 1942-1951 (Standard drawing No. 1600-130/131, Training aids, A. A. range miniature, 14 October 1942, TM 9-855, Targets, target material and training course layouts, 1 November 1951, pp 28, 29).

Embankments/trenches/etc.

None.

Buildings

No buildings are mentioned in the standard plans for these ranges. However, a range may have had a control tower, latrine, target storage building, ammunition storage building, other storage sheds, and administrative/maintenance buildings supporting general range functions. The range may also have been part of a larger installation range complex that contained these buildings.

Moving target range

Soldiers were trained on these ranges to fire at moving targets. "An early version of the range had the target moving on a track that zigzagged either toward or away from the firing position. A later version used a target moving on a triangular shaped track" ("RO-1" 29). This later version also had a continuous line of firing points, embankments in front of the target track with a built in target repair pit, a latrine, and an ammunition point (Figure 273). Soldiers fired at targets mounted on carrying cars moving along the tracks.

Typical layout and danger area

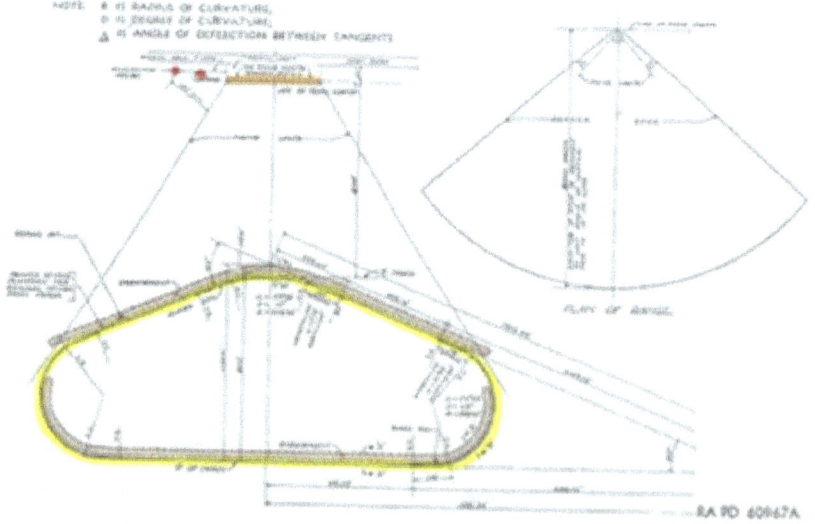

Figure 273. Moving target range, circa 1951 (TM 9-855, Targets, target material and training course layouts, 1 November 1951, p 35).

Firing lines

A continuous line of firing points ran perpendicular to the triangular target track.

Targets

Targets moved on a carrying car along a triangular target track.

Embankments/trenches/etc.

Embankments ran in between the firing line and the triangular target track to protect and partially conceal the track.

Buildings

A latrine and ammunition point are mentioned in the standard plans for these ranges. A range may also have had a control tower, target storage building, other storage sheds, and administrative/maintenance buildings supporting general range functions. The range may also have been part of a larger installation range complex that contained these buildings.

Night firing range

Soldiers trained on these ranges to fire under night conditions. The range consisted of firing lines established at 0, 25, and 50 meters (some fitted with foxholes), E and F-type silhouette pop-up targets, and light posts (see Figures 275 and 276). Soldiers took turns firing at pop-up targets from firing lines.

Layouts

Danger area

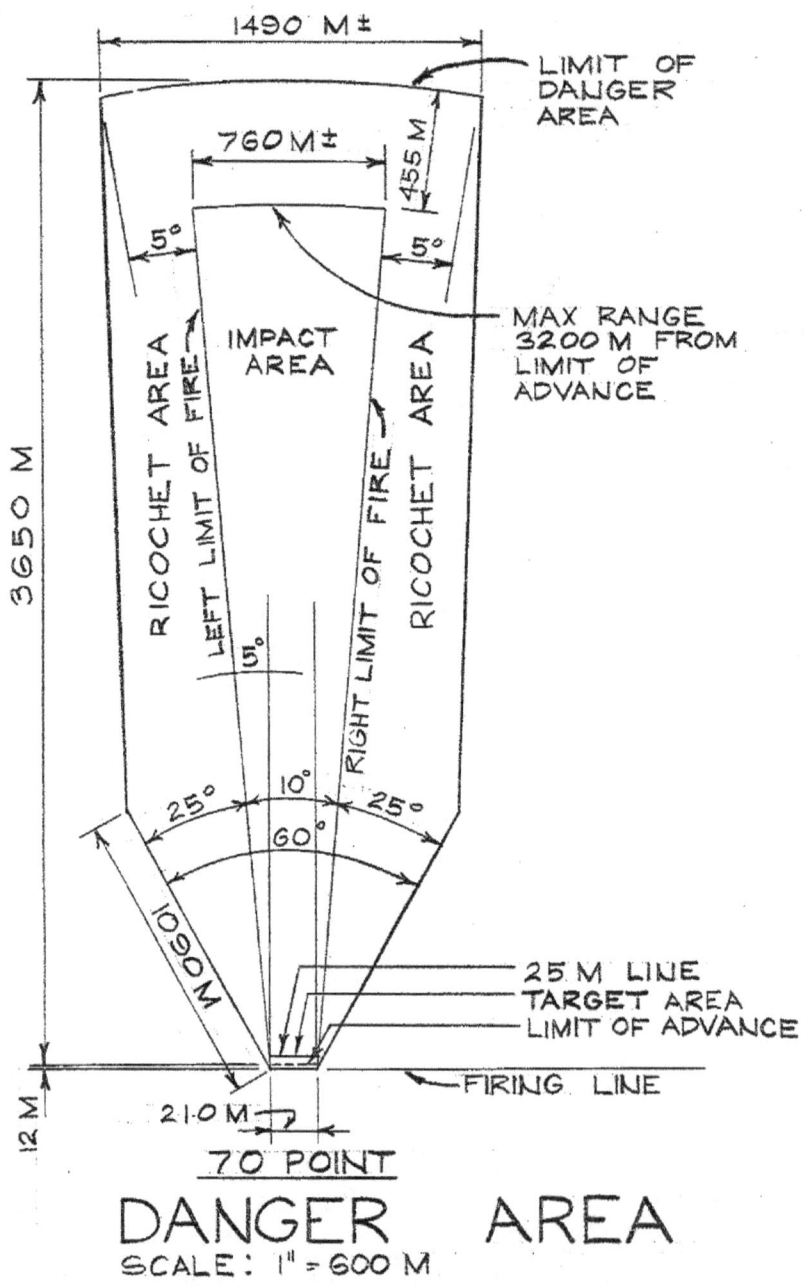

Figure 274. ATC facilities, night firing range "N", danger area, Fort Bragg, NC, 1966 (Standard drawing 28-13-117 sheet 8, Construction of ranges, phase 1, U.S. ATC Facilities, Fort Bragg, NC, range "N", night firing range, 28 June 1966).

Typical layouts

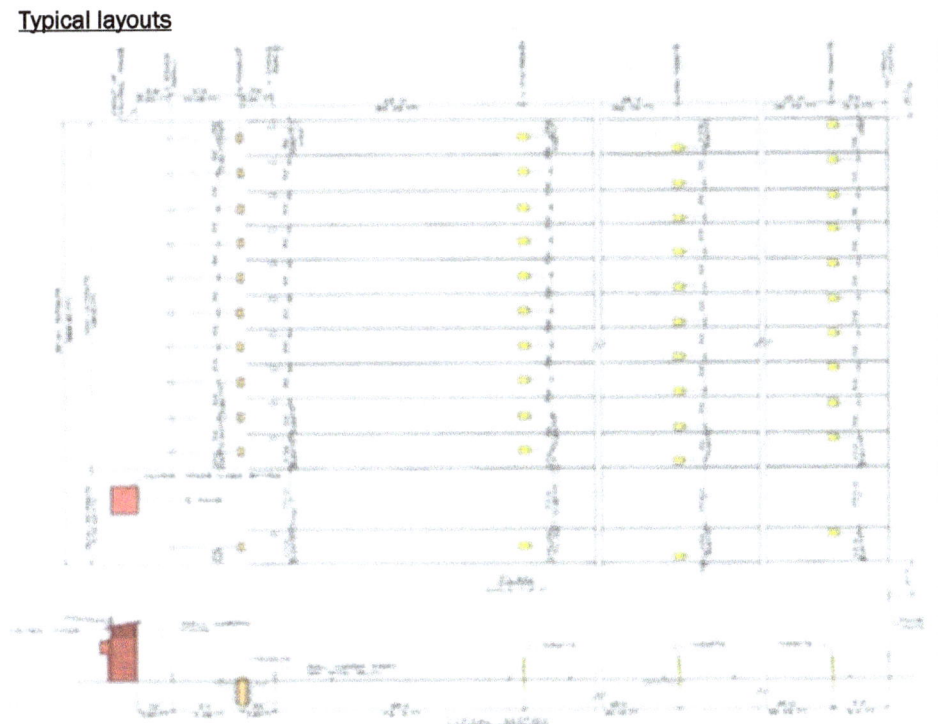

Figure 275. ATC facilities, night firing range "N", plan and typical section, Fort Bragg, NC, 1966 (Standard drawing 28-13-117 sheet 8, Construction of ranges, phase 1, U.S. ATC Facilities, Fort Bragg, NC, range "N", night firing range, 28 June 1966).

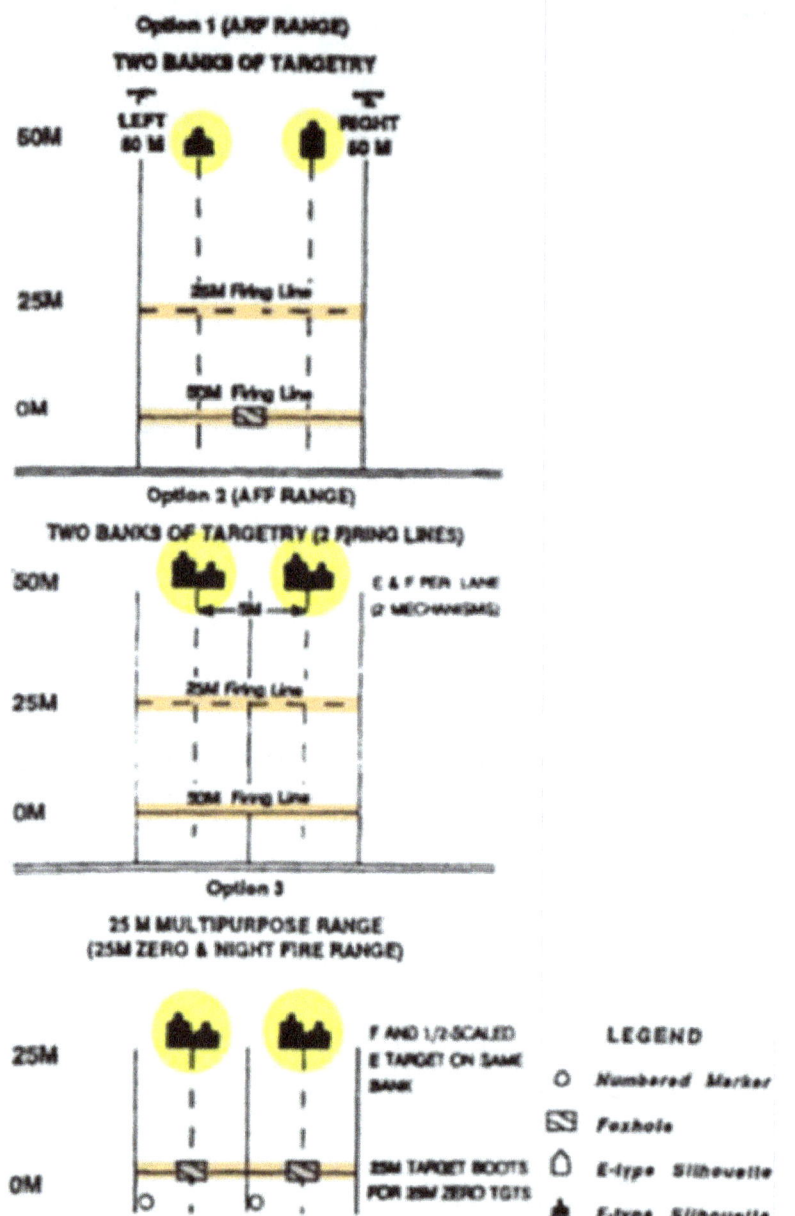

Figure 276. Night firing range, circa 1992 (TC 25-8, Training ranges, 25 February 1992, p 6-14).

Firing line

"Soldiers fired from lines established at 0 meters (m), 25m, and 50m. Firing on this range began at the 25m firing line" ("RO-1" 31). Some firing lines were fitted with concrete pipe foxholes (see Figure 277).

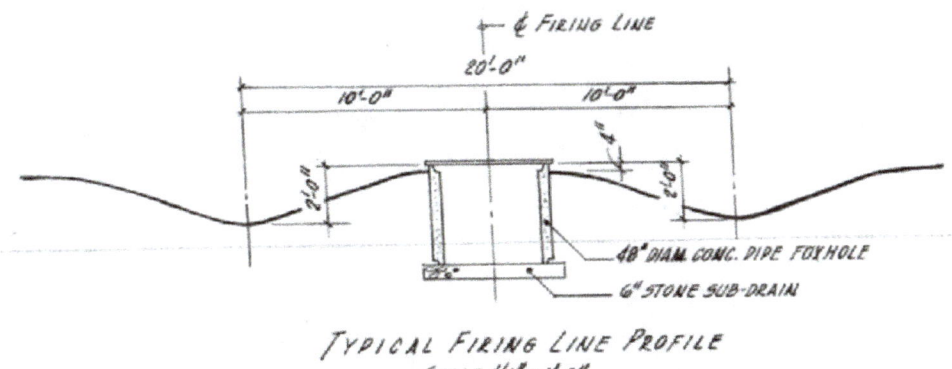

Figure 277. ATC facilities, night firing range "N", firing line, Fort Bragg, NC, 1966 (Standard drawing 28-13-117 sheet 9, Construction of ranges, phase 1, U.S. ATC Facilities, Fort Bragg, NC, range "N", night firing range, 28 June 1966).

Targets

These ranges used E- and F-type silhouette pop-up targets.

Embankments/trenches/etc.

None.

Buildings

Control towers, bleachers, ammunition storage buildings, target storage buildings, and latrines are mentioned in the plans for these ranges. A range may also have had other storage sheds and administrative/maintenance buildings supporting general range functions. The range may also have been part of a larger installation range complex that contained these buildings.

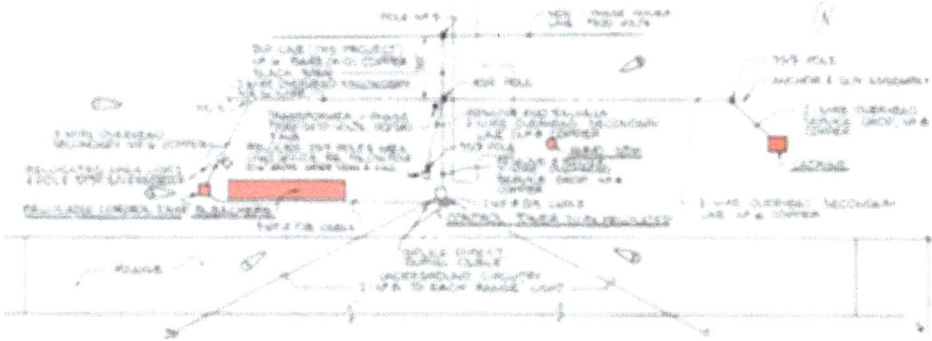

Figure 278. Night fire range electrical layout, Fort Bragg, NC, 1966 (Standard drawing 28-13-117 sheet 28, Electrical distribution, 28 June 1966).

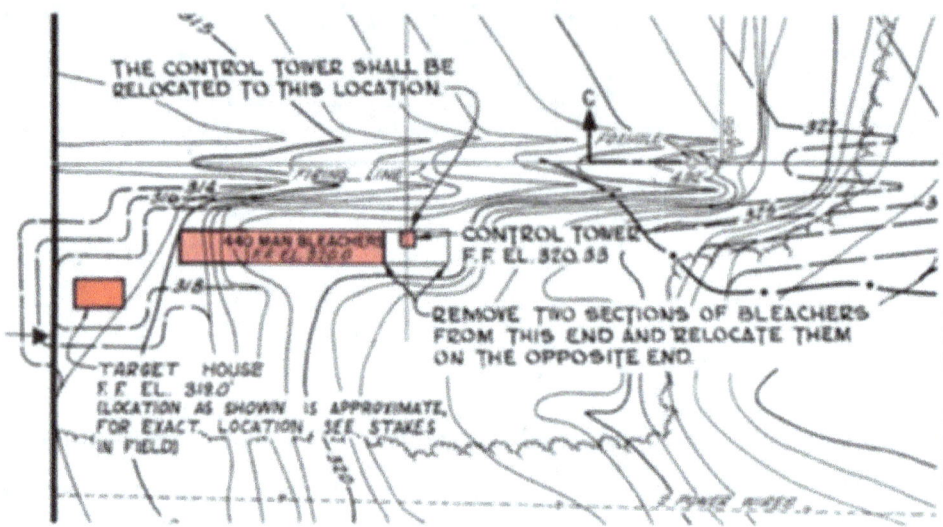

Figure 279. ATC facilities, night firing range "N", range extension plan, Fort Bragg, NC, undated (Standard drawing 28-13-117 sheet 9A, construction of ranges, phase 1, U.S. ATC facilities, Fort Bragg, NC, range "N", night firing range extension (70 points to 110 points), undated).

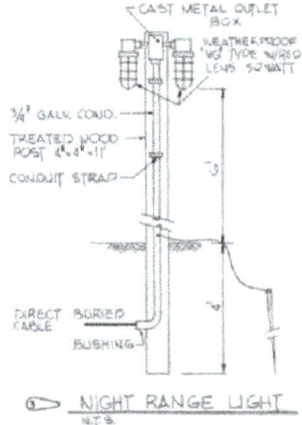

Figure 280. Night range light, Fort Bragg, NC, 1966 (Standard drawing 28-13-117 sheet 28, Electrical distribution, 28 June 1966).

Skeet range

Although Skeet Ranges (also known as basic deflection ranges) were constructed on several types of installations, the Army Air Force and Air Force were primary users. This range was used to train gunners how to lead aerial targets, and to wing axis fighter planes in combat. The range consisted of a firing line in the form of a half circle (sometimes fitted with firing towers), a control tower, and high and low houses out of which high and low flying pigeon targets were shot (Figure 281). As pigeons were fired from high and low houses away from soldiers, they took turns leading and firing at the aerial targets. A variation of the range used clay pigeons, thrown toward the firing position for gunners to practice attacking an approaching target. ("RO-1" 28).

Layouts

Typical layout

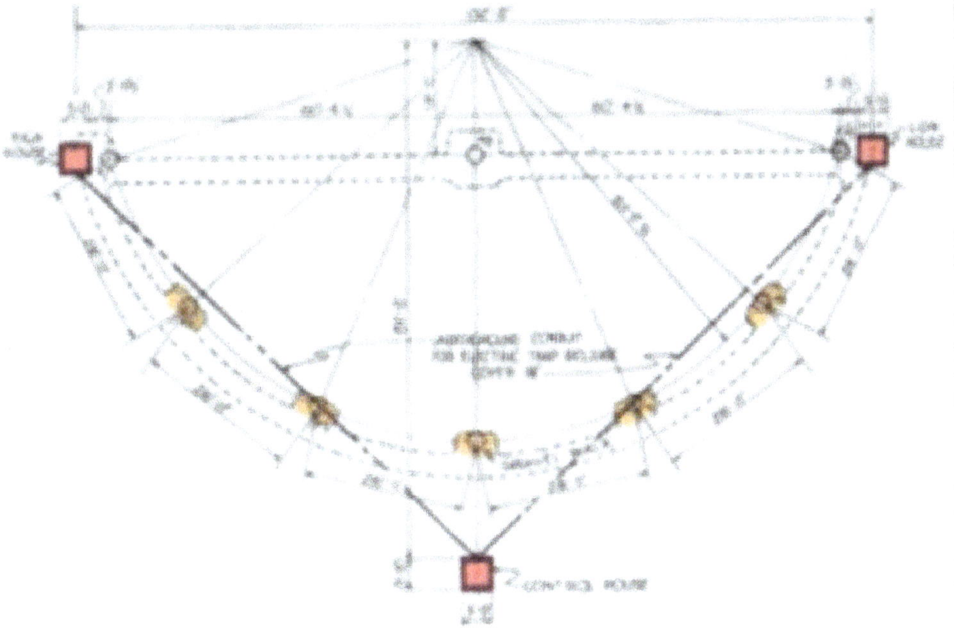

Figure 281. Skeet range, circa 1945 (AAF 85-0-1, Army Air Forces gunnery and bombardment ranges, 15 June 1945, p 2-1-3).

Figure 282. On one of a dozen skeet ranges, aerial gunnery students learn to fire rapidly and to lead their target at Fort Myers Army Air Field, FL, December 1942 (NARA College Park, RG 342-FH, box 2207, photo 4A-18414).

Figure 283. Skeet target practice for cadets at NAS Corpus Christi, TX, 23 July 1941 (NARA College Park, RG 80-G, box 1978, photo 463690).

Firing lines

Firing positions were spaced evenly along a half circle firing line. Some firing lines had shooting towers to simulate firing from aircraft (see Figures 284 and 285).

Figure 284. Close-up of tower on the 7-mile skeet range at Geiger Field (near Spokane), WA, 3 March 1942 (NARA College Park, RG 342-FH, box 2202, photo 4A-18382).

Targets

Skeet range targets were clay pigeons.

Embankments/trenches/etc.

None.

Buildings

A typical skeet range had a control tower, and high and low houses from which clay pigeons were released at high or low angles (see Figures 281 and 285). Additionally, some ranges had towers at firing points to simulate firing from an aircraft (see Figures 284 and 285). A range may also have had a latrine, target storage building, ammunition storage building, other storage sheds, and administrative/maintenance buildings supporting general range functions. The range may also have been part of a larger installation range complex that contained these buildings.

Figure 285. Mile skeet range at Geiger Field (near Spokane), WA, 4 March 1942 (NARA College Park, RG 342-FH, box 2207, photo 4A-18423).

Submarine target range

Sailors trained on a submarine target range to fire at moving targets that simulated submarines in water. The range consisted of a firing line (appears to be covered and combined with a range building in Figure 286), targets moving on a track, and embankments behind the target line and in front of the target track assembly. Soldiers fired at the targets as they moved along the track.

Layouts

Figure 286. Overall view of submarine target range at Camp Wissahickon, NJ, undated (circa 1918) (NARA College Park, RG 24-TC, box 1, Folder P).

Firing lines

Firing lines likely consisted of a line of cleared firing points, and may have been covered (see Figure 286).

Targets

Targets were silhouette panels mounted on dollies that moved up and down along a target track to simulate movement in water (Figure 287). They were controlled by an operator behind the embankment wall (Figure 288).

Figure 287. Looking down on target track on submarine target range at Camp Wissahickon, NJ, undated (circa 1918) (NARA College Park, RG 24-TC, box 1, folder K).

Figure 288. Sailor operating targets on track at submarine target range at Camp Wissahickon, NJ, undated (circa 1918) (NARA College Park, RG 24-TC, box 1, folder K).

Embankments/trenches/etc.

A dirt and log embankment was constructed in front of the moving target track assembly and an embankment wall was constructed behind the target line (Figure 289).

Figure 289. End of target track on submarine target range at Camp Wissahickon, NJ, undated (circa 1918) (NARA College Park, RG 24-TC, box 1, folder K).

Buildings

Only a combination of covered firing line and range building is shown in Figure 286. However, a range may have also had a control tower, latrine, target storage building, ammunition storage building, other storage sheds, and administrative/maintenance buildings supporting general range functions. The range may also have been part of a larger installation range complex that contained these buildings.

Moving firing points and fixed targets

Class B or combat range

"The WWI version of a Combat Range was known as a Class 'B' Range (Figure 290). The WWII version was known as a Combat Range (Figure 291). Soldiers trained on these ranges for combat conditions, firing on the move at surprise targets. The ranges consisted of starting points on a firing line, and targets partially concealed by embankments, trenches, or the natural terrain. The Class 'B' Range used hidden targets and the Combat Range used cable operated pop-up targets" ("RO-1" 33). Soldiers moved down the range and engaged hidden targets as they became visible.

Layouts

Figure 290. Class "B" range, circa 1913 (War Department Document No. 422, Small arms firing manual – 1913 (corrected to 15 April 1917), 28 February 1913, p 199).

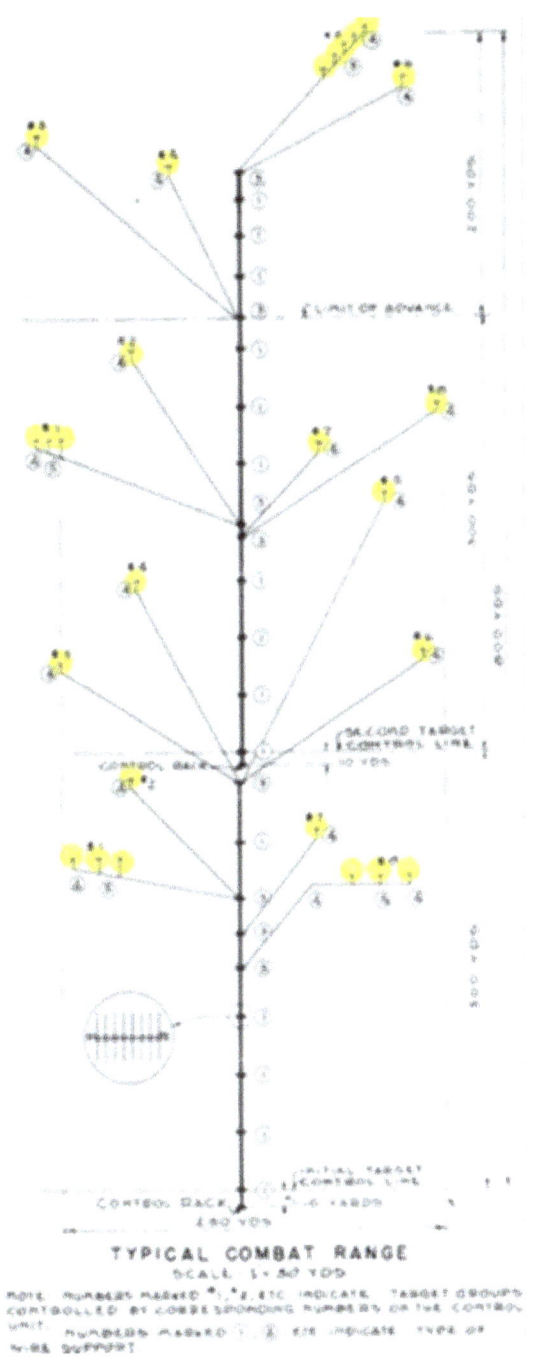

Figure 291. Typical combat range, Fort Bragg, NC, 1951 (Standard drawing 28-13-18 sheet 1 of 1, Range, field target, small arms, plan and details, 21 November 1951).

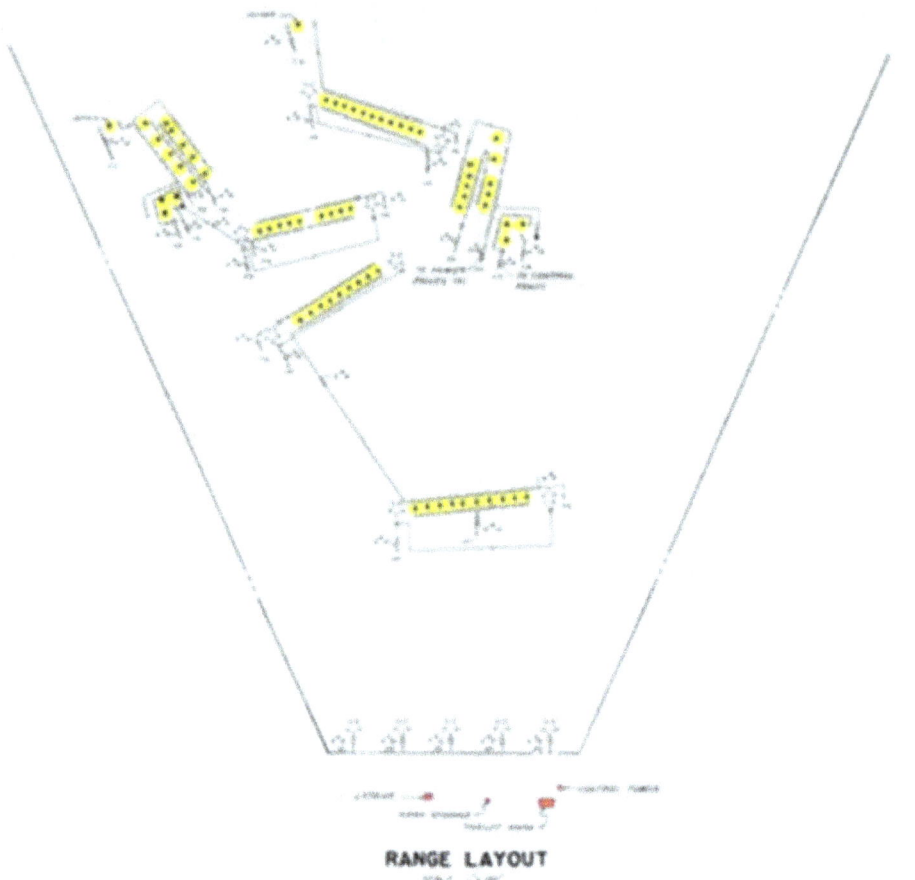

Figure 292. Combat firing range electrical layout, Fort Bragg, NC, 1966 (Standard drawing 28-13-117 sheet 32, Electrical distribution, 28 June 1966).

Firing lines

Soldiers moved forward and engaged targets as they became visible.

Targets

The Class "B" Range used hidden targets, and the Combat Range used cable operated pop-up targets (Figures 293, 294, and 295) ("RO-1" 33).

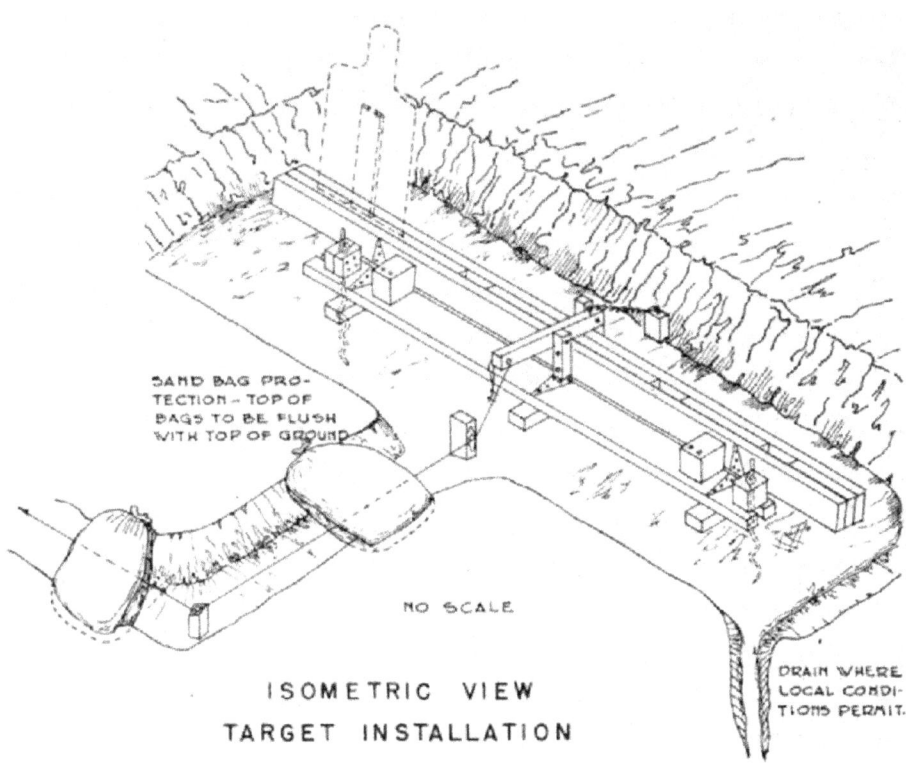

Figure 293. Small arms range, field target, isometric view of target installation, Fort Bragg, NC, 1951 (Standard drawing 28-13-18 sheet 1 of 1, Range, field target, small arms, plan and details, 21 November 1951).

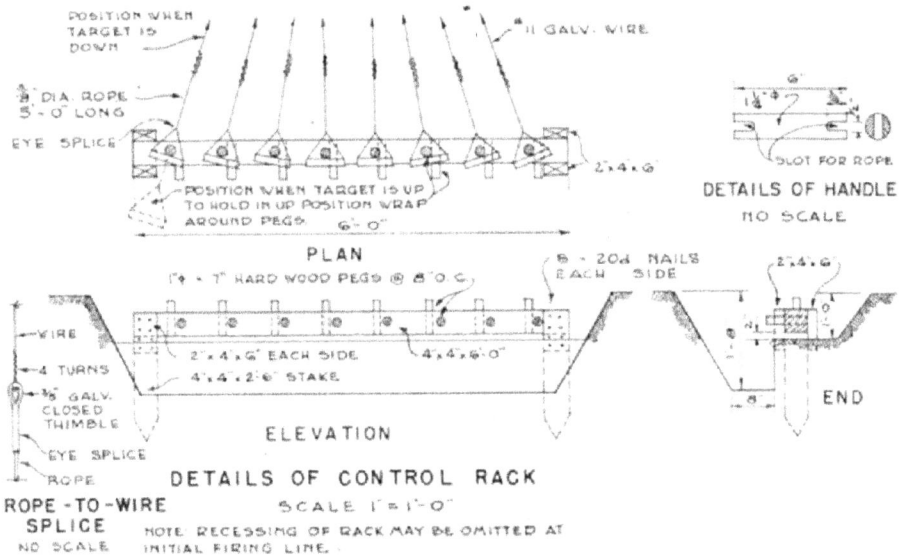

Figure 294. Small arms range, field target, details of control rack, Fort Bragg, NC, 1951 (Standard drawing 28-13-18 sheet 1 of 1, Range, field target, small arms, plan and details, 21 November 1951).

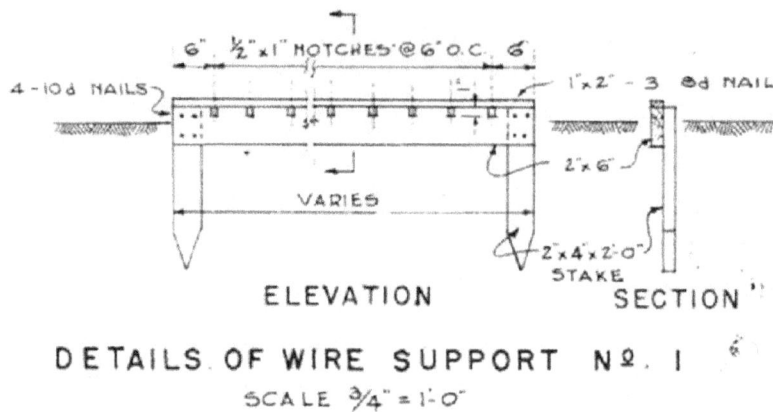

Figure 295. Small arms range, field target, details of wire support, Fort Bragg, NC, 1951 (Standard drawing 28-13-18 sheet 1 of 1, Range, field target, small arms, plan and details, 21 November 1951).

Embankments/trenches/etc.

Embankments may have been built in some locations to partially conceal targets and make the terrain more realistic of combat conditions. Pop-up targets were also partially placed in a trench or behind an embankment for concealment.

Buildings

Control towers, latrines, ammunition storage buildings, target houses, and bleachers are mentioned in the plans for these ranges (see Figures 296 and 297). A range may also have had other storage sheds and administrative/maintenance buildings supporting general range functions. The range may also have been part of a larger installation range complex that contained these buildings.

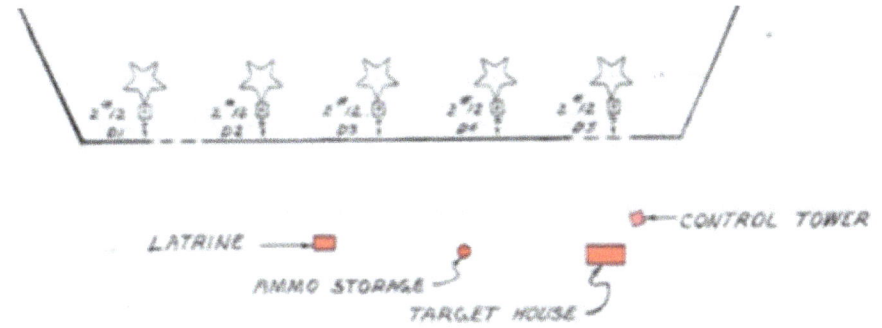

Figure 296. Combat firing range electrical layout, Fort Bragg, NC, 1966 (Standard drawing 28-13-117 sheet 32, Electrical distribution, 28 June 1966).

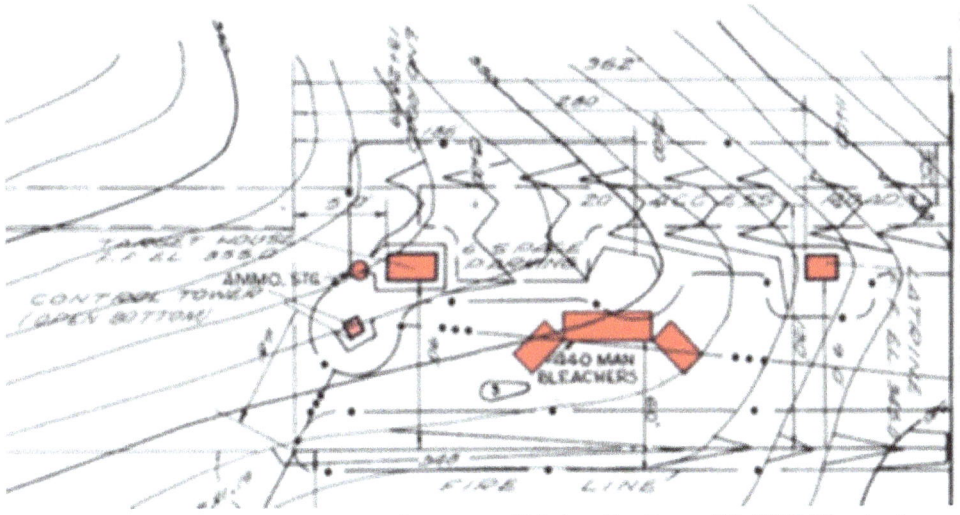

Figure 297. ATC facilities, combat firing range "Q" plan, Fort Bragg, NC, 1966 (Standard drawing 28-13-117 sheet 10, Construction of ranges, phase 1, U.S. ATC facilities, Fort Bragg, NC, range "Q", combat firing range, 28 June 1966).

Dismounted submachine gun practice course

Submachine gunners were trained on these ranges to engage stationary targets while moving ("RO-1" 34). The range consisted of a stationary firing point (phase A); a semicircular firing line in three phases (phases B, C, and D); tape lines to mark divisions between phases; red and green flags to direct the starting and stopping of firing; E, F, and M silhouette type stationary, pivot, hinge, and sled targets; and a control stand or tower for the officer in charge (see Figure 299). Gunners first fired from a stationary point at an M-type stationary target, and then fired in three phases between red flags as they moved along a curved firing line at pivot, hinge, and sled targets. Training was completed with .45 caliber submachine guns.

Layouts

Danger area

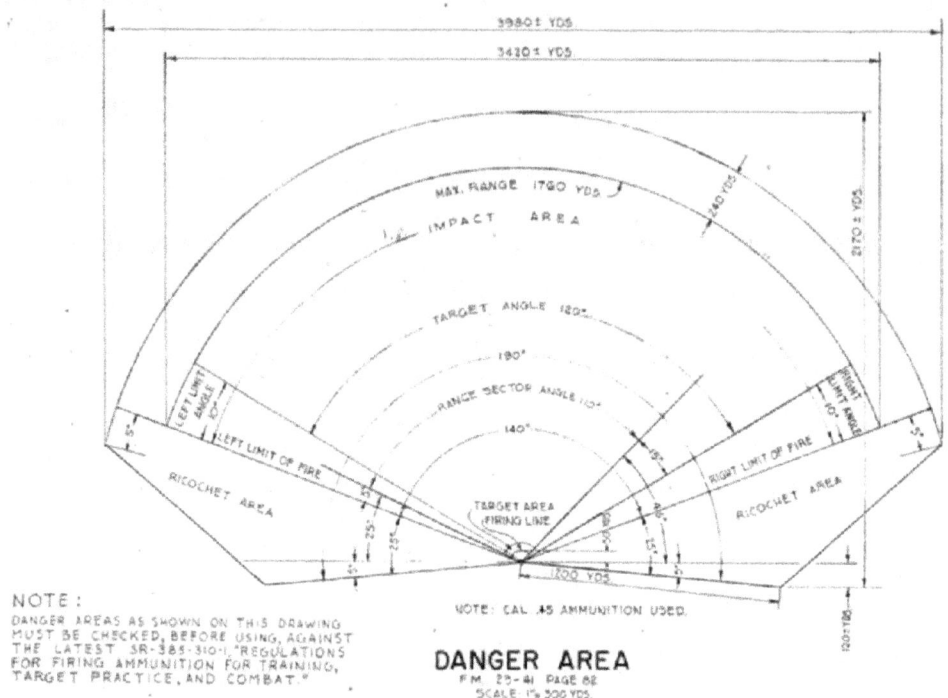

Figure 298. Dismounted submachine gun practice course, danger area, Fort Bragg, NC, 1951 (Standard drawing 28-13-13, sheet 1 of 2, Range, submachine gun, dismounted practice course, plans and details, 7 December 1951).

Typical layout

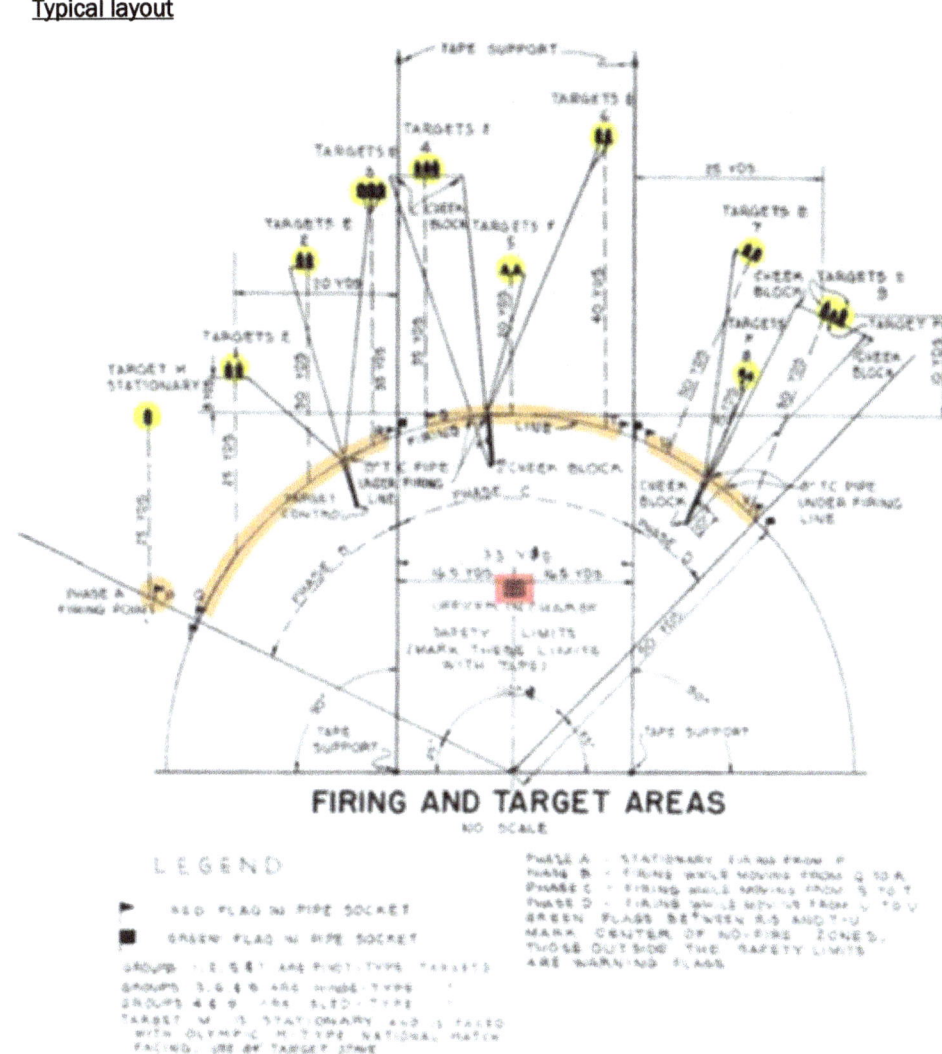

Figure 299. Dismounted submachine gun practice course, firing and target areas, Fort Bragg, NC, 1951 (Standard drawing 28-13-13, sheet 1 of 2, Range, submachine gun, dismounted practice course, plans and details, 7 December 1951).

Firing lines

Firing lines on this range included an initial firing point (phase A), and three phases of firing on the move along a curved firing line (phases B, C, and D). Red flags in pipe sockets marked the beginning and ending of firing in each phase, and green flags marked the center of no fire zones along the firing line (see Figure 299).

Targets

Soldiers began firing at a stationary M-type target (faced with Olympic M-type national match facing, and mounted on an 84" target stave) in phase A. In phase B, they fired at two groups of two pivot-type E silhouette targets and one group of three hinge-type E silhouette targets. In phase C, they fired at a group of three sled-type E silhouette targets, a group of two pivot-type F silhouette targets, and a group of two hinge-type E silhouette targets. In phase D, they fired at a group of two hinge-type E silhouette targets, a group of two hinge-type F silhouette targets, and a group of three sled-type (two E and one F silhouettes) targets.

<u>Pivot targets</u>

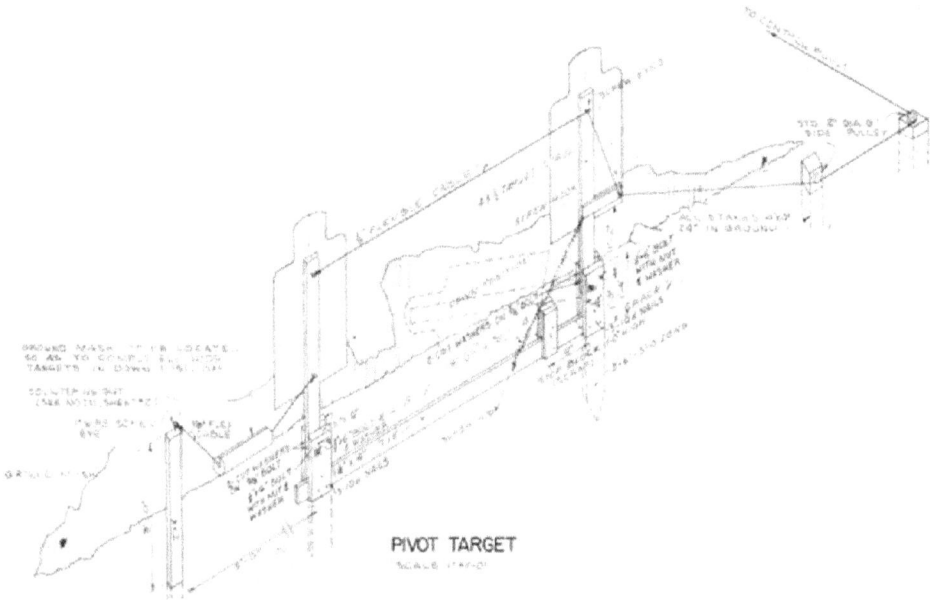

Figure 300. Dismounted submachine gun practice course, pivot target, Fort Bragg, NC, 1951 (Standard drawing 28-13-13 sheet 1 of 2, Range, submachine gun, dismounted practice course, plans and details, 7 December 1951).

Hinge targets

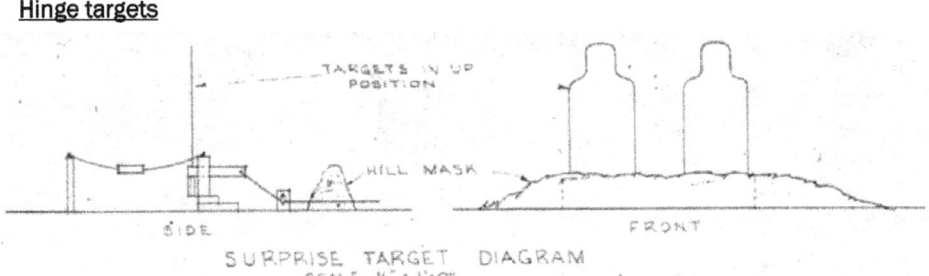

Figure 301. Dismounted submachine gun practice course, surprise target diagram, Fort Bragg, NC, 1951 (Standard drawing 28-13-13 sheet 2 of 2, Range, submachine gun, dismounted practice course, details, 7 December 1951).

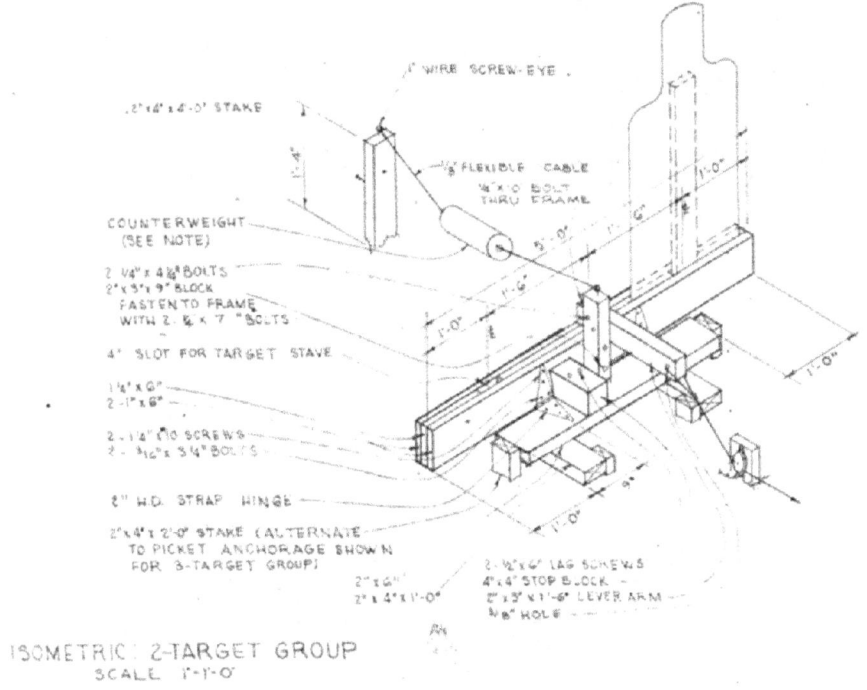

Figure 302. Dismounted submachine gun practice course, surprise 2-target group isometric, Fort Bragg, NC, 1951 (Standard drawing 28-13-13 sheet 2 of 2, Range, submachine gun, dismounted practice course, details, 7 December 1951).

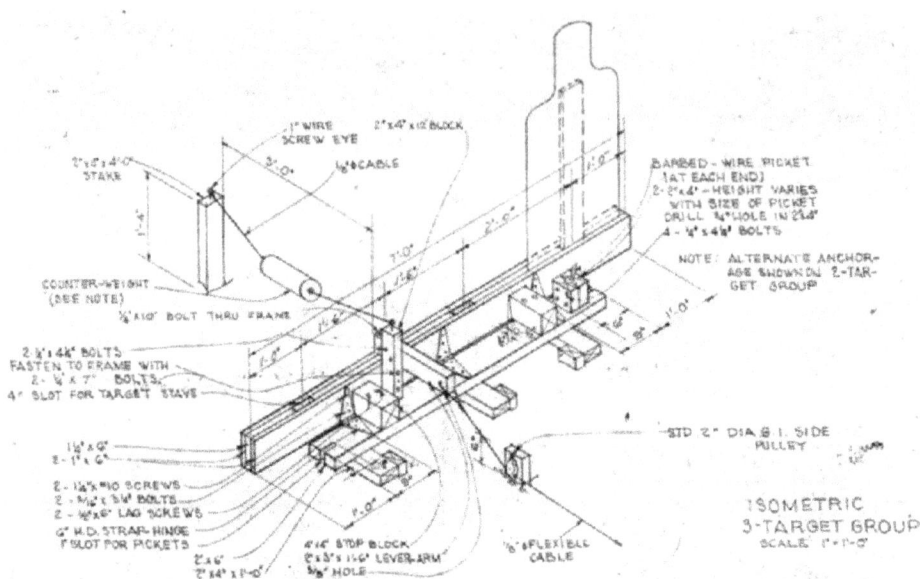

Figure 303. Dismounted submachine gun practice course, surprise 3-target group isometric, Fort Bragg, NC, 1951 (Standard drawing 28-13-13 sheet 2 of 2, Range, submachine gun, dismounted practice course, details, 7 December 1951).

Sled targets

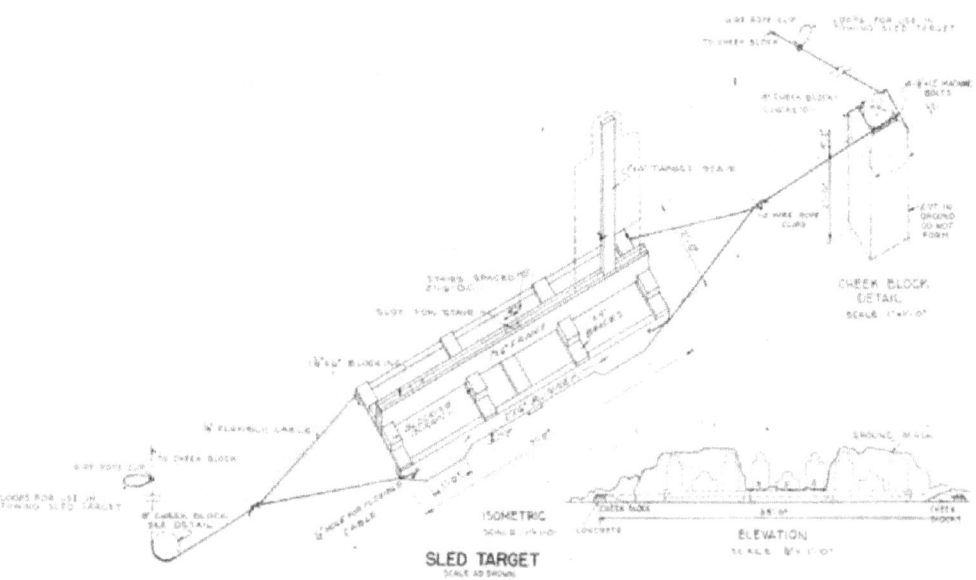

Figure 304. Dismounted submachine gun practice course, sled target, Fort Bragg, NC, 1951 (Standard drawing 28-13-13 sheet 1 of 2, Range, submachine gun, dismounted practice course, plans and details, 7 December 1951).

Embankments/trenches/etc.

Embankments were often built in front of surprise and moving target assemblies (see Figures 301 and 304).

Buildings

A control tower or stand was placed behind and in the middle of phases B, C, and D firing lines for the officer in charge of training. A range may also have had a latrine, target storage building, ammunition storage building, other storage sheds, and administrative/maintenance buildings supporting general range functions. The range may also have been part of a larger installation range complex that contained these buildings.

Mounted pistol range

Soldiers were trained on these ranges to fire pistols from horseback. The range consisted of a riding course with stationary targets to the right and left of the direction the horse and rider were traveling (see Figure 305). As the rider approached targets set at various angles from the course, he fired on them with his pistol.

Layouts

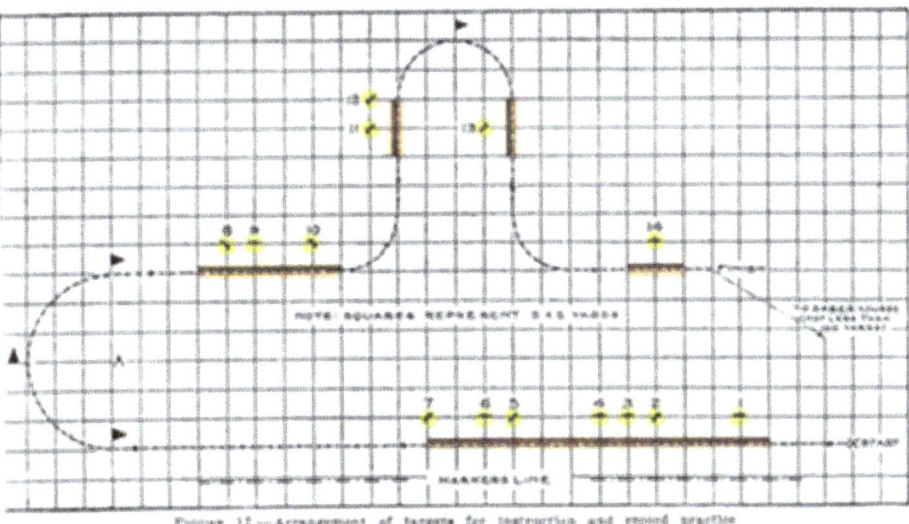

Figure 305. Mounted pistol course, circa 1932 (Basic field manual, volume III – Basic weapons, Part one rifle company, Chapter 3 – Automatic pistol marksmanship, 5 April 1932, p 40).

Firing lines

Firing lines were the stretches along the riding course where targets became visible.

Targets

Stationary targets were set at 90- and 45-degree angles 5 yds away from the riding course on either side.

Embankments/trenches/etc.

None.

Buildings

No buildings are mentioned in the standard plans for these ranges. However, a range may have had a control tower, latrine, target storage building, ammunition storage building, other storage sheds, and administrative/maintenance buildings supporting general range functions. The range may also have been part of a larger installation range complex that contained these buildings.

Moving vehicle machine gun range

Vehicle crewmen trained on these ranges to fire machine guns mounted on moving vehicles ("RO-1" 36). The range consisted of a firing runway (fitted with obstacles for full-track vehicles) that approached a target line of kneeling silhouette, standing silhouette, and 5' by 8' panel targets (see Figure 306). As the vehicle crew traveled down the runway in wheeled, half-track, or full-track vehicles they fired at the targets from moving and halted (wheeled and half-track vehicles only) firing positions. Training was completed with .50 caliber machine guns (see Figure 307).

Layouts

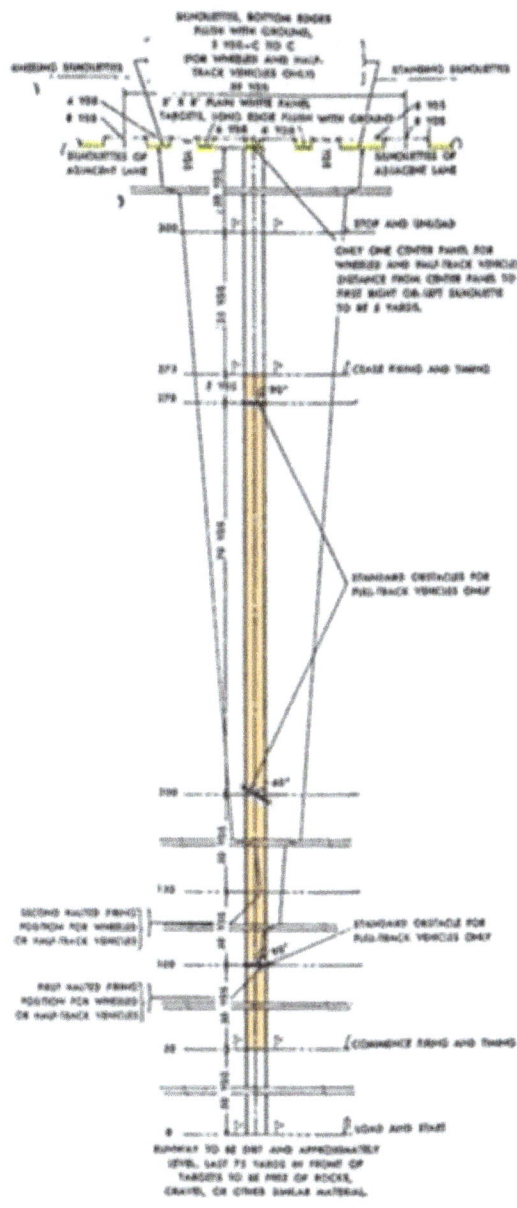

Figure 306. Moving vehicle range, machine gun, circa 1942 (Standard drawing No. 1600-135, Moving vehicle ranges, 23 October 1942).

Firing lines

Firing was performed from moving and halted (for wheeled and half-track vehicles only) positions on a 225-yd stretch of an approximately level dirt runway (with driving obstacles for full-track vehicles only) (see Figure

306). After moving 50 yds in their vehicles, the crews began timed firing. Twenty-five yards after the end of the 225-yd firing stretch, the crews stopped their vehicles and unloaded.

Figure 307. Trainees firing the 50 caliber machine gun mounted in the half-track at Fort Knox, KY, August 1942 (NARA College Park, RG 111-SC WWII, box 85, photo SC114295).

Figure 308. Machine gun firing range at Casper Army Air Field, WY, 5 May 1943 (NARA College Park, RG 342-FH, box 2202, photo 4A-17283).

Figure 309. Students operate machine guns from an E-9 gunnery truck at Harvard Army Air Field, NE, 22 April 1945 (NARA College Park, RG 342-FH, box 2202, photo 4A-17254).

Targets

Targets consisted of kneeling silhouettes, standing silhouettes, and 5' x 8' plain white panel targets placed flush with the ground on a target line at the end of the firing runway (Figure 310). Only one center target panel was used for wheeled and half-track vehicles (see Figure 311).

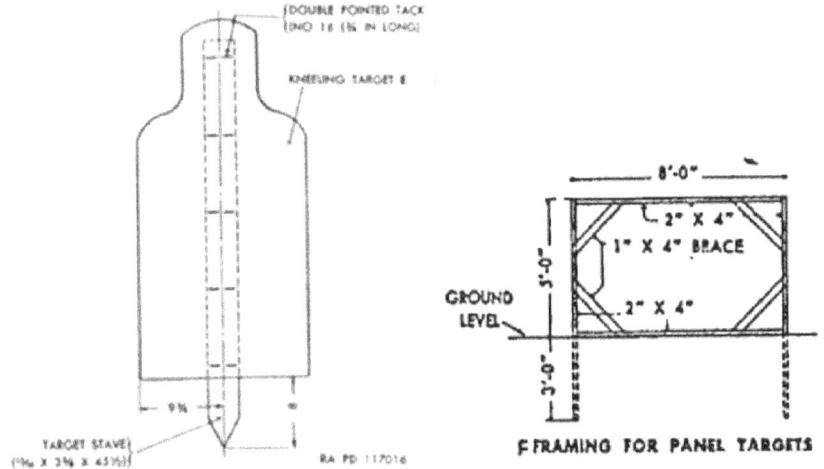

Figure 310. Kneeling pasteboard target E, circa 1951, and panel target, circa 1942 (TM 9-855, Targets, target materials, and rifle range construction, 17 November 1951, pp 174, 176, "Framing for panel targets" RO-1 fig 22 pg. 36, Standard drawing No. 1600-135, Moving vehicle ranges, 23 October 1942).

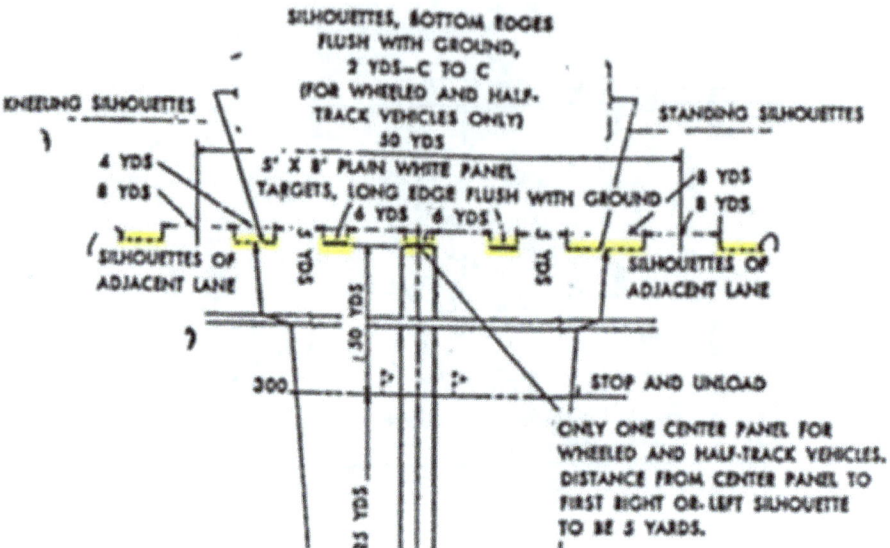

Figure 311. Moving vehicle range, machine gun, circa 1942 (Standard drawing No. 1600-135, Moving vehicle ranges, 23 October 1942).

Embankments/trenches/etc.

Obstacles placed on the firing runway (for full-track vehicles only) helped simulate battle road conditions for firing (see Figure 312).

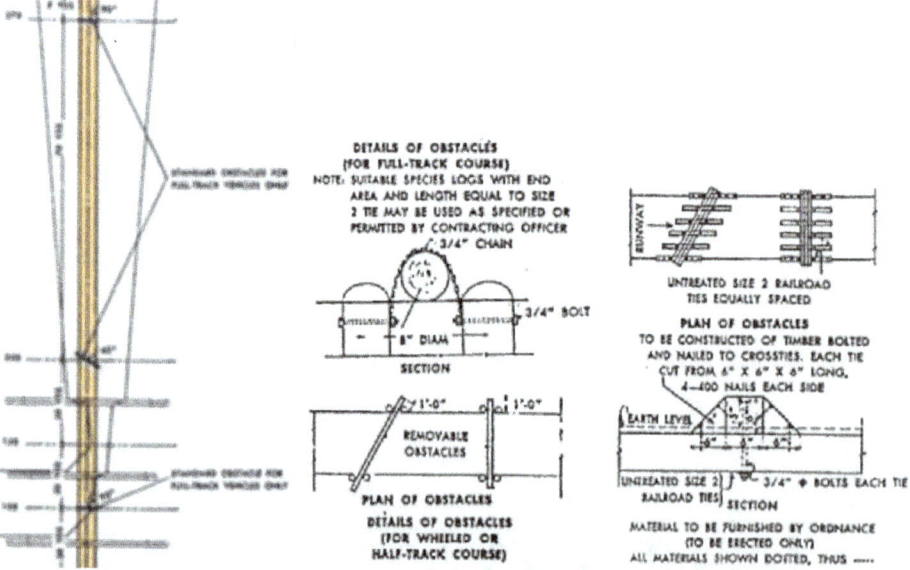

Figure 312. Moving vehicle range, machine gun, obstacles, circa 1942 (Standard drawing No. 1600-135, Moving vehicle ranges, 23 October 1942).

Buildings

No buildings are mentioned in the standard plans for these ranges. However, a range may have had a control tower, latrine, target storage building, ammunition storage building, other storage sheds, and administrative/maintenance buildings supporting general range functions. The range may also have been part of a larger installation range complex that contained these buildings.

Moving vehicle submachine gun range

Armored vehicle crews trained on these ranges to fire submachine guns from a moving vehicle. "A submachine gun was assigned to armored vehicle crews for use when vehicle armament was not" ("RO-1" 34). The range consisted of a rectangular firing road, three groups of fixed E- and F-type targets, red flags in between which firing was commenced, and a white flag that signaled the end of the course (see Figure 315). As the vehicle moved down the firing road, they reached their first red flag, which signified the commencement of firing and timing. They fired at group one until they reached the second red flag. They then traveled to the third red flag where they fired at group two until they reached the fourth red flag. They traveled down the road to the fifth red flag where they halted for fifteen seconds, reversed their gun, and then commenced firing at targets behind the vehicle until they reached the sixth red flag, where their timing was stopped. They then traveled to the seventh (white) flag, where they halted and cleared their guns. Training was completed with submachine guns.

Layouts

Danger area

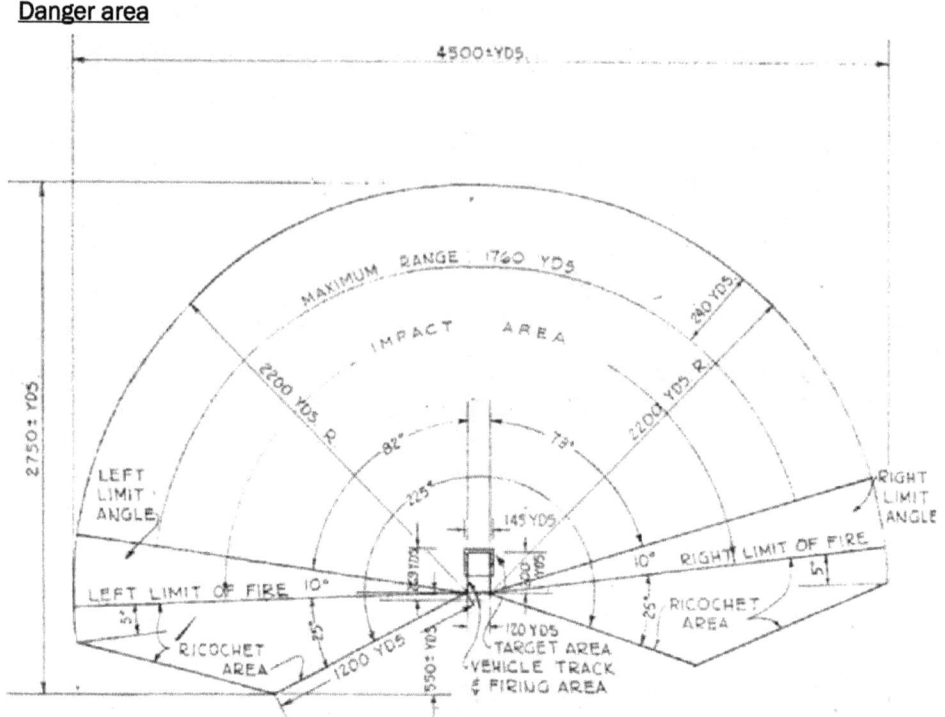

Figure 313. Submachine gun moving vehicle range, danger area, Fort Bragg, NC, 1951 (Standard drawing 28-13-21 sheet 1 of 1, Range, moving vehicle, submachine gun, plan and details, 21 November 1951).

Typical layout

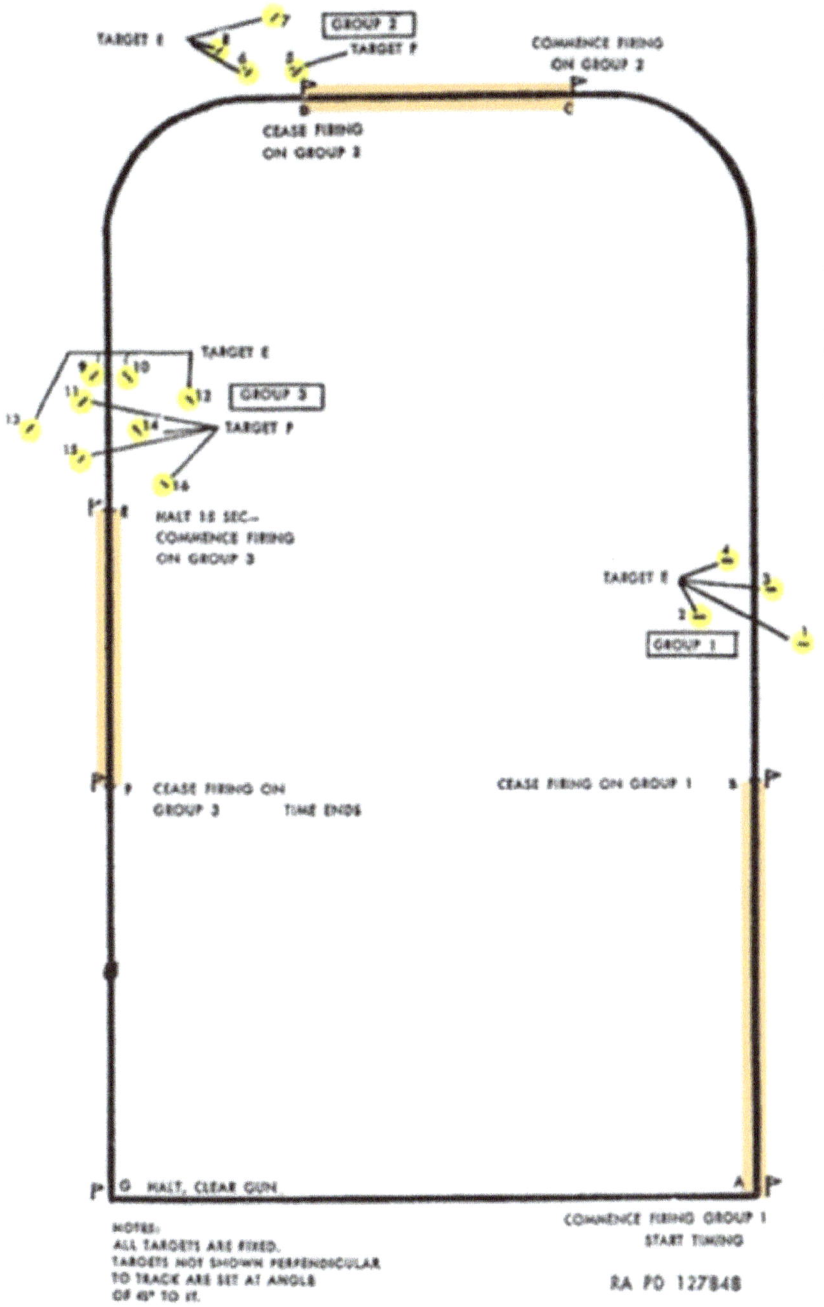

Figure 314. Moving Vehicle range, submachine gun, circa 1942 to 1951 (Standard drawing No. 1600-135, Moving vehicle ranges, 23 October 1942).

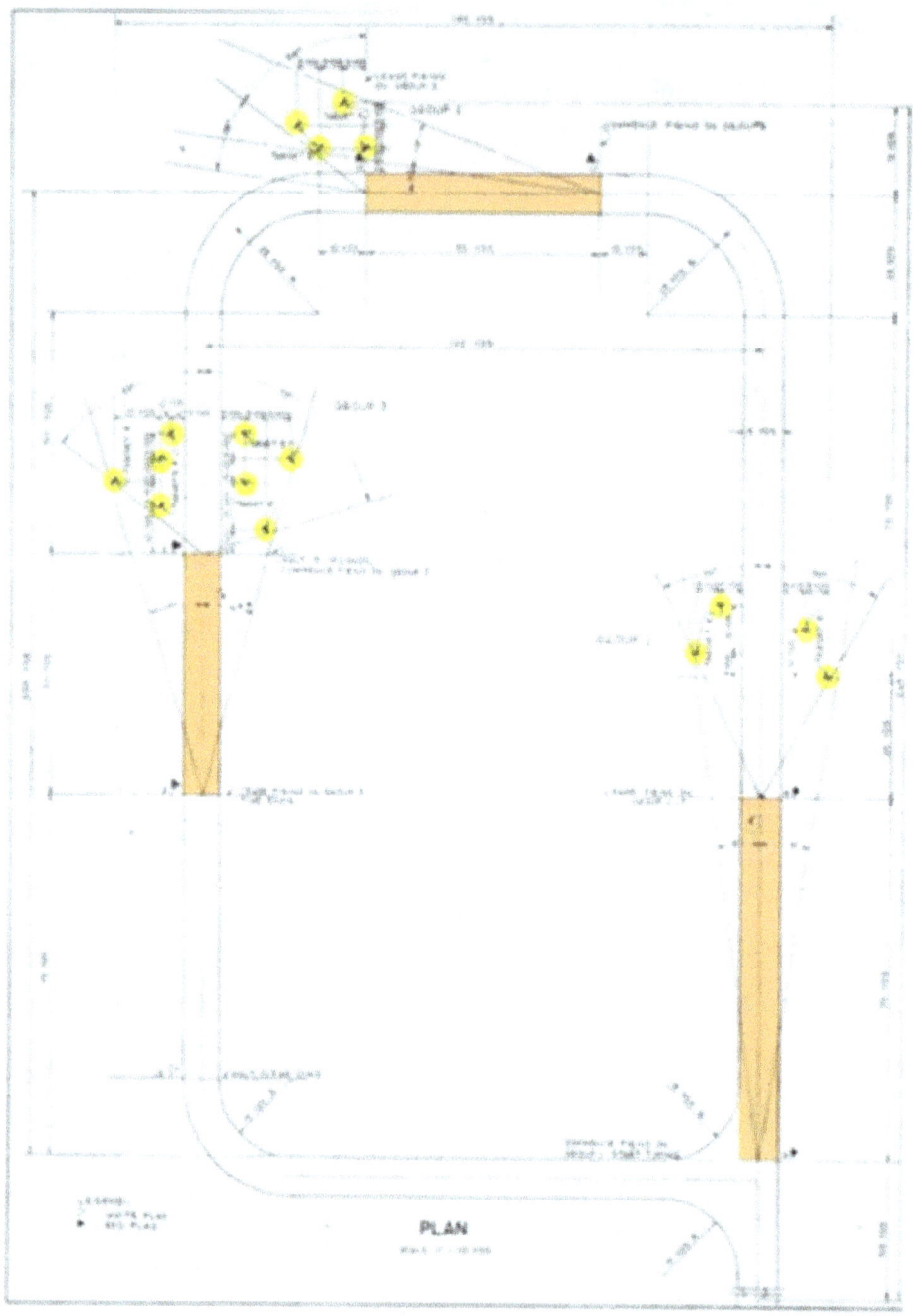

Figure 315. Submachine gun moving vehicle range, plan, Fort Bragg, NC, 1951 (Standard drawing 28-13-21 sheet 1 of 1, Range, moving vehicle, submachine gun, plan and details, 21 November 1951).

Firing lines

Soldiers fired from a moving vehicle on three sections of the road marked by red flags at the beginning and end of the section. In the first section, personnel fired for 75 yds at targets on both sides of the road. In the second section, personnel fired for 50 yds at targets on the right side of the road. In the third section, personnel fired for 50 yds at targets behind the vehicle on both sides of the road (see Figure 315).

Targets

Three groupings of E- and F-type silhouette targets were used on the course. The first group consisted of four E-type targets set at 90-degree angles from the road. The second group consisted of two E- and two F-type targets set at a 45-degree angle from the road. The third group consisted of four E- and four F-type targets set at a 45-degree angle from the road (see Figures 314 and 315).

Embankments/trenches/etc.

None.

Buildings

No buildings are mentioned in the standard plans for these ranges. However, a range may have had a control tower, latrine, target storage building, ammunition storage building, other storage sheds, and administrative/maintenance buildings supporting general range functions. The range may also have been part of a larger installation range complex that contained these buildings.

Transition range

Individual soldiers were trained on these ranges to fire on the move at targets that appeared suddenly. The range consisted of lanes with an initial firing point (sometimes fitted with a foxhole, window frame, rubble pile, stump, barricade, ditch, roof top, log, or prone position support), pop up or moving targets hidden by various embankments, and target control racks or control pits (see Figure 318). As a soldier moved down his firing lane, he fired at pop-up targets as they appeared, similar to a field or combat range ("RO-1" 37). Training was completed with .30 caliber rifles (see Figure 316).

Layouts

Danger areas

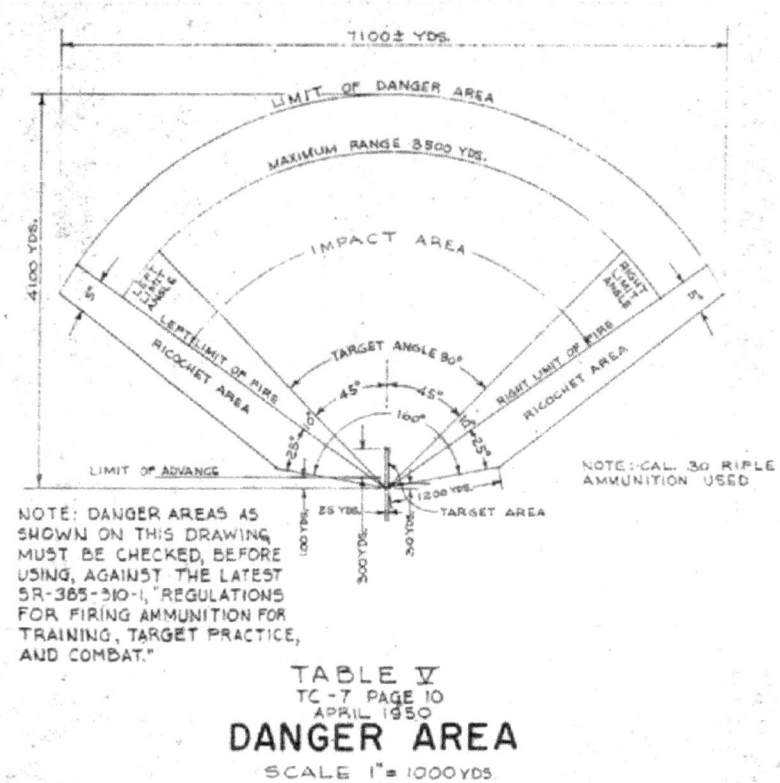

Figure 316. Transition range danger area table V, Fort Bragg, NC, 1955 (Standard drawing 28-13-04 sheet 1 of 1, Range, transition, details, 6 June 1955).

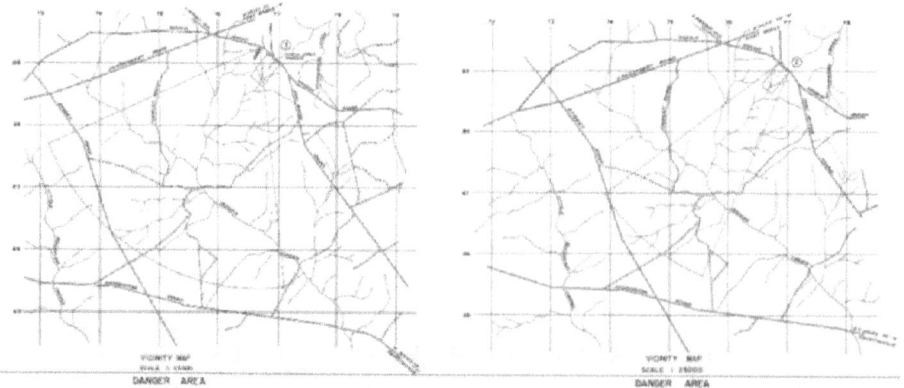

Figure 317. Superimposed transition range, danger areas, Fort Bragg, NC, 1954 (Standard drawing 28-13-106 sheet 1 of 1, Superimposed transition range, table VIII, 22 March 1954, Standard drawing 28-13-107 sheet 1 of 1, Superimposed transition range, table VII, 28 March 1954).

Typical layouts

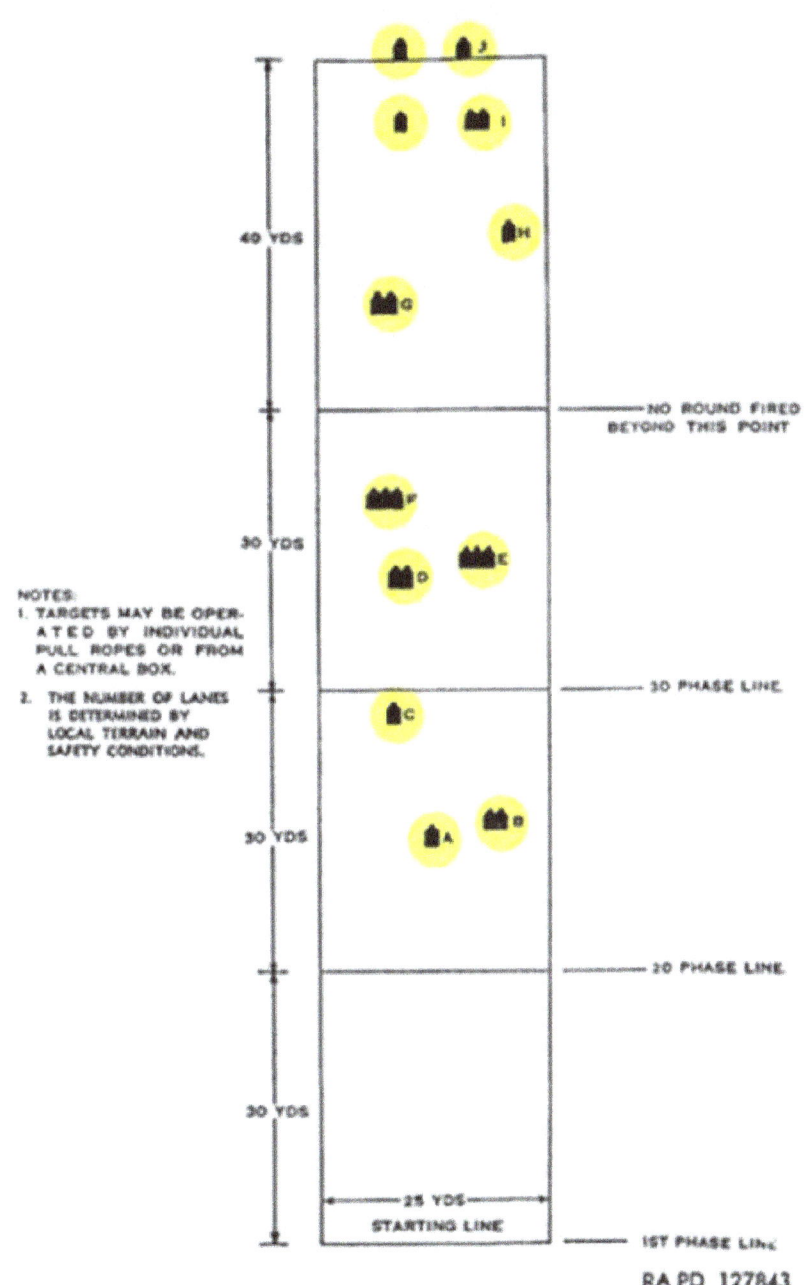

Figure 318. Typical rifle transition range target layout, circa 1943 to 1952 (Standard drawing No. 1600-200, Transition firing course, 18 August 1943).

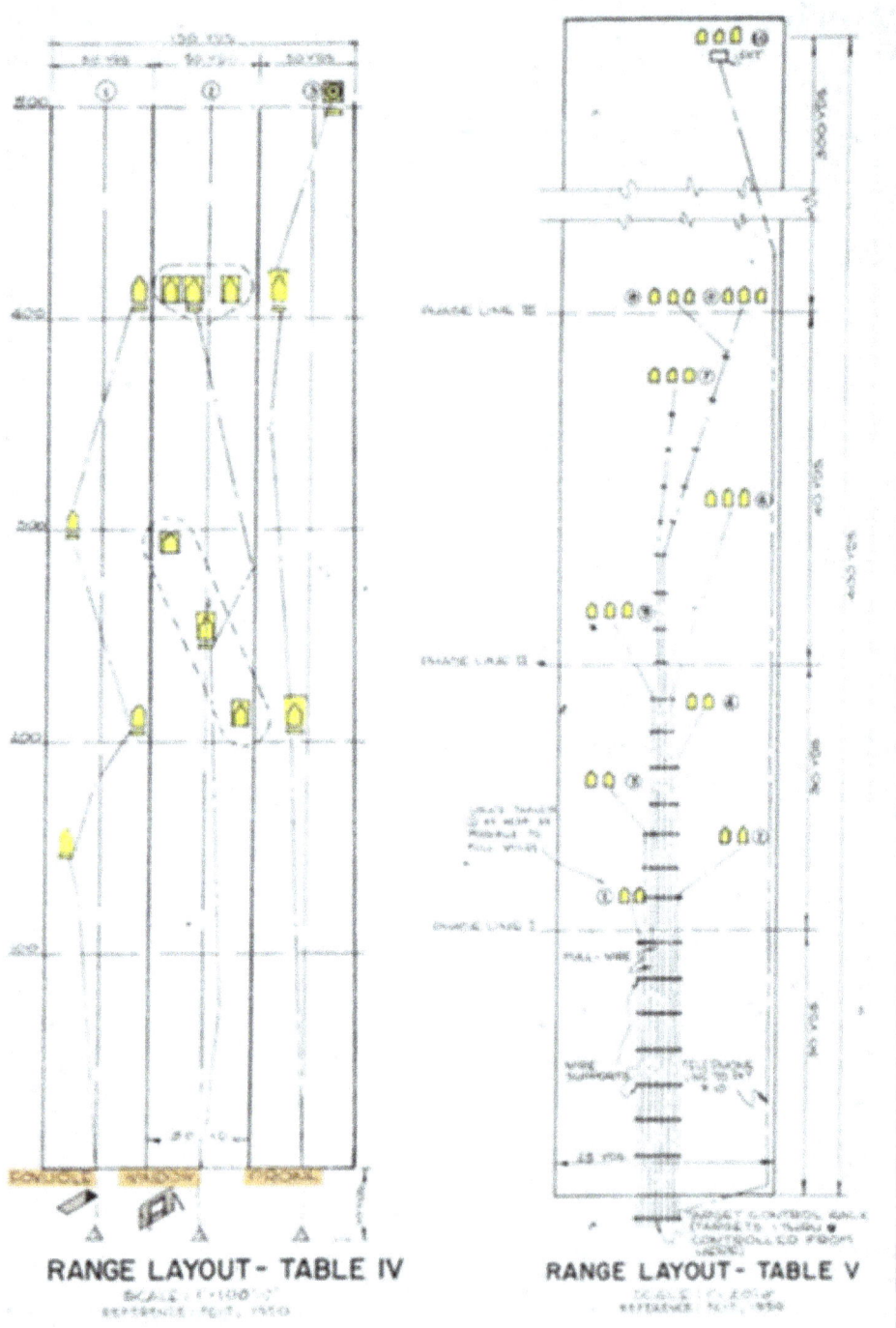

Figure 319. Transition range layouts, table IV and V, Fort Bragg, NC, 1955 (Standard drawing 28-13-04 sheet 1 of 1, Range, transition, details, 6 June 1955).

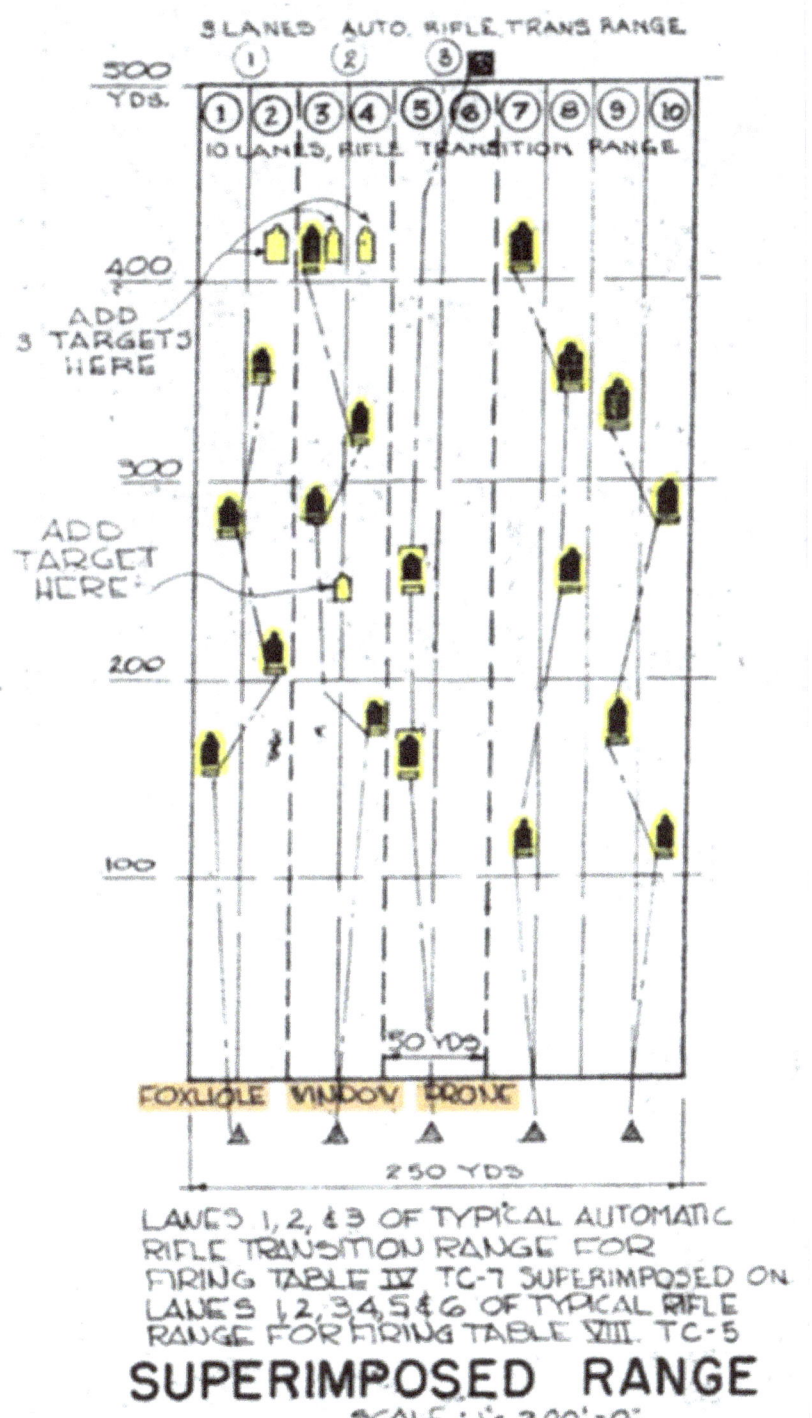

Figure 320. Superimposed transition range, Fort Bragg, NC, 1955 (Standard drawing 28-13-04 sheet 1 of 1, Range, transition, details, 6 June 1955).

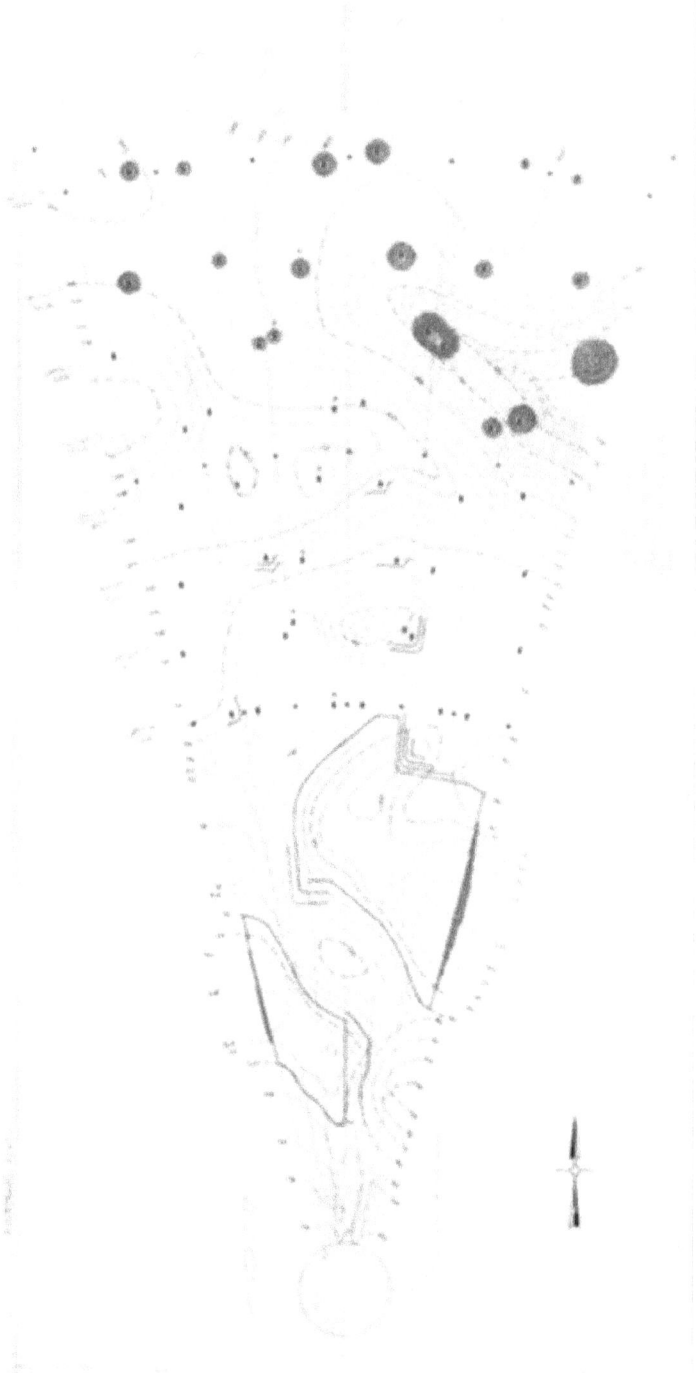

Figure 321. Transition firing range, machine gun range 53, overall site plan, Fort Bragg, NC, 1978 (Standard drawing DFE-3278 sheet 3 of 10, Transition firing range, machine gun-range 53, overall site plan, 19 January 1978).

Figure 322. Aerial photograph of Heins Rifle Transition Range, Fort Knox, KY, undated (Training 3a, vol. 1, Stock Shot #296, Fort Knox, KY, Heins Rifle Transition Range, Fort Knox, photographer Unk., undated, USAARMC Photo Branch, Fort Knox, KY).

Firing lines

At the starting line of the range an initial firing point may have included a foxhole, window frame, rubble pile, stump, barricade, ditch, roof top, log, or prone position support (see Figure 323). This initial firing point may have been built up as a mound (see Figures 324 and 325). As soldiers moved down their individual lanes, they fired from various positions at targets when they appeared.

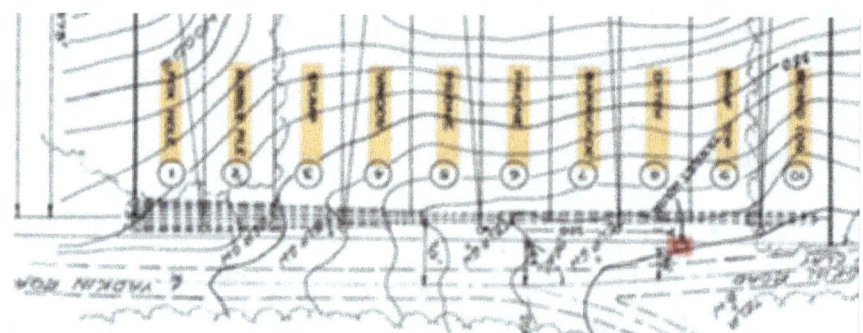

Figure 323. Superimposed transition range, firing line, Fort Bragg, NC, 1954 (Standard drawing 28-13-107 sheet 1 of 1, Superimposed transition range, Table VII, 28 March 1954).

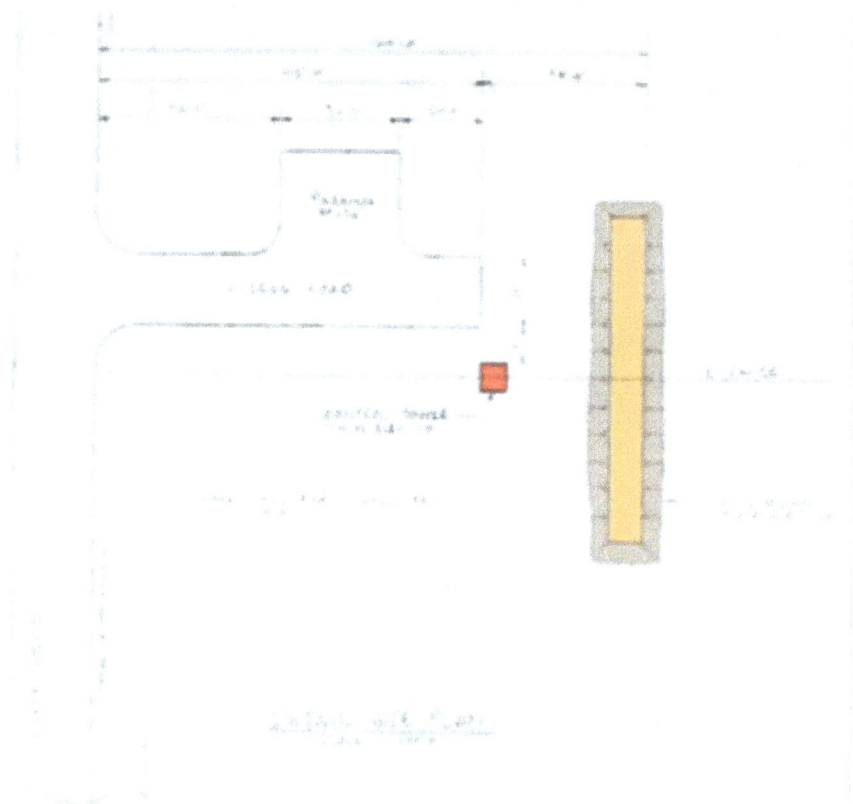

Figure 324. Transition firing range, machine gun range 53, Detail site plan, Fort Bragg, NC, 1978 (Standard drawing DFE-3278 sheet 6 of 10, Transition firing range, machine gun-range 53, range details, 19 January 1978).

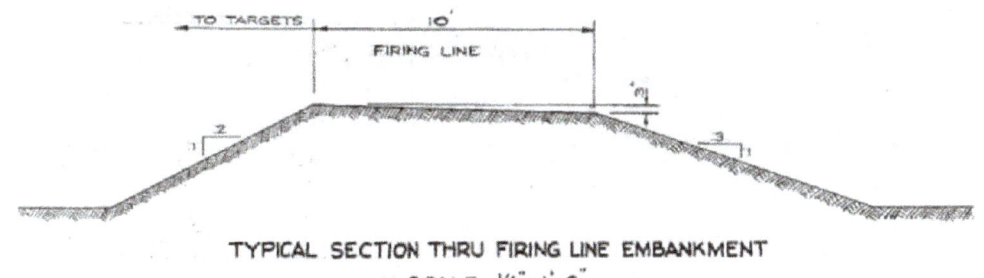

Figure 325. Superimposed transition range, firing line, Fort Bragg, NC, 1954 (Standard drawing 28-13-107 sheet 1 of 1, Superimposed transition range, table VII, 28 March 1954).

Figure 326. Marines at the transition range at MCB Camp Lejeune, NC, 1961 (NARA College Park, RG 127-GG-958, photo 341782).

Figure 327. A drill instructor teaches recruits the proper method of firing from inside a building at MCRD San Diego (Camp Pendleton), CA, December 1972 (NARA College Park, RG 127-GG-936, photo 230641).

Targets

Transition ranges used pop-up or moving E-type silhouette targets, B-type targets, or machine gun panel targets with silhouettes as aiming points (see Figures 328 through 330). Targets were made to pop-up or move through a control rack system, or a control pit next to the target from which individual targets were moved (see Figure 331). Targets raised and lowered individually, simultaneously, or simultaneously in groups. They were often concealed by a variety of embankments, including above ground moving target pits (Figure 332), above ground deep pits (Figure 333), shell holes (Figure 334), foxholes (Figure 335), vertical target pits (Figure 336), and concrete M-30 pop-up target emplacements (Figure 337). A mock window frame also may have concealed them (see Figure 338).

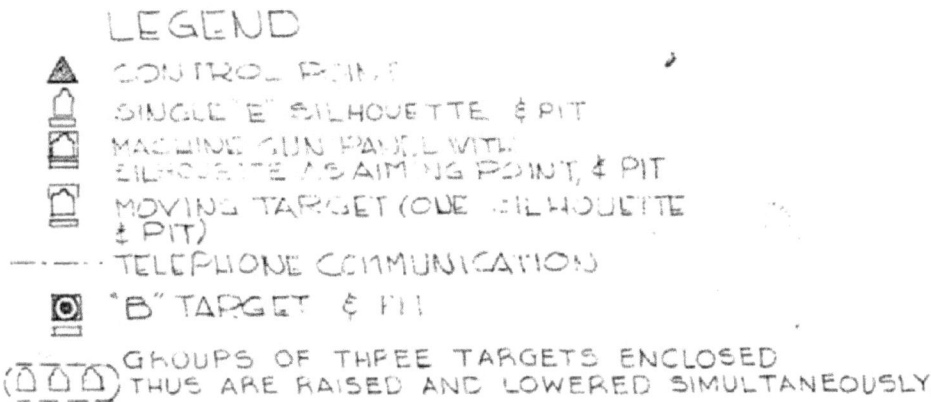

Figure 328. Superimposed transition range, Fort Bragg, NC, 1955 (Standard drawing 28-13-04 sheet 1 of 1, Range, transition, details, 6 June 1955).

Machine gun panel target

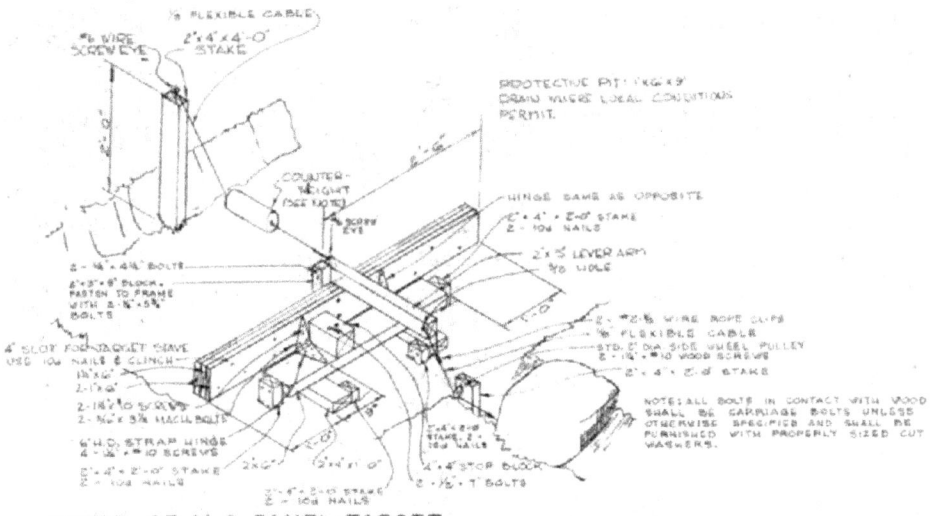

Figure 329. Transition range detail of M.G. Panel target, Fort Bragg, NC, 1955 (Standard drawing 28-13-04 sheet 1 of 1, Range, transition, details, 6 June 1955).

Moving target

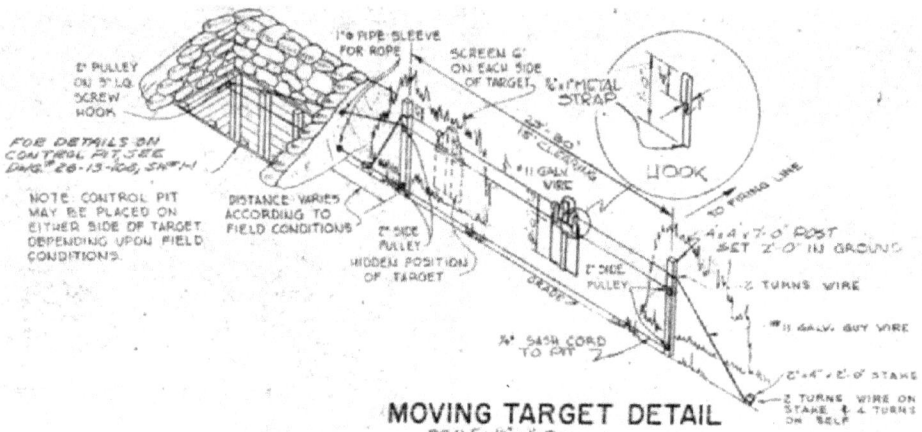

Figure 330. Transition range moving target detail, Fort Bragg, NC, 1955 (Standard drawing 28-13-04 sheet 1 of 1, Range, transition, details, 6 June 1955).

Target control pit

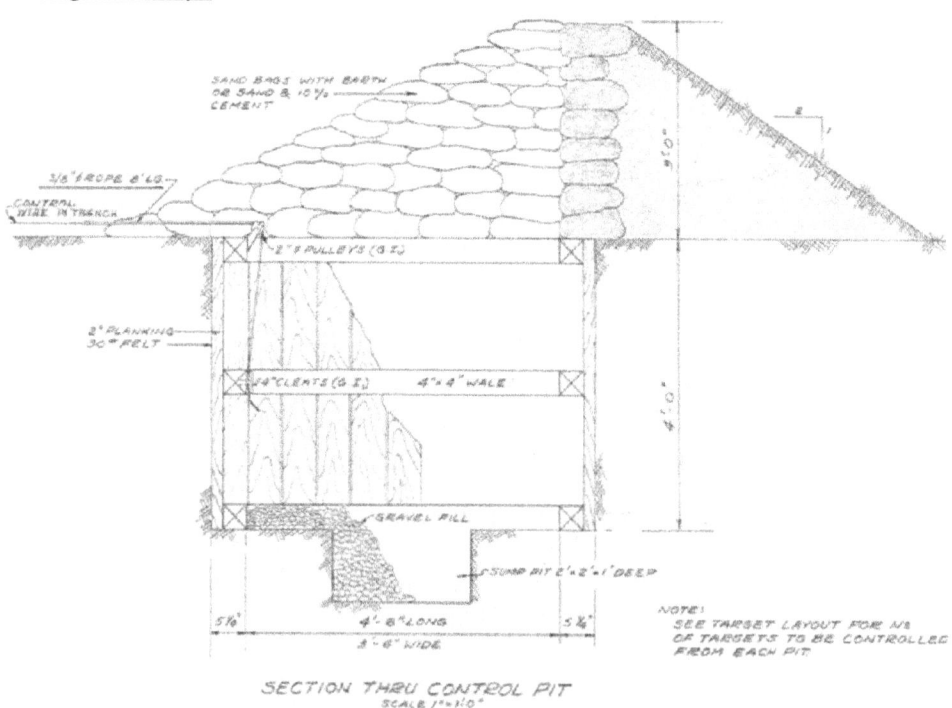

Figure 331. Superimposed transition range, section thru control pit, Fort Bragg, NC, 1954 (Standard drawing 28-13-108 sheet 1 of 1, Superimposed transition range, table VIII, 22 March 1954).

Above ground moving target pit

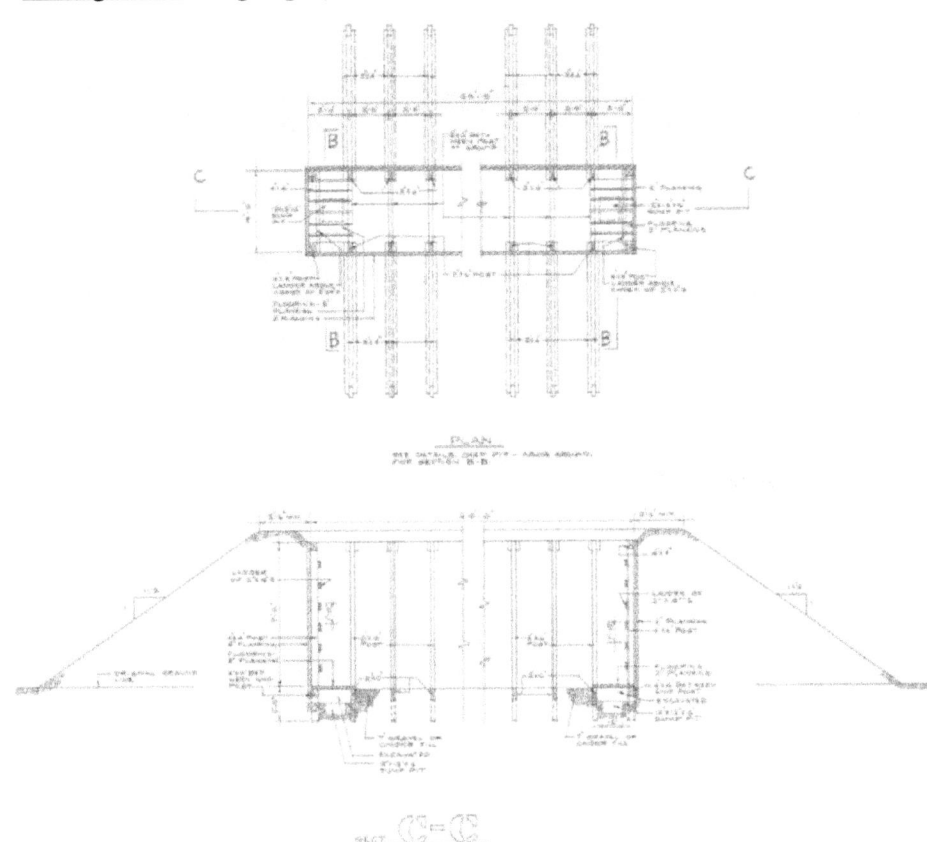

Figure 332. Above ground moving target pit, section and plan, Fort Bragg, NC, 1955 (Standard drawing 28-13-05 sheet 1 of 1, Transition ranges, M1 rifle and MG, above ground pits, construction details, 6 June 1955).

Above ground deep pit

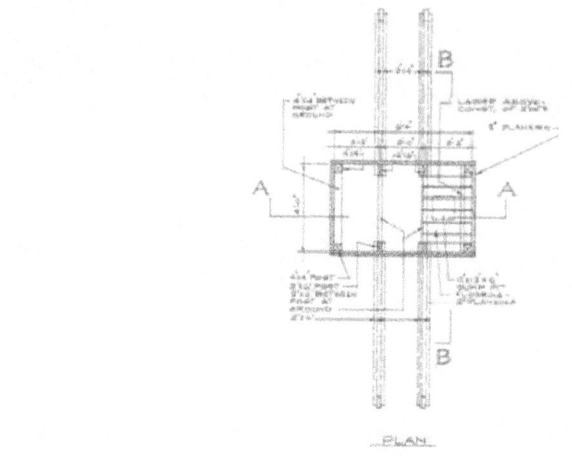

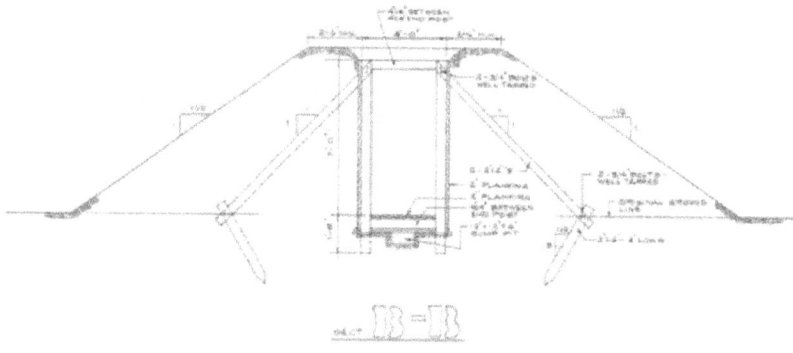

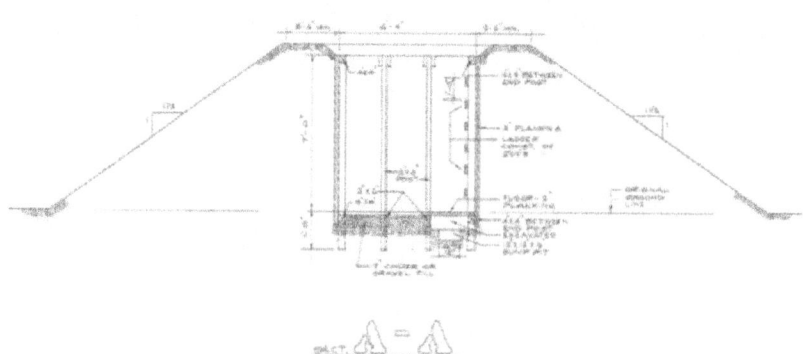

Figure 333. Above ground deep pit, sections and plan, Fort Bragg, NC, 1955 (Standard drawing 28-13-05 sheet 1 of 1, Transition ranges, M1 Rifle and MG, above ground pits, construction details, 6 June 1955).

Shell hole

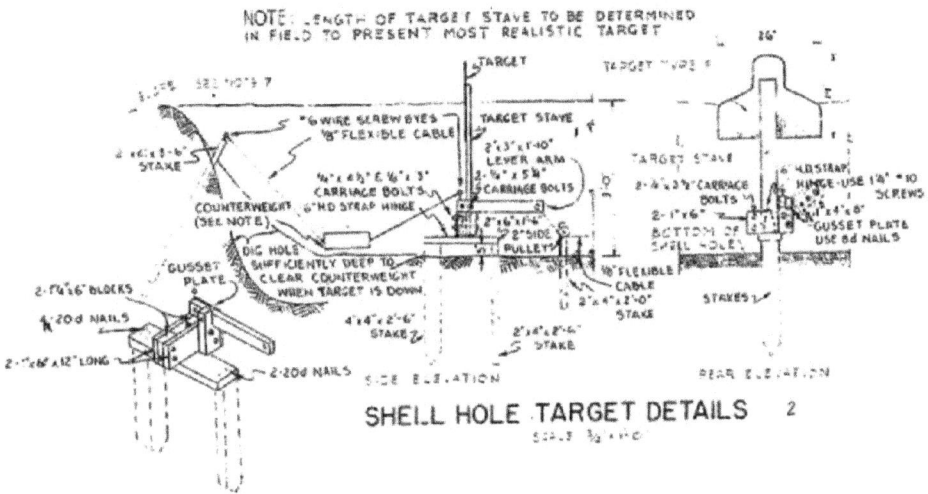

Figure 334. Shell hole target details, Fort Bragg, NC, 1957 (Standard drawing 28-13-05, close combat course, plan and details, 8 September 1957).

Fox hole

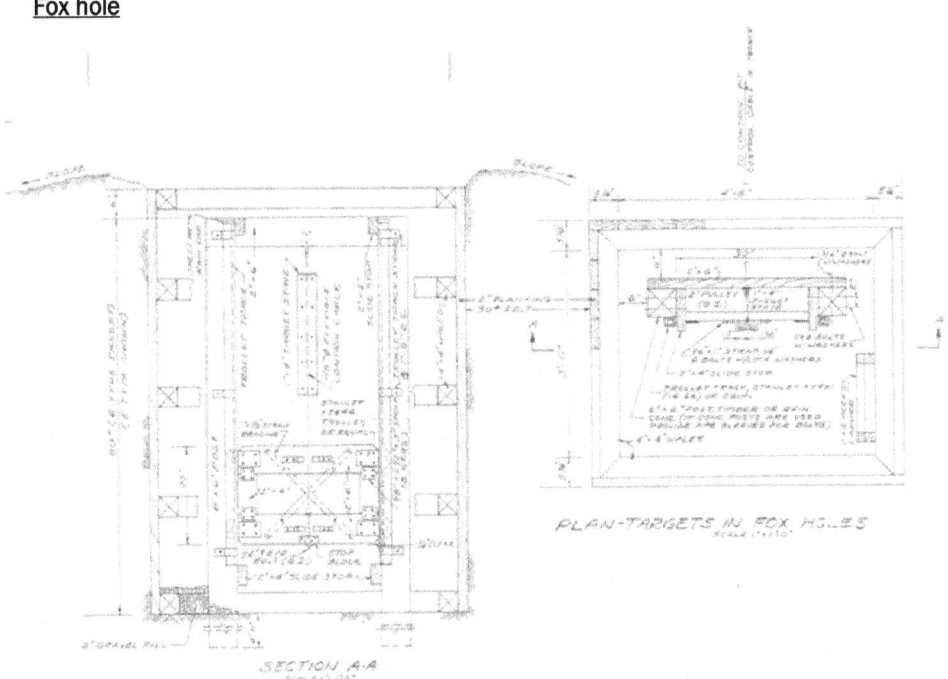

Figure 335. Superimposed transition range, plan of targets in foxholes, and section, Fort Bragg, NC, 1954 (Standard drawing 28-13-108 sheet 1 of 1, Superimposed transition range, table VIII, 22 March 1954).

Vertical target pit

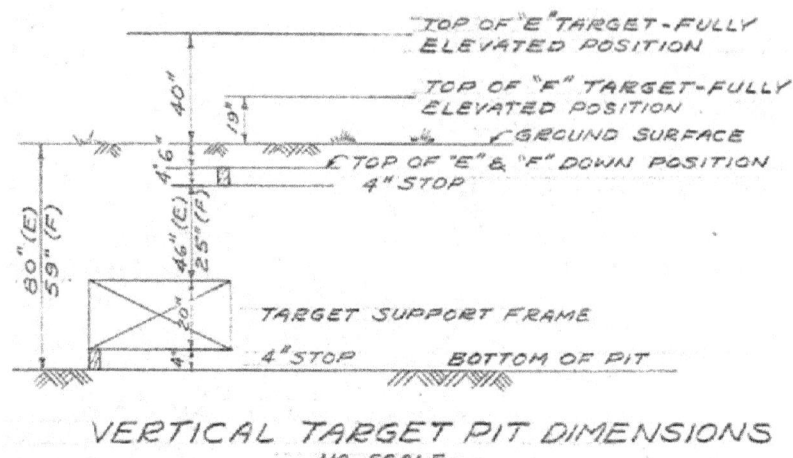

Figure 336. Superimposed transition range, vertical target pit dimensions, Fort Bragg, NC, 1954 (Standard drawing 28-13-108 sheet 1 of 1, Superimposed transition range, table VIII, 22 March 1954).

Concrete M-30 pop-up target emplacement

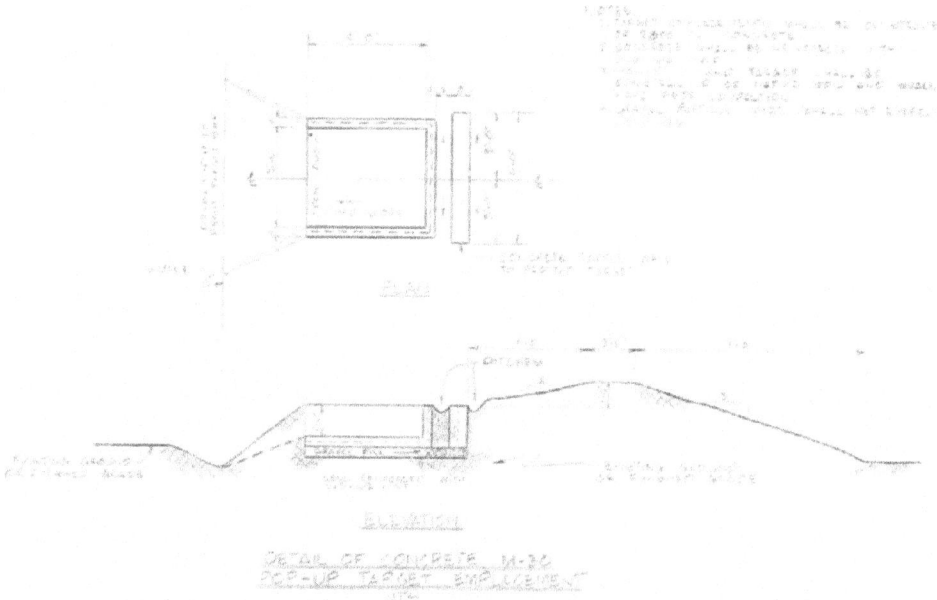

Figure 337. Transition firing range, machine gun range 53, detail of concrete M-30 pop-up target emplacement, Fort Bragg, NC, 1978 (Standard drawing DFE-3278 sheet 7 of 10, Transition firing range, machine gun-range 53, range details, 19 January 1978).

Window target

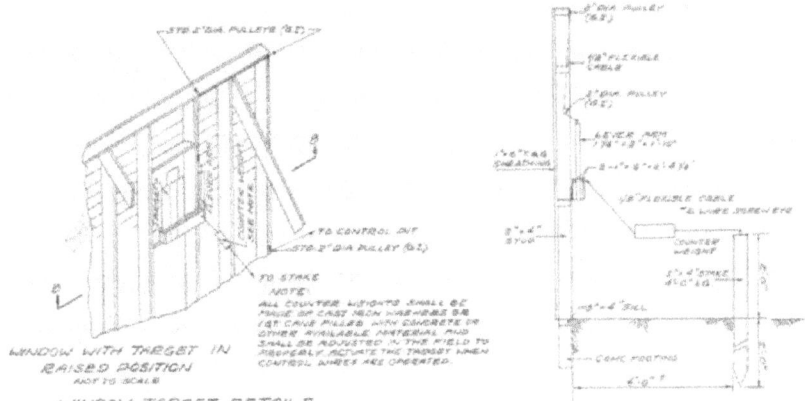

Figure 338. Superimposed transition range, target layout plan and window target details, Fort Bragg, NC, 1954 (Standard drawing 28-13-108 sheet 1 of 1, Superimposed transition range, table VIII, 22 March 1954).

Embankments/trenches/etc.

Embankments and trenches were built for both firing lines (see Figures 323 through 325) and targets (see Figures 332 through 338) as shown above. The centerline profiles of firing lanes below show the terrain created by the embankments.

Figure 339. Transition firing range, machine gun range 53, centerline profiles for lanes 1 to 3, Fort Bragg, NC, 1978 (Standard drawing DFE-3278 sheet 4 of 10, transition firing range, machine gun-range 53, centerline profiles, lanes 1 through 3, 19 January 1978).

Figure 340. Transition firing range, machine gun range 53, centerline profiles for lanes 4 to 6, Fort Bragg, NC, 1978 (Standard drawing DFE-3278 sheet 5 of 10, Transition firing range, machine gun-range 53, centerline profiles, lanes 4 through 6, 19 January 1978).

Buildings

No buildings other than a target house are mentioned in the plans for these ranges. However, a range may also have had a control tower, latrine, ammunition storage building, other storage sheds, and administrative/maintenance buildings supporting general range functions. The range may also have been part of a larger installation range complex that contained these buildings.

Moving firing points & moving targets

Moving base range

Aircraft gunners were trained on these ranges to fire from a moving position at moving targets. "The range consisted of a roadway with multiple trap houses. As gunners were driven down a road, they shot at clay pigeons launched from trap houses" ("RO-1" 39). All firing was performed when trap houses were in line with fixed targets (see Figure 341).

Layouts

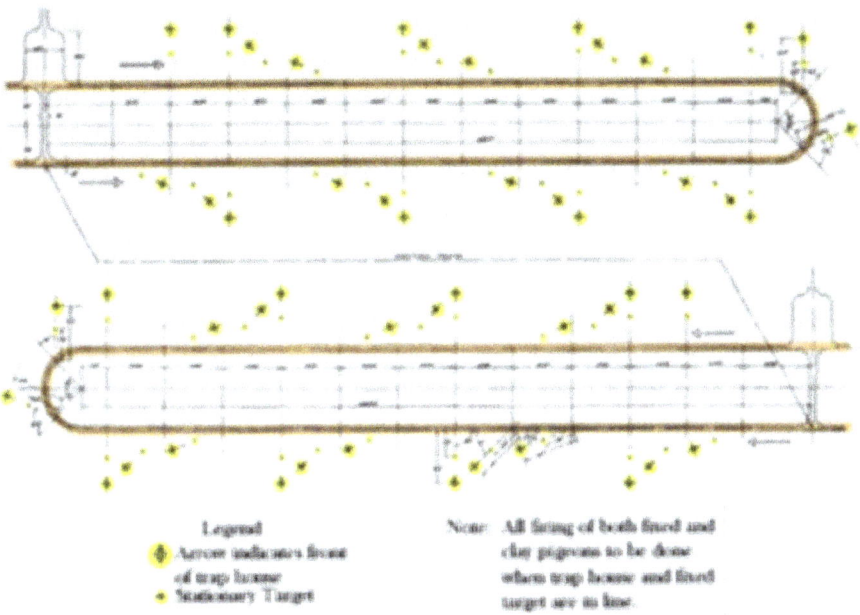

Figure 341. Moving base range, circa 1945 (AAF Manual 85-0-1, Army Air Forces gunnery and bombardment ranges, 15 June 1945, pp 2-2-1 to 2-2-3).

Firing lines

Aircraft gunners shot from moving vehicles at points on the road when trap houses were in line with fixed targets.

Targets

Targets were clay pigeons and fixed targets.

Embankments/Trenches/Etc.

None.

Buildings

No buildings are mentioned in the standard plans for these ranges. However, a range may have had a control tower, latrine, target storage building, ammunition storage building, other storage sheds, and administrative/maintenance buildings supporting general range functions. The range may also have been part of a larger installation range complex that contained these buildings.

Moving target submachine gun range

Soldiers were trained on these ranges to fire submachine guns at moving targets from moving vehicles. The range consisted of a road on which moving vehicles were driven, a target track upon which 5 x 8-ft targets moved, an embankment in front of the target track, a winch house and counterweight tower, and parapets with target repair and wing pits (see Figures 343 and 344). Soldiers fired from vehicles moving along the road at targets moving along the track ("RO-1" 38). Training was completed with .45 caliber submachine guns.

Layouts

Danger area

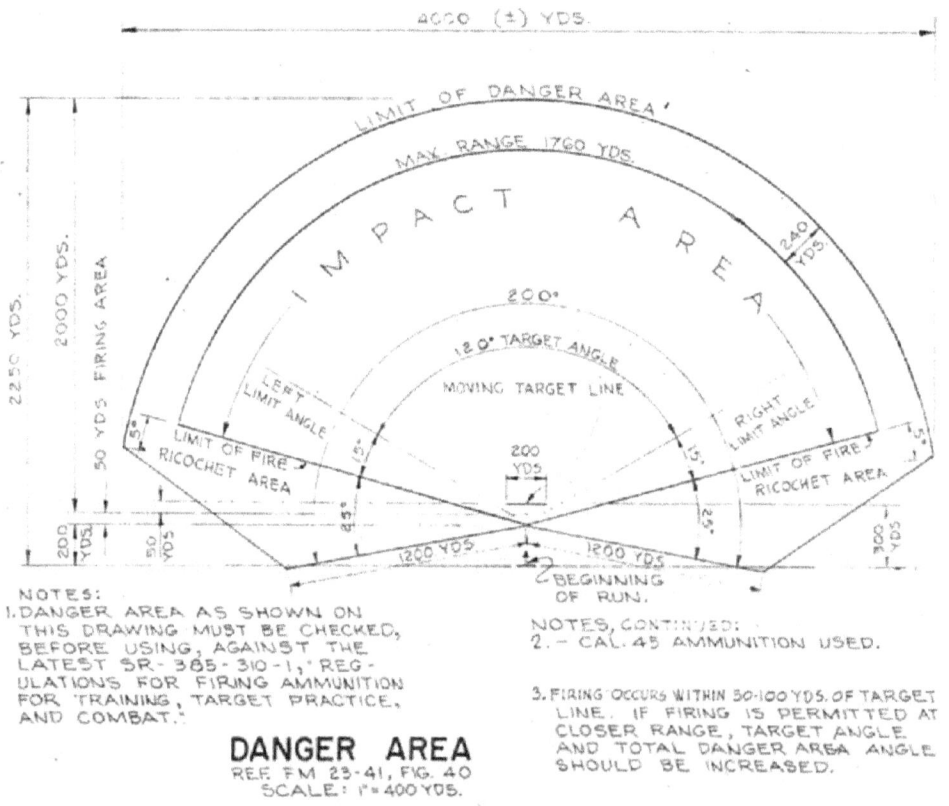

Figure 342. Moving target submachine gun range, danger area, Fort Bragg, NC, 1952 (Standard drawing 28-13-14 sheet 1 of 3, Range, moving target, submachine gun, plans and details, 20 June 1952).

Typical layouts

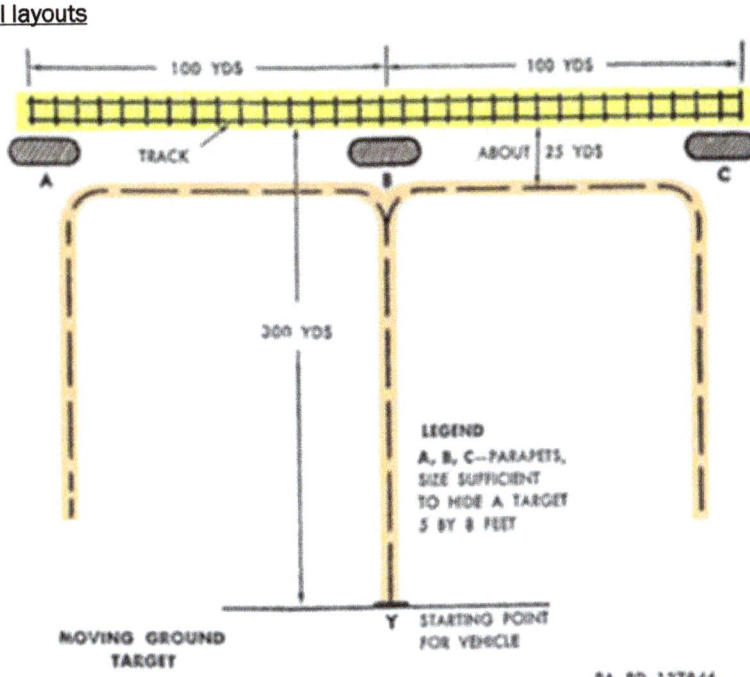

Figure 343. Submachine gun moving target range, circa 1952 (Standard drawing No. 28-13-14, Range, moving target, submachine gun, sheets 1 to 3, 20 June 1952).

Figure 344. Moving target submachine gun range, plan, Fort Bragg, NC, 1952 (Standard drawing 28-13-14 sheet 1 of 3, Range, moving target, submachine gun, plans and details, 20 June 1952).

Firing lines

Soldiers fired from moving vehicles on a road that circled past a moving target track (see Figure 343). Firing occurred within 50-100 yds of the target line.

Targets

Figure 345 shows sketches of 5 by 8-ft panel targets that moved along the track.

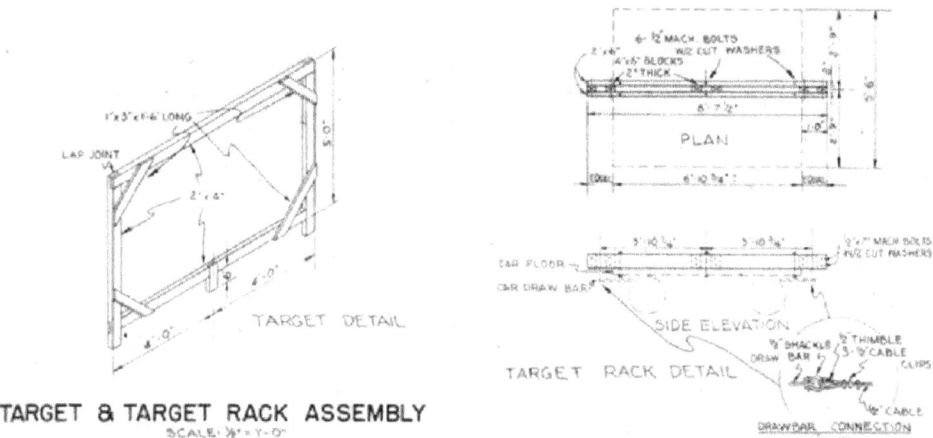

Figure 345. Moving target submachine gun range, target and target rack assembly, Fort Bragg, NC, 1952 (Standard drawing 28-13-14 sheet 1 of 3, Range, moving target, submachine gun, plans and details, 20 June 1952).

Embankments/trenches/etc.

An embankment was placed in front of the target track (Figure 346), and included parapets with target repair and wing pits (see Figures 347 through 349).

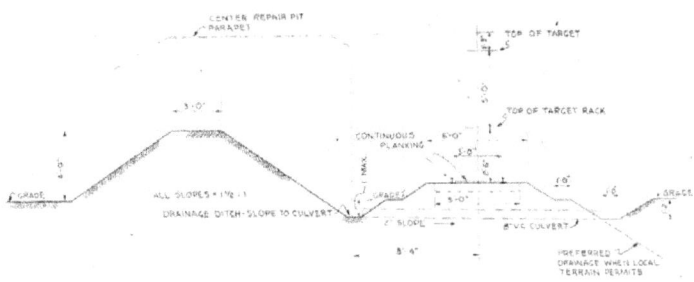

Figure 346. Moving target submachine gun range, section thru firing area, Fort Bragg, NC, 1952 (Standard drawing 28-13-14 sheet 2 of 3, Range, moving target, submachine gun, repair pit and parapet details, 20 June 1952).

Repair pits and wing walls

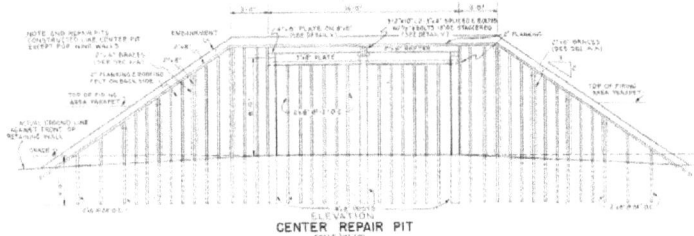

Figure 347. Moving target submachine gun range, center repair pit elevation, Fort Bragg, NC, 1952 (Standard drawing 28-13-14 sheet 2 of 3, Range, moving target, submachine gun, repair pit and parapet details, 20 June 1952).

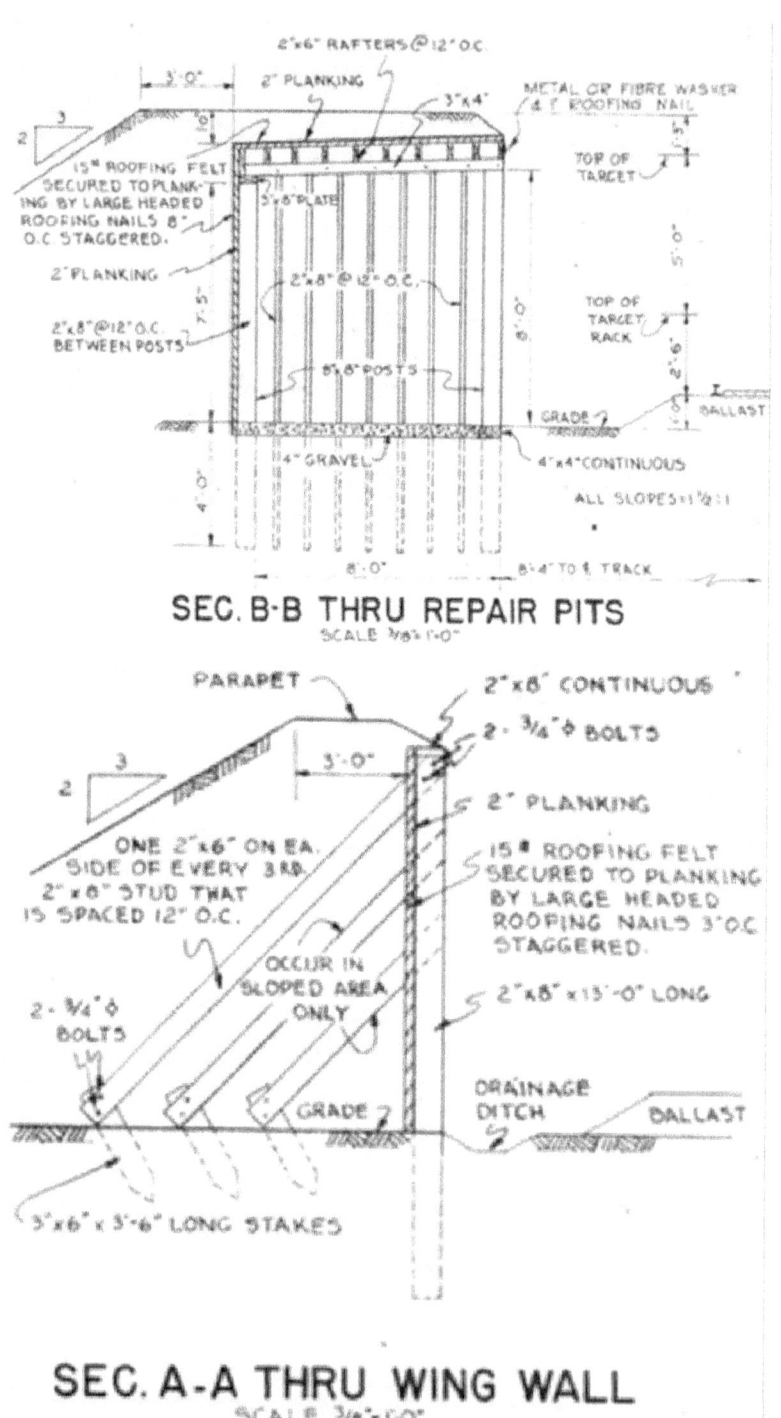

Figure 348. Moving Target submachine gun range, sections thru repair pits and wing walls, Fort Bragg, NC, 1952 (Standard drawing 28-13-14 sheet 2 of 3, Range, moving target, submachine gun, repair pit and parapet details, 20 June 1952).

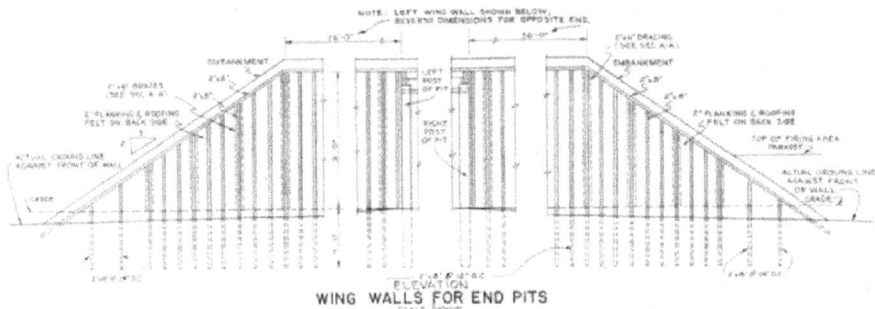

Figure 349. Moving target submachine gun range, wing walls for end pits, Fort Bragg, NC, 1952 (Standard drawing 28-13-14 sheet 2 of 3, Range, moving target, submachine gun, repair pit and parapet details, 20 June 1952).

Buildings

No buildings other than a winch house and counterweight tower (see Figures 350 through 354) are mentioned in the standard plans for these ranges. However, a range may have had a control tower, latrine, target storage building, ammunition storage building, other storage sheds, and administrative/maintenance buildings supporting general range functions. The range may also have been part of a larger installation range complex that contained these buildings.

Winch house and counterweight tower

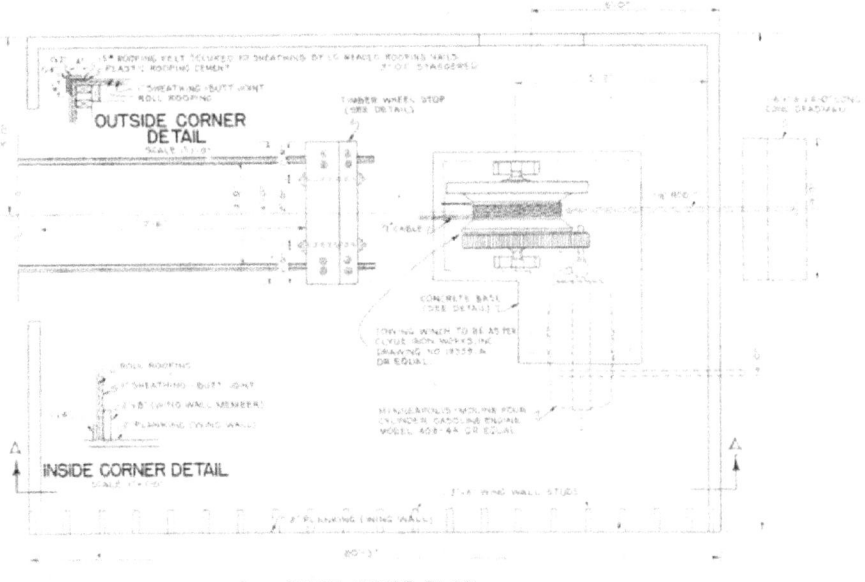

Figure 350. Moving target submachine gun range, winch house plan, Fort Bragg, NC, 1952 (Standard drawing 28-13-14 sheet 3 of 3, Range, moving target, submachine gun, winch house, 20 June 1952).

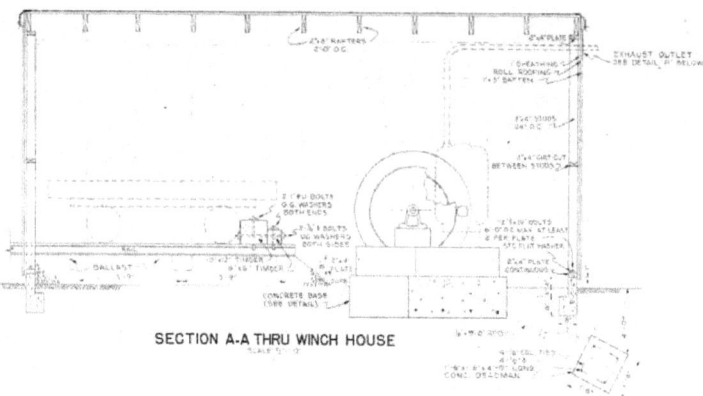

Figure 351. Moving target submachine gun range, winch house section, Fort Bragg, NC, 1952 (Standard drawing 28-13-14 sheet 3 of 3, Range, moving target, submachine gun, winch house, 20 June 1952).

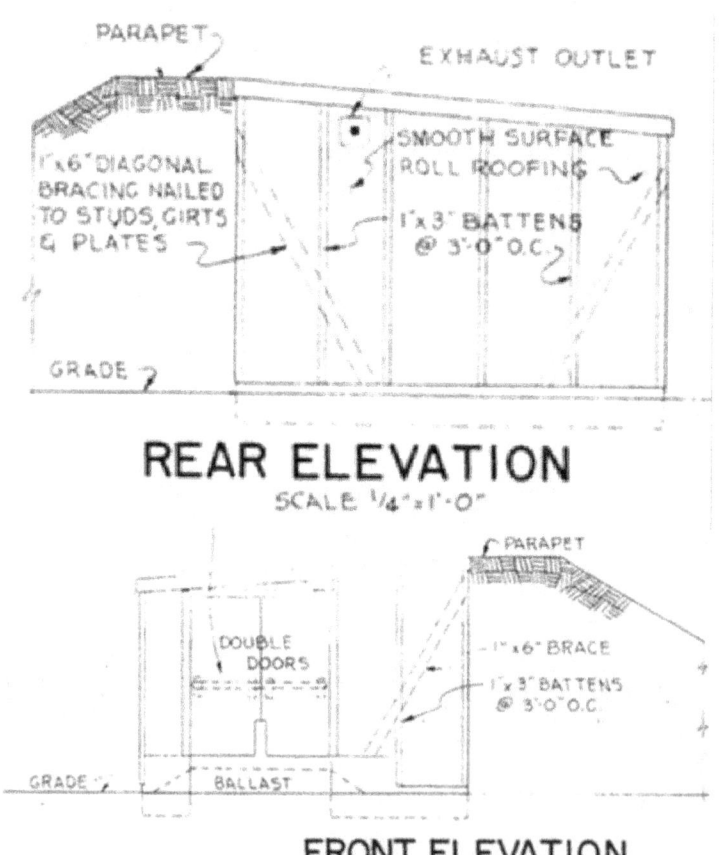

Figure 352. Moving target submachine gun range, winch house rear and front elevations, Fort Bragg, NC, 1952 (Standard drawing 28-13-14 sheet 3 of 3, Range, moving target, submachine gun, winch house, 20 June 1952).

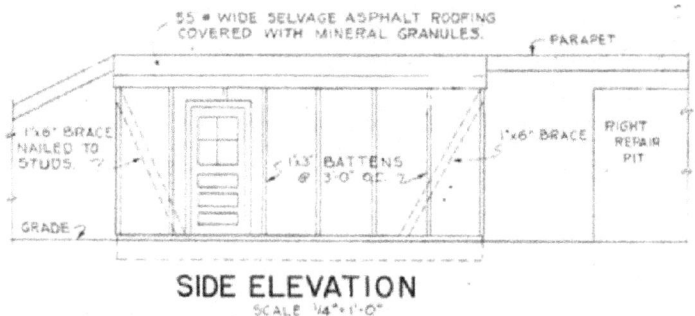

Figure 353. Moving target submachine gun range, winch house side elevation, Fort Bragg, NC, 1952 (Standard drawing 28-13-14 sheet 3 of 3, Range, moving target, submachine gun, winch house, 20 June 1952).

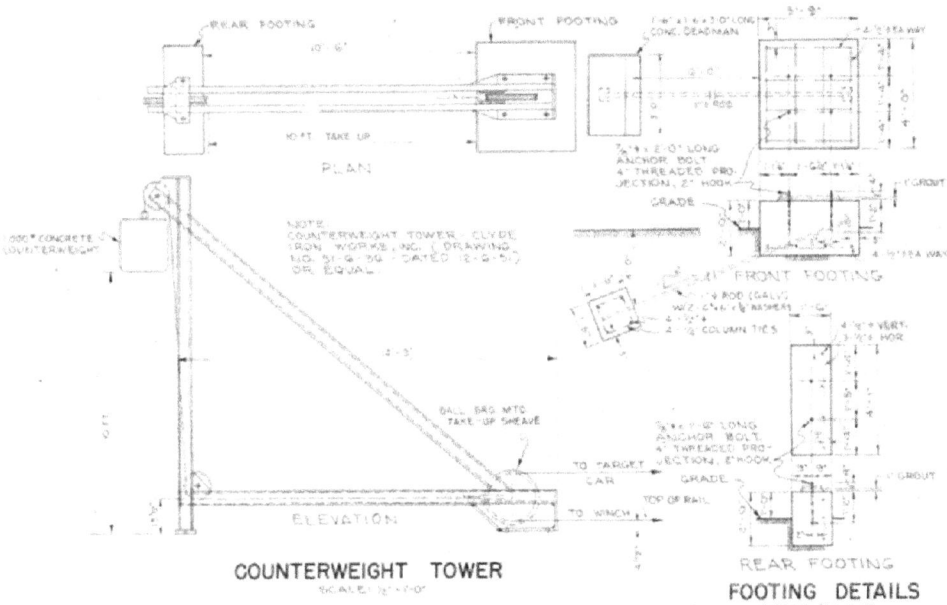

Figure 354. Moving target submachine gun range, counterweight tower, Fort Bragg, NC, 1952 (Standard drawing 28-13-14 sheet 1 of 3, Range, moving target, submachine gun, plans and details, 20 June 1952).

Hand and rifle grenade ranges

Hand grenade ranges

"Hand grenade ranges were located at Ground Forces Training Centers and at other installations where training of ground forces was the primary mission. Hand grenades were authorized for virtually all types of units" ("RO-14" 12-13). "Hand grenade ranges could have been stand-alone or shared with other ranges. However, most hand grenade ranges were normally stand-alone and, because of their size, did not require integration into a combined range impact area. An infantry division training installation required at least three practice grenade courts and a live grenade court" ("RO-14" ES2).

"Formal training with hand grenades consisted of several courses to teach troops how to use hand grenades in various situations. However, only one course involved live grenades, the remaining courses involved practice grenades. The range for the hand grenade confidence course consisted of a practice grenade court for throwing the practice grenades and a live grenade course for throwing fragmentation grenades. These two courses were constructed alike and close together in order to allow easy movement from one to the other. Investigations performed at numerous closed and active installations indicate that, ranges where only practice grenades were used required a small area and were often located near the cantonment area. Sometime after 1954, hand grenade courses could be modified to support the unit mission essential task list, terrain, and commander's intent" ("RO-14" 15-16).

Weapons

"Although hand grenades had been employed since the 15th century, the U.S. Army did not begin designing hand grenades until 1917" ("RO-14" 2). "Hand grenade range operations involved high explosive, smoke, chemical, practice, fragmentation, frangible, and dummy grenades" ("RO-14" ES2). "Each hand grenade was a standalone weapon and not a part of another system" ("RO-14" 2).

"Explosive, practice, and dummy grenades were used on the range. Exploding white phosphorus smoke grenades were also used at grenade ranges. Burning type grenades were not associated with grenade ranges but were used for training at other ranges and maneuver areas. Investigations performed at numerous closed and active installations indicate that a variety of other grenades including burning grenades (in small quantities) might have been used for demonstration purposes on hand grenade ranges" ("RO-14" 9).

"A key ingredient in determining whether a grenade course was restricted to practice grenades was its location. Investigations performed at numerous closed and active installations indicate that grenade courts located in a cantonment area were restricted to dummy and practice grenades because they did not meet the safety requirements established for live grenade ranges" ("RO-14" 17).

training Procedures

Weapons (if required) were drawn from the armory and munitions were drawn from the ammunition supply point (ASP). Both were normally transported to the range by the unit using organic (integral) vehicles. Some practice, smoke, fragmentation, and lachrymatory hand grenades also required assembly before use. An igniting fuse was screwed into the grenade body and a charge of black powder was placed through a filling hole in the base of the grenade. An assembly area was established for use as a briefing point and ammunition issue point. Training involved throwing practice or live grenades as determined by training requirements. Prior to 1940, if a hand grenade failed, it was recovered and placed in a deep stream or other body of water or buried in the ground. By 1940, destruction of grenade duds was done singly or in bulk using either TNT blocks or fire. Troops then policed the area, taking trash to an offsite disposal area, or burying trash in a fox hole if they were on their way to do more training ("RO-14" 9-12).

Layouts

Danger areas

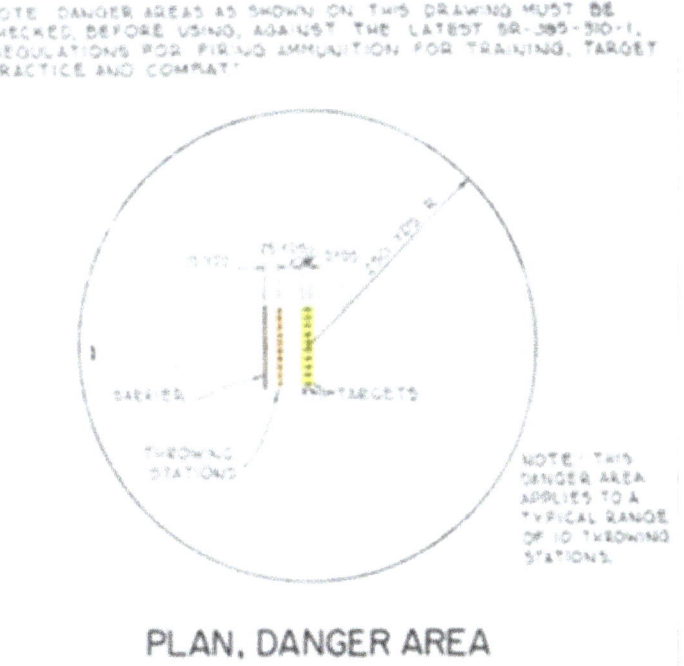

Figure 355. Live grenade course, plan, danger area, Fort Bragg, NC, 1951 (Standard drawing 28-13-44 sheet 1 of 1, Live grenade course, plans and details, 21 November 1951).

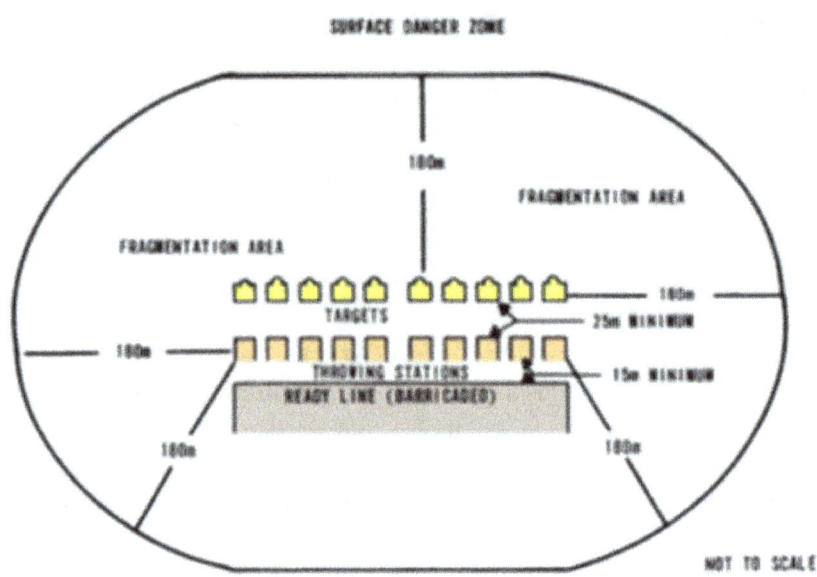

Figure 356. Fragmentation and offensive hand grenade range surface danger area, circa 1943-1986 (AR 385-63, Regulations for firing ammunition for training, target practice, and combat, 17 June 1968, Chapter 8, AR 385-63, Safety regulations for firing ammunition for training, target practice, and combat, 28 February 1973, p 8-3, AR 385-63/MCO P3570.1, Policies and Procedures for firing ammunition for training, target practice, and combat, 22 February 1978, p 7-3).

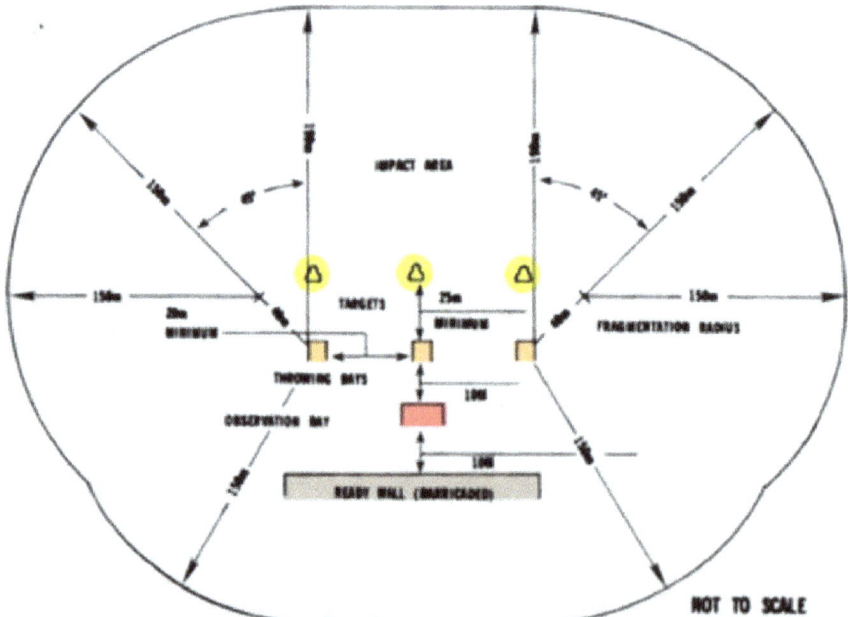

Figure 357. Fragmentation and offensive hand grenade range modified surface danger area, circa 1983-1986 (FM 23-30, AR 385-63/MCO P3570.1A, Policies and procedures for firing ammunition for training, target practice, and combat, 15 October 1983, p 7-3).

Angle court

"An angle court is illustrated in Figure 358 below. A line 18 ft long was constructed connecting points A and B. From the center point of this line, point C was established at a distance 50 yds to the perpendicular. Taped or outlined trenches were constructed 3 ft wide and centered 5 yds apart to serve as targets. The grenade thrower stood with their leading foot at point C. This course was in use from WWI until WWII. By 1944, this type of court was no longer in use. The layout indicates it was a practice court, restricted to dummy and practice grenades" ("RO-14" 21).

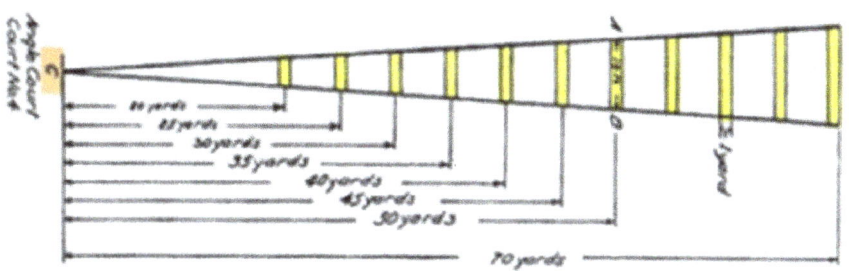

Figure 358. Angle court, circa 1927-1944 (War Department Document No. 918, Volume III, Manual of basic training and standards of proficiency for the national guard, 1927, p 436, FM 23-30, Grenades, 15 June 1942, p 20, FM 23-30, Hand and rifle grenades, rocket, AT, HE, 2.36-in., 14 February 1944, pp 30-34, Ibid., pp 31-34, FM 23-30, Grenades, January 1940, p 15, FM 23-30, Grenades, 15 June 1942, p 19).

Crater court

"Crater courts were constructed by placing a strip of canvas 4 ft long and 2 in. wide on the ground. At a distance 25 yds in front, a circular pit 3 yds in diameter and a minimum of 18 in. deep was dug. Grenade throwing was conducted with the toe of the leading foot placed on the tape. The crater court was designed as a practice range for using practice and frangible hand grenades. However, frangible grenades appear to have been instituted around 1943 and fallen out of use by 1949. The crater court is depicted in Figure 359 below. The crater court depicted in Figure 359 was in use from WWI until the end of WWII" ("RO-14" 20).

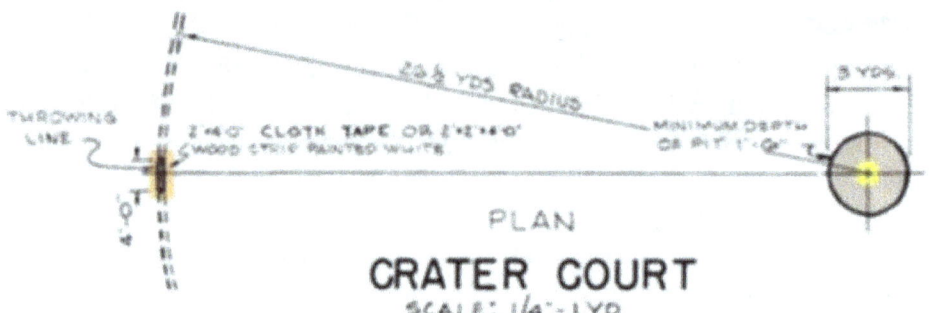

Figure 359. Practice grenade course, crater court, Fort Bragg, NC, 1951 (Standard drawing 28-13-43 sheet 1 of 1, Practice grenade course layout and details, 21 November 1951).

Grenade assault course

"Grenade assault courses were preferably built on rough, sparsely wooded terrain and varied in length from 150 to 200 yds. At installations where suitable terrain was limited and a combat reaction course existed, it was permissible to incorporate a grenade assault course into the combat reaction course. The course provided situations where a soldier could conduct a grenade attack from various positions. Both practice and frangible grenades were used on this course. The suggested assault course is outlined in Table 3 below" ("RO-14" 29). "A 1988 variation of the assault course is depicted in Figure 360 below" ("RO-14" 30).

Table 3. Suggested Assault Course Table, 1944 (FM 23-30, Hand and Rifle Grenades, Rocket, AT, HE, 2.36-in., 14 February 1944, pp 39 – 41).

Station No.	Protection	Target	Distance (yards)	Grenade
1	Logs	Machine gun emplacement.	25	Training hand grenade.
2	Wall	Window	10	Training hand grenade.
3	Foxhole	Bobbing targets from trench.	25	Training hand grenade.
4	Knoll	Mortar emplacement.	20	Training hand grenade.
5	Logs or other natural barrier.	Prone dummies pulled by rope behind bush.	25	Training hand grenade.

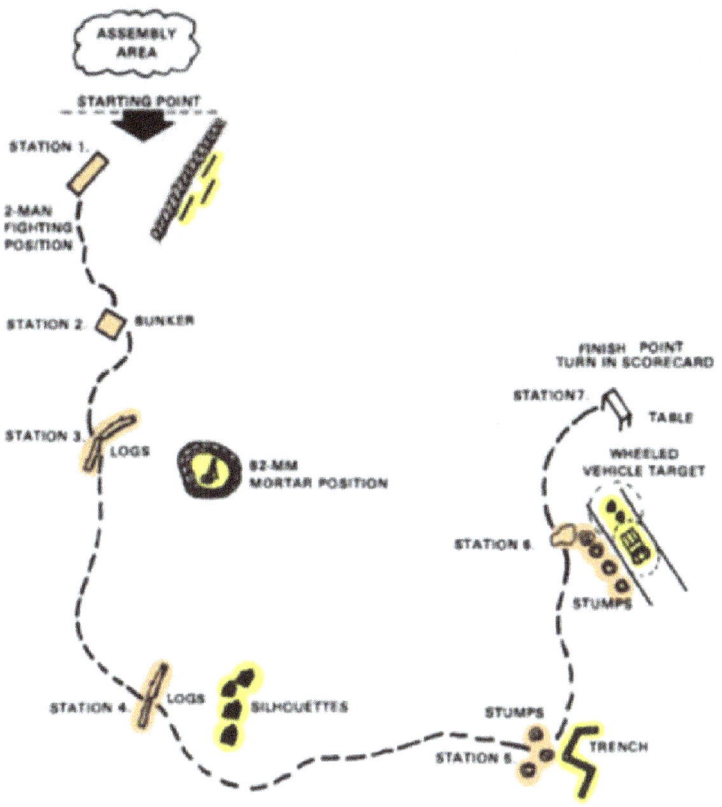

Figure 360. Assault course layout, circa 1988 (FM 23-30, Hand and rifle grenades, rocket, AT, HE, 2.36-in., 14 February 1944, pp 39 – 41, FM 23-30, Hand and rifle grenades, 14 April 1949, pp 42-44).

Live grenade practice course

"The live grenade practice course was used exclusively for throwing live hand grenades. A typical 6-bay hand grenade course required an area of approximately 18,000 sq ft (150 x 120 ft). The course included individual throwing bays or a trench, with targets and an impact area approximately 75 ft in front of the throwing line. The course was laid out with a ready line located behind a barrier at least 5-ft high and a throwing area located a minimum of 45 ft in front of the barrier. Throwing bays were constructed from sandbags or concrete. Targets consisted of a circular outline, a crater, and/or a foxhole. A schematic plan and elevation view of a 5-bay live grenade practice course is depicted in Figure 361 below" ("RO-14" 25). "By 1988, the live grenade practice court appears to have been modified to the layout depicted in Figure 363 below" ("RO-14" 26). "The surface danger area for a live grenade practice course was established by 1968 and is illustrated in Figure 356 above. This type of course evolved from the trenches grenade court when it was recognized that protection to the thrower was necessary, even though there was no longer a need to train in trench warfare. This course was used from WWII until sometime prior to 1988 and was restricted to live, HE grenades" ("RO-14" 27). "The surface danger area for fragmentation and offensive hand grenade ranges was modified slightly in 1983 as shown in Figure 357 above. This modified danger area was still in effect in 1986" ("RO-14" 29).

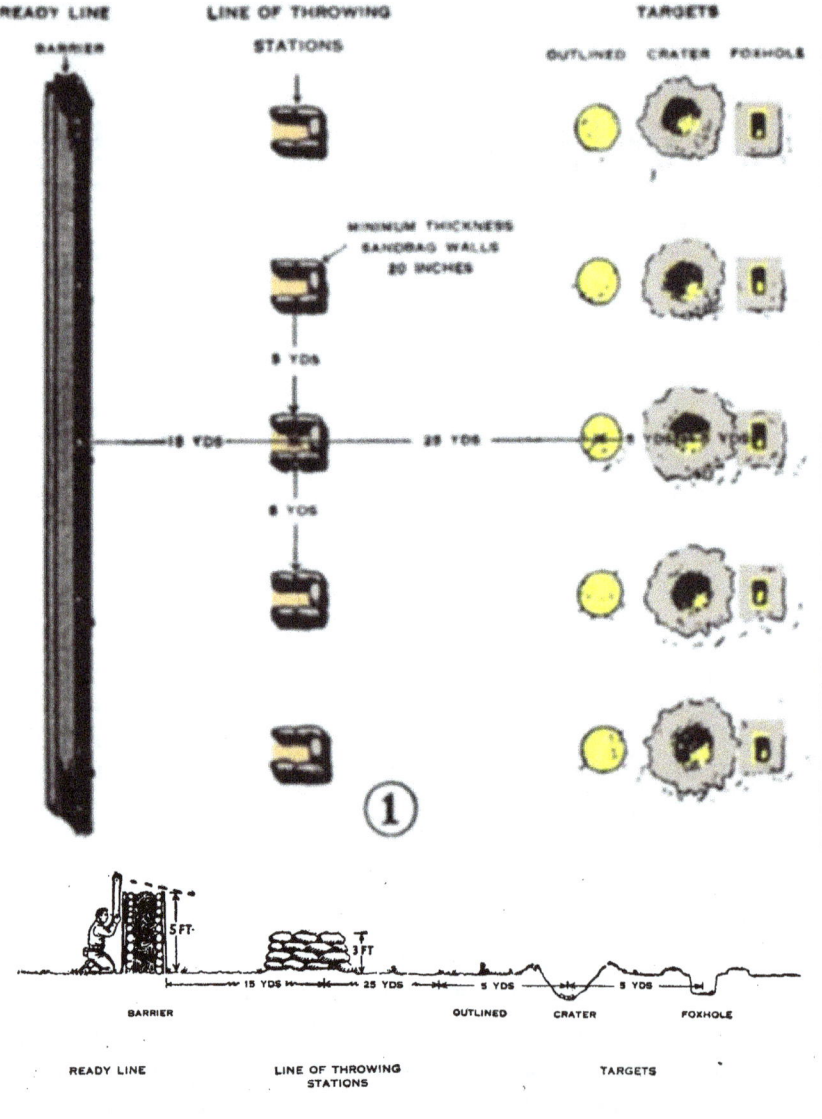

Figure 361. Live grenade practice course, plan and profile, circa 1944-1949 (FM 23-30, Hand and rifle grenades, rocket, AT, HE, 2.36-in., 14 February 1944, pp 35, 36, FM 23-30, Hand and rifle grenades, 14 April 1949, pp 40, 41).

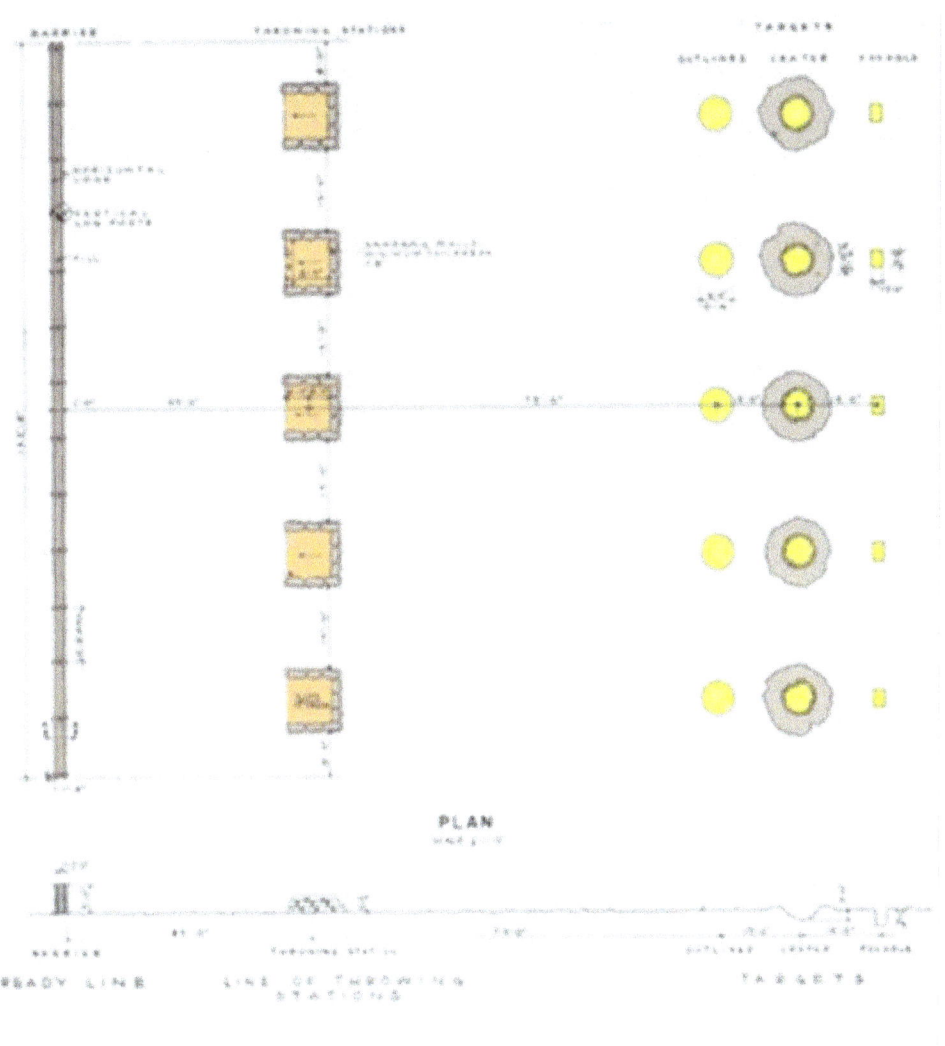

Figure 362. Live grenade course, plan and profile, Fort Bragg, NC, 1951 (Standard drawing 28-13-44 sheet 1 of 1, Live grenade course, plans and details, 21 November 1951).

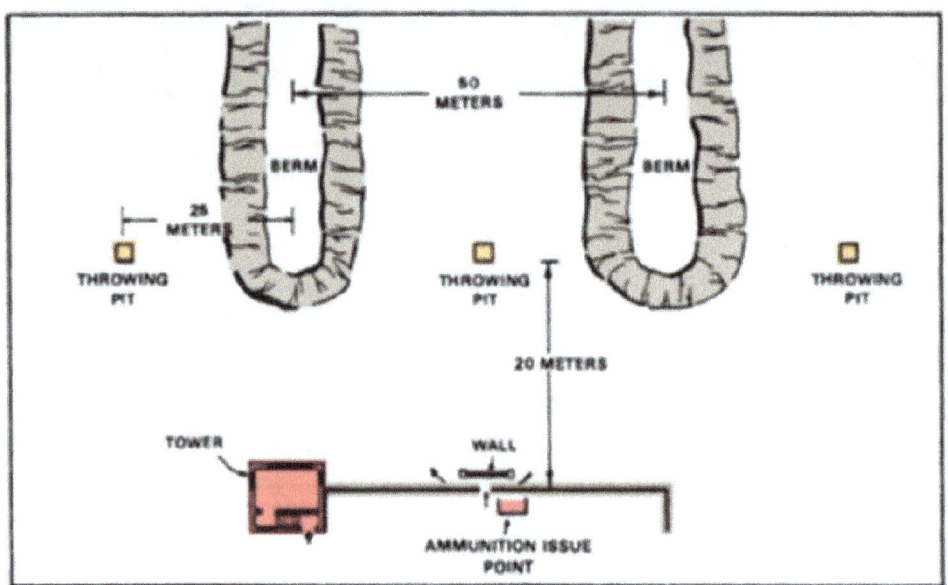

Figure 363. Live grenade pit layout, circa 1988 (FM 23-30, Grenades and pyrotechnic signals, 27 December 1988, p 4-6).

Figure 364. Periscopes, throwing bays, and impact area at Remagen Hand Grenade Range at Fort Jackson, SC, 1 November 1961 (NARA College Park, RG 111-SC post-1955, box 371, photo SC592064).

Main grenade court

"Main grenade courts consisted of five rows of lines as indicated in Figure 365 below. Each row consisted of two lines of canvas or target cloth placed 1 yd apart to indicate a trench. Each court was 40 yds wide and 35 or 50 yds long. Four spider webs were constructed at one end (the target end). Each spider web had three concentric circles that were 1, 2, and 3 yds in diameter, respectively, and outlined with cloth strips. Small stakes with pieces of cloth nailed to the top were placed in the center of each web to serve as targets" ("RO-14" 18). "Investigations performed at numerous closed and active installations indicate that the main grenade court depicted in Figure 365 was in use from WWI to the Vietnam time period. In 1944, the main grenade court was identified as a practice grenade court. Use of this court was restricted to dummy ammunition in 1940, and frangible and practice grenades in 1944. However, frangible grenades appear to have been instituted around 1943 and fallen out of use by 1949" ("RO-14" 18, 19).

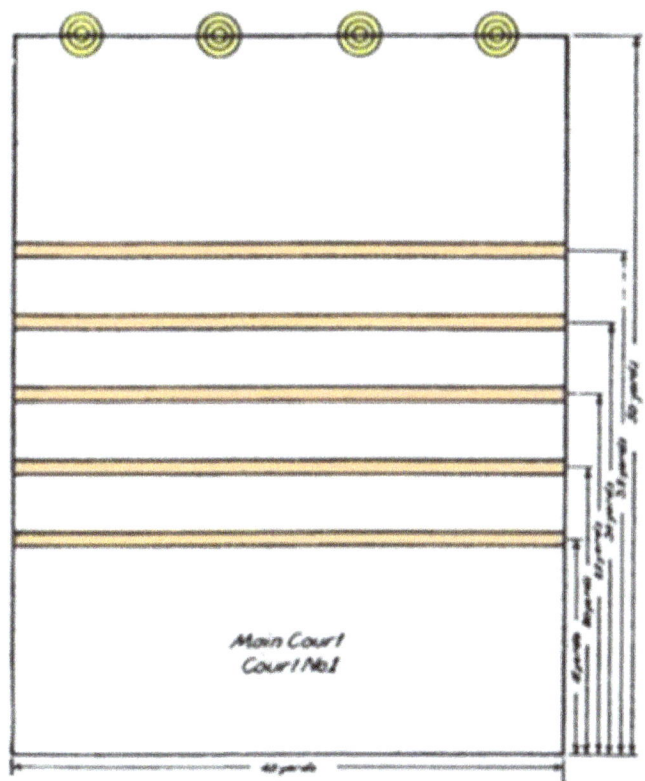

Figure 365. Main grenade court, circa 1927-1949 (FM 23-30, Grenades, January 1940, p.16, FM 23-30, Hand and rifle grenades, rocket, AT, HE, 2.36-in., 14 February 1944, p 30, 31).

Throwing pit (fox hole court)

"Throwing pits were 8 ft long, 4 ft wide, and 7 ft deep. Twenty-five yards away, along the longitudinal axis, a pit 18 ft long, 2 ft wide, and 1 ft deep was dug with its longitudinal axis perpendicular. At distances of 20 yds, along a 45° angle, pits 18 ft long, 2 ft wide, and 1 ft deep were dug so that the longitudinal axis of each was perpendicular to the 45° lines. This layout is depicted in Figure 366 below" ("RO-14" 19). "The grenade-throwing pit shown in Figure 366 was in use from WWI until the end of WWII. During WWII, it was referred to as a Fox Hole Court. The throwing pit was designed as a practice range for using practice and frangible grenades. However, frangible grenades appear to have been instituted about 1943 and fallen out of use by 1949" ("RO-14" 20).

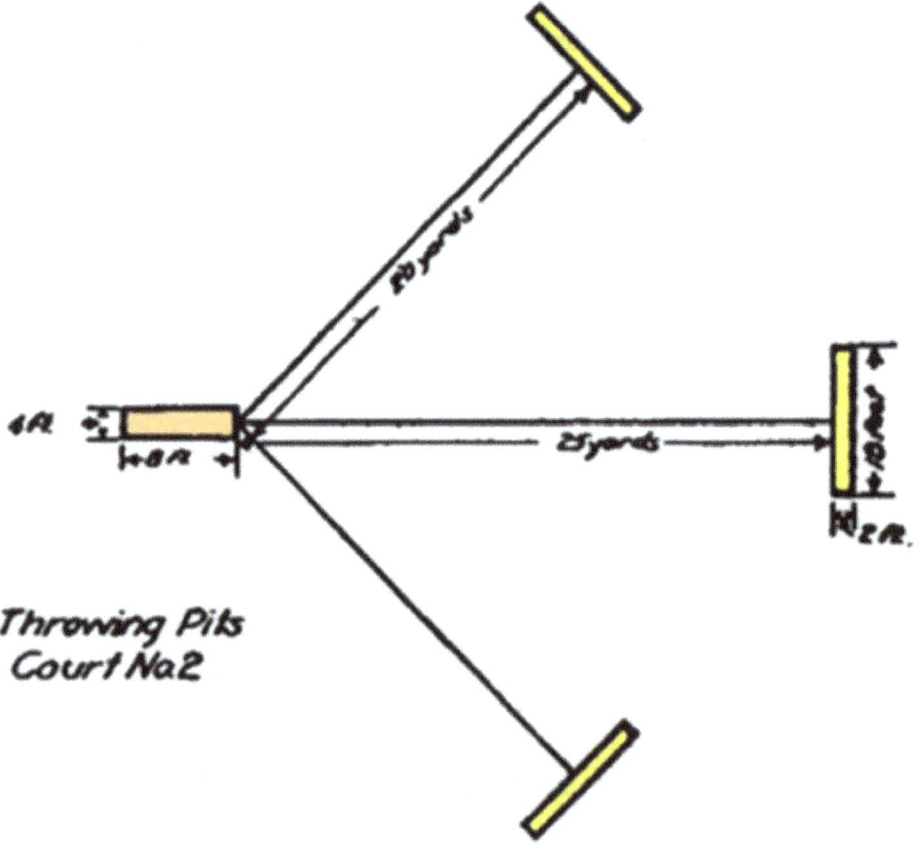

Figure 366. Throwing pits course, circa 1942 (FM 23-30, Grenades, 15 June 1942, p 23).

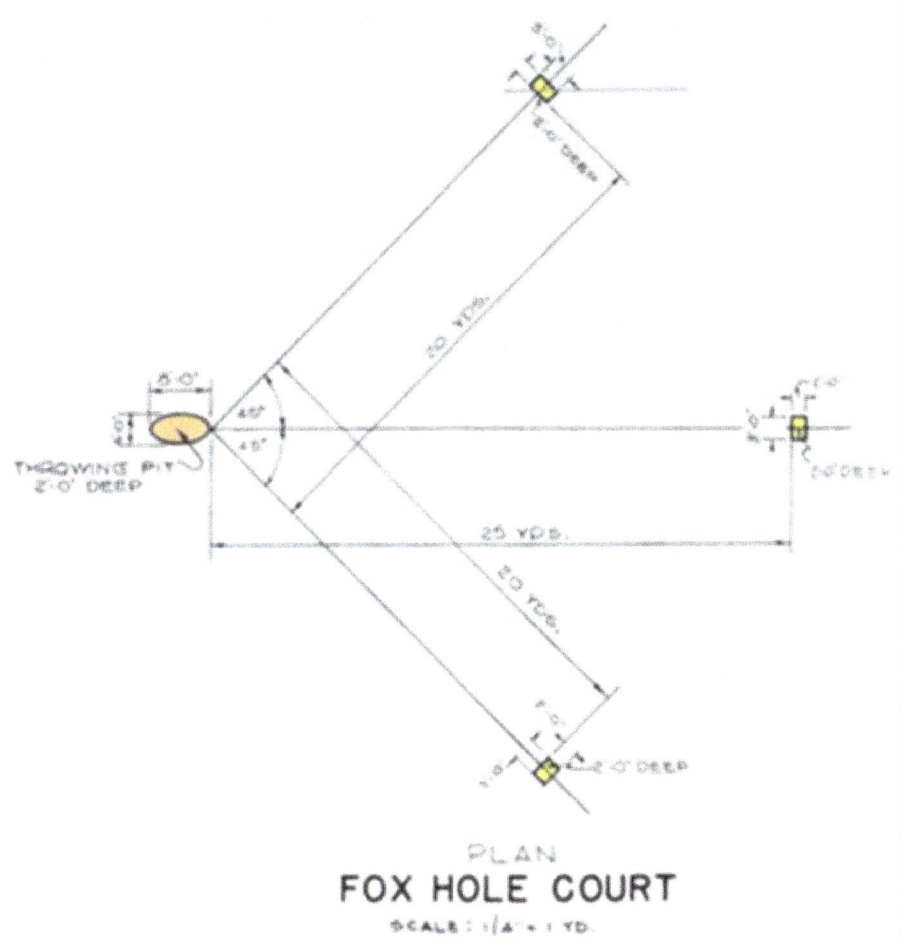

Figure 367. Practice grenade course, foxhole court, Fort Bragg, NC, 1951 (Standard drawing 28-13-43 sheet 1 of 1, Practice grenade course layout and details, 21 November 1951).

Trench court

"Trenches were constructed according to the specifications for Type A fire trenches. For a class of 150 men, 12 fire bays and 12 traverse bays were constructed. Three masks (sandbag walls 7 ft high and 12 ft long built on the circumference of a 12-ft circle) were placed along the trenches about 10 yds behind the parados (the parados is the dirt piled behind the trench to protect from reverse fire) to protect officers in charge of the firing line. This placed the officers in charge above the trenches, which was the best position from which to control throwing. The trench court is depicted in Figure 368 below" ("RO-14" 22). "During WWI, personnel required training in trench warfare. In order to meet this need, training centers developed extensive trench training areas. Between 1942 and 1944, it appears trench courts were no longer in use. In-ground trenches were replaced by aboveground-simulated trenches using sandbags. Fragmentation, practice, and live grenades were used on the trench court" ("RO-14" 22).

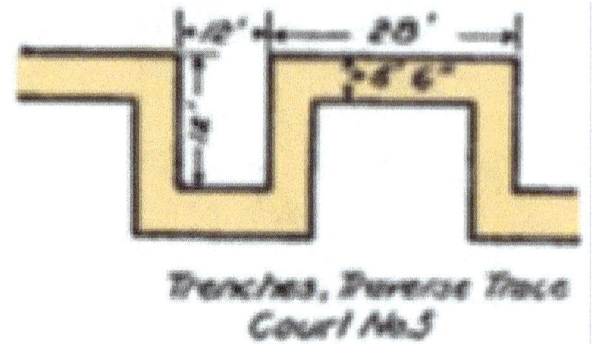

Figure 368. Trench court, circa 1927-1942 (War Department Document No. 918, Volume III, Manual of basic training and standards of proficiency for the National Guard, 1927, p 437, FM 23-30, Grenades, 15 June 1942, pp 22, 23).

Figure 369. Throwing live fragmentation hand grenades at the dummy targets at Fort Jackson, SC, 1 November 1943 (NARA College Park, RG 111-SC WWII, box 681, photo SC324452).

Vertical target court

"Vertical targets were constructed in groups of four to represent windows at various heights above the ground. The vertical targets court shown in Figure 370 below was used from WWI until at least 1951 and was intended for use with dummy and practice grenades, although training with burning and frangible grenades was permitted" ("RO-14" 23).

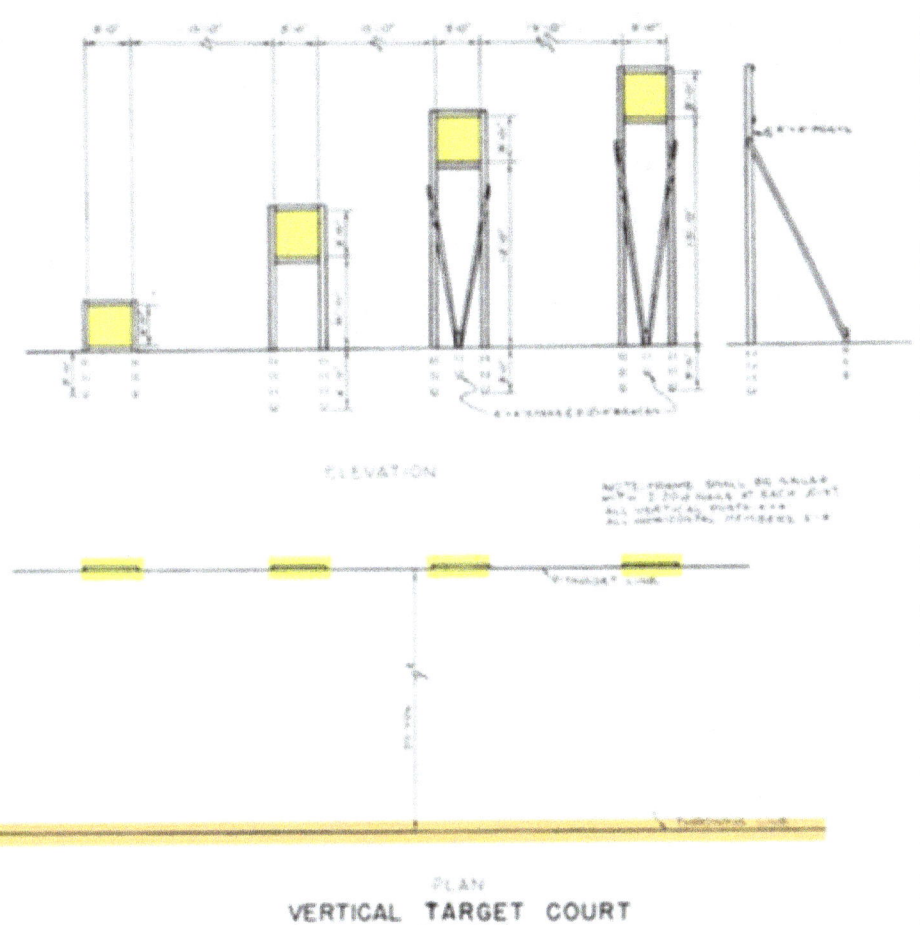

Figure 370. Practice grenade course, vertical target court, Fort Bragg, NC, circa 1927 to 1951 (Standard drawing 28-13-43 sheet 1 of 1, Practice grenade course layout and details, 21 November 1951).

Woods court

"The only mention of this court in the available documentation is in the 1945 change to the 1944 edition of FM 23-30 and the 1949 edition of FM 23-30. The woods court was constructed to simulate restrictions placed on the hand grenade thrower by dense underbrush or low-hanging trees. The court consisted of a length of chicken wire stretched over frames or stakes 3 to 4 ft above the ground. Single strands of wire, rope, netting, or other covering material could be substituted for chicken wire. The wire was covered with leaves and branches. The foxhole, crater, or similar type targets were placed to the front within the range of an underhand throw (approximately 12 yds). Practice grenades were used on the woods court, although live grenades were allowed on the woods court in 1945. This court is depicted in Figure 371 below" ("RO-14" 24).

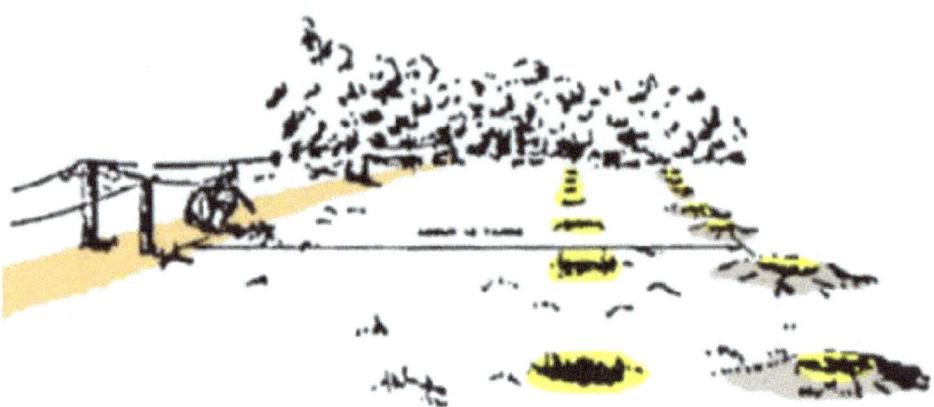

Figure 371. Woods Court, circa 1945-1949 (change 2 to FM 23-30, Hand and rifle grenades, rocket, AT, HE, 2.36-in., 18 October 1945, pp 13, 14, FM 23-30, Hand and rifle grenades, 14 April 1949, pp 34-39).

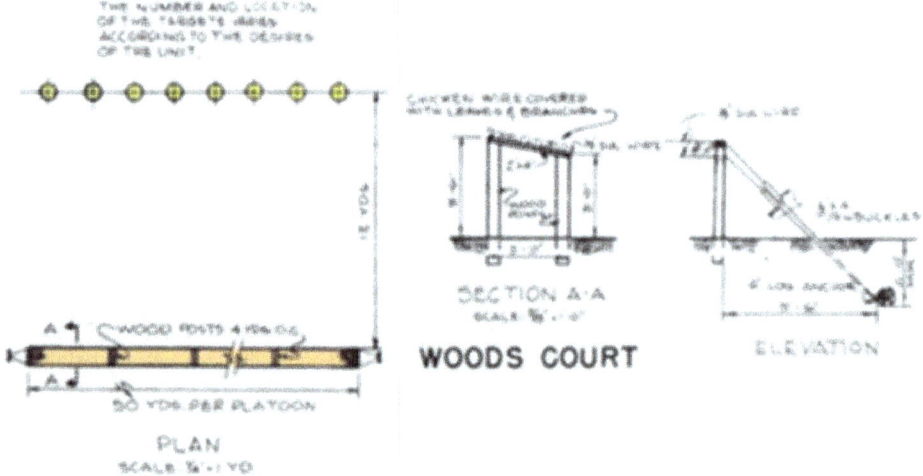

Figure 372. Practice grenade course, Woods Court, Fort Bragg, NC, 1951 (Standard drawing 28-13-43 sheet 1 of 1, Practice grenade course layout and details, 21 November 1951).

Firing lines

Firing positions on hand grenade ranges may have consisted of tape lines, foxholes, logs, stumps, bunkers, trenches, sandbag throwing stations, concrete throwing pits, throwing bays, etc (see examples below).

Sandbag throwing stations

Figure 373. Wearing helmets and barricaded behind sand bags as a precaution against flying grenade splinters, a recruit assumes a perfect stance as he lets fly his live grenade at MCRD Parris Island, SC, 11 June 1946 (NARA College Park, RG 127-GC, box 35, photo A16066).

Concrete throwing pits

Figure 374. The straining on the faces of these recruits show that they want to give the grenade a prodigious heave to give them time to duck behind the concrete barrier before the missile explodes at MCRD Parris Island, SC, 28 April 1949 (NARA College Park, RG 127-GC, box 35, photo 19159).

Figure 375. A recruit lies in a grenade pit just after he has thrown a grenade, there is an instructor in each pit at MCRD Parris Island, SC, 6 December 1951 (NARA College Park, RG 127-GC-586, box 35, photo A60683).

Throwing bays

Figure 376. Live bays on the Remagen Hand Grenade Range at Fort Jackson, SC, 25 March 1965 (NARA College Park, RG 111-SC post-1955, box 390, photo SC615409).

Targets

Targets on hand grenade ranges may have included silhouettes, tape or cloth circle outlines, craters, foxholes, trenches, mortar positions, wheeled vehicle towed targets, and simulated windows. "The various targets used on a hand grenade court were constructed of strips of canvas, target cloth, white tape, engineer's tape, or wood" ("RO-14" 56). Boxes were often set in the ground to hold the sides of targets.

Embankments/trenches/etc.

Where possible, throwing areas and ready lines were separated from target areas by steel/concrete/wooden/sandbag revetments, walls, or earthen berms to lessen the effect of high velocity/low angle fragments (e.g., 50m long and 1.8m high). Some were fitted with periscopes to allow soldiers to observe the impact of grenades (see Figure 379).

Figure 377. Grenade bursting in front of pits at MCRD Parris Island, SC, 6 December 1951 (NARA College Park, RG 127-GC-586, box 35, photo A60641).

Figure 378. Recruits lying behind wall waiting their turn to throw grenades at MCRD Parris Island, SC, 6 December 1951 (NARA College Park, RG 127-GC-586, box 35, photo A60662).

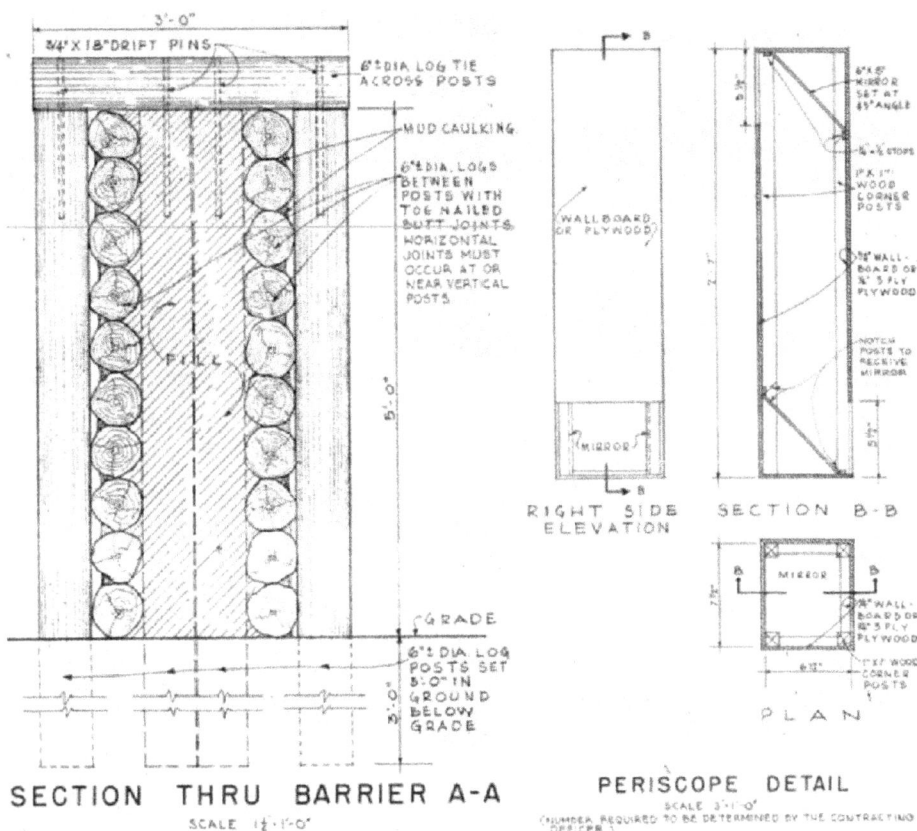

Figure 379. Live grenade course, section thru barrier and periscope detail, Fort Bragg, NC, 1951 (Standard drawing 28-13-44 sheet 1 of 1, Live grenade course, plans and details, 21 November 1951).

Buildings

"An assembly area was established for use as a briefing point and ammunition issue point" ("RO-14" 10). "Sometime after 1954, observation pits were constructed of polycarbonate resin sheets and laminated windowpanes in order to enable the Officer in Charge to better observe and control the throwing pits" ("RO-14" 15). A range may also have had a latrine, target storage building, ammunition storage building, other storage sheds, and administrative/maintenance buildings supporting general range functions. The range may also have been part of a larger installation range complex that contained these buildings.

Figure 380. Aerial photograph of Millcreek Grenade Range, Fort Knox, KY, 1974 (Training 15, vol. 1, 3-T-321-4/AH-74 Stock Shot # 1075, Millcreek grenade range, photo by: Mr. Maurice Monday, 6 June 1974, photo sac. A-V Sys Br. TASO, DPT, Fort Knox, KY 40121).

Rifle grenade ranges

"Rifle grenade ranges were located at Ground Forces Training Centers and at other installations where training of ground forces was the primary mission. Rifle grenades were authorized for most units authorized rifles" ("RO-14" 12-13). "Rifle grenade ranges could have been stand-alone or shared with other ranges. A typical rifle grenade range might have been integrated into a range layout at division training sites with overlapping cell boundaries with other ranges. An infantry division training installation required at least three practice grenade courts, a live grenade court, and one moving target range for anti-tank weapons, including anti-tank rifle grenades" ("RO-14" ES2). "A 37-mm antitank range could also be utilized for field firing of AT practice rifle grenades by constructing several individual prone shelters or standing type one-man foxholes. Therefore, every Ground Forces Training Center would likely have had an antitank gun range" ("RO-14" 12-13).

"Rifle grenade ranges were initially used to train personnel to fire anti-personnel grenades at area targets. However, these ranges were modified to accommodate training to attack tanks with anti-tank grenades. Formal training with rifle grenades consisted of several courses to teach troops how to use rifle grenades in various situations. However, only one course involved live grenades, the remaining courses involved practice grenades. Investigations performed at numerous closed and active installations indicate that, ranges where only practice grenades were used required a small area and were often located near the cantonment area" ("RO-14" 15-16).

Weapons

"The U.S. Army did not begin designing rifle grenades until 1917. Rifle grenades were developed to fill in the gap between the hand grenade and the light trench mortar. The types of rifle grenade initially used by the U.S. Army during World War I were the V.B. fragmentation rifle grenade designed by the French and the Babbitt grenade" ("RO-14" 2). "Rifle grenade range operations involved high explosive, smoke, chemical, practice, fragmentation, frangible, and dummy grenades" ("RO-14" ES2). "The rifle grenade was fired from a launcher attached to a .30 caliber rifle or carbine" ("RO-14" 2).

"Explosive, practice, and dummy grenades were used on the range. Exploding white phosphorus smoke grenades were also used at grenade ranges. Burning type grenades were not associated with grenade ranges but were used for training at other ranges and maneuver areas. Investigations performed at numerous closed and active installations indicate that a variety of other grenades, including burning grenades (in small quantities), might have been used for demonstration purposes rifle grenade ranges" ("RO-14" 9).

Training procedures

Weapons (if required) were drawn from the armory and munitions were drawn from the ammunition supply point (ASP). Both were normally transported to the range by the unit using organic (integral) vehicles. The grenade launcher was attached to the rifle barrel by means of a wing nut and clamp, or a latch and valve screw. The grenade projection adapter was rotated onto the rifle barrel until the safety lever and arming clip were down. Some practice, smoke, fragmentation, and lachrymatory rifle grenades also required assembly before use. An igniting fuse was screwed into the grenade body and a charge of black powder was placed through a filling hole in the base of the grenade. An assembly area was established for use as a briefing point and ammunition issue point. Training involved firing practice or live rifle grenades as determined by training requirements. After firing, the crew inspected the rifle and performed after firing first echelon (Operator) maintenance, including cleaning and lubricating. If the crew was returning to garrison immediately after firing, maintenance of the rifle may have been postponed until they arrived at their destination. Prior to 1940, if a grenade failed, it was recovered and placed in a deep stream or other body of water, or buried in the ground. By 1940, destruction of grenade duds was done singly or in bulk using either TNT blocks or fire." Troops then policed the area, taking trash to an offsite disposal area, or burying trash in a fox hole if they were on their way to do more training ("RO-14" 9-12).

Layouts

Danger areas

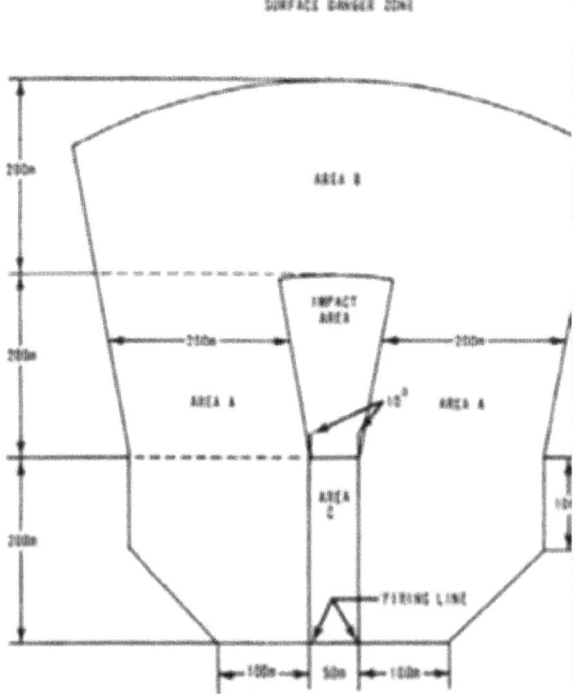

Figure 381. Rifle grenade surface danger area, circa 1968-1973 (AR 385-63, Regulations for firing ammunition for training, target practice and combat, 17 June 1968, p 8-4, AR 385-63, Safety regulations for firing ammunition for training, target practice, and combat, 28 February 1973, p 8-5, AR 385-63/MCO P3570.1, Policies and procedures for firing ammunition for training, target practice, and combat, 22 February 1978, p 7-5, AR 385-63/MCO P3570.1A, Policies and procedures for firing ammunition for training, target practice, and combat, 15 October 1983, p 7-5).

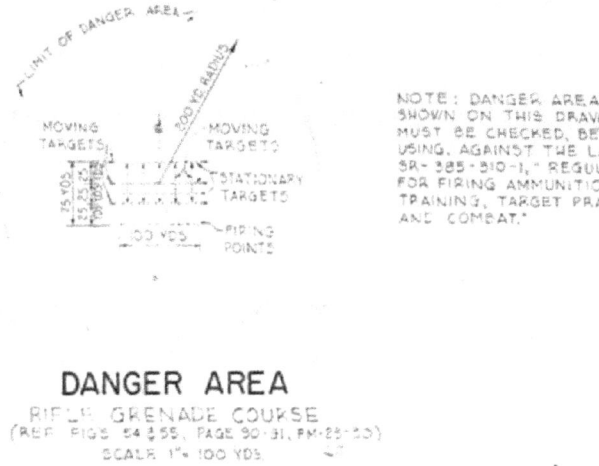

Figure 382. Practice rifle grenade course, danger area, Fort Bragg, NC, 1952 (Standard drawing 28-13-45 sheet 1 of 2, Practice rifle grenade course plans, 30 January 1952).

Typical layouts

There were three types of rifle grenade ranges: 1) antipersonnel marksmanship (stationary), 2) antitank marksmanship (stationary and moving target), and 3) antitank field firing. One or all of these types could have been located on the rifle grenade range" ("RO-14" 30).

Antipersonnel marksmanship rifle grenade Range

"The High Trajectory Fire Court encompassed an area 200 yds long and approximately 40 yds wide with targets located at the far end. Firing points were established at ranges of 50, 100, 150, and 200 yds" ("RO-14" 31). "The Flat Trajectory Fire Court was 75 yds long and had 6 targets at the far end. Sandbags were placed at the 25-yd mark, shell craters dug at the 50-yd mark, and stakes indicated other firing points at the 75-yd mark" ("RO-14" 31).

Pre-WWII rifle grenade range

"In 1927, fragmentation and live rifle grenades were fired from kneeling, prone, and trench positions on what appeared to be a separate court from a hand grenade court. The firer placed himself 150 yds from the target. A description of the court is not provided in the reference document. The earliest depiction of a rifle grenade court is in a 1932 Basic Field Manual (BFM) and is reproduced in Figure 383 below. Three circles, each with a 10-yd radius, were outlined on the ground at distances of 150, 200, and 250 yds from a firing line. Two E targets (placed at an angle) or another visible object were placed in the center of each circle. Two firing points were used. One firing point was open. The second firing point was masked so that the target could not be seen when the firer was in a kneeling position. Use of this range appears to have been discontinued prior to 1942" ("RO-14" 32).

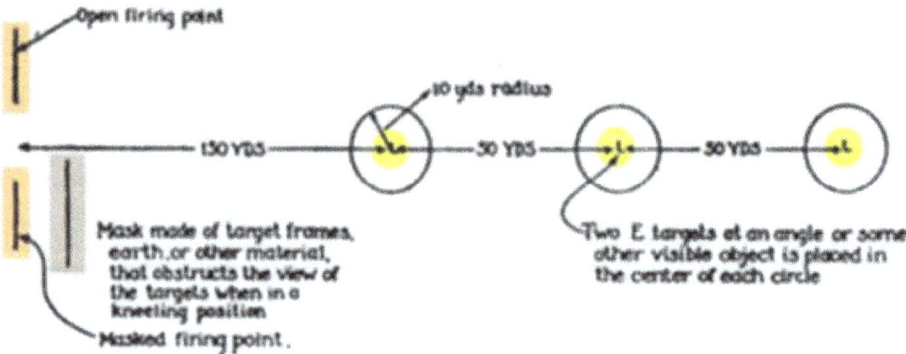

Figure 383. Rifle grenade court, circa 1927-1942 (Basic Field Manual, Instruction with hand and rifle grenades, 14 July 1932, p 30).

Stationary target rifle grenade course

"A stationary target rifle range course is depicted in Figure 384 below. Only practice rifle grenades were fired on this course. Training was done in an area not occupied by other troops and a clear distance of 200-yds was established behind the target" ("RO-14" 33). "The stationary target rifle grenade course was used from WW II to 1986 and accommodated live or practice grenade training. The stationary target rifle grenade course surface danger area is depicted in Figure 381 above" ("RO-14" 34).

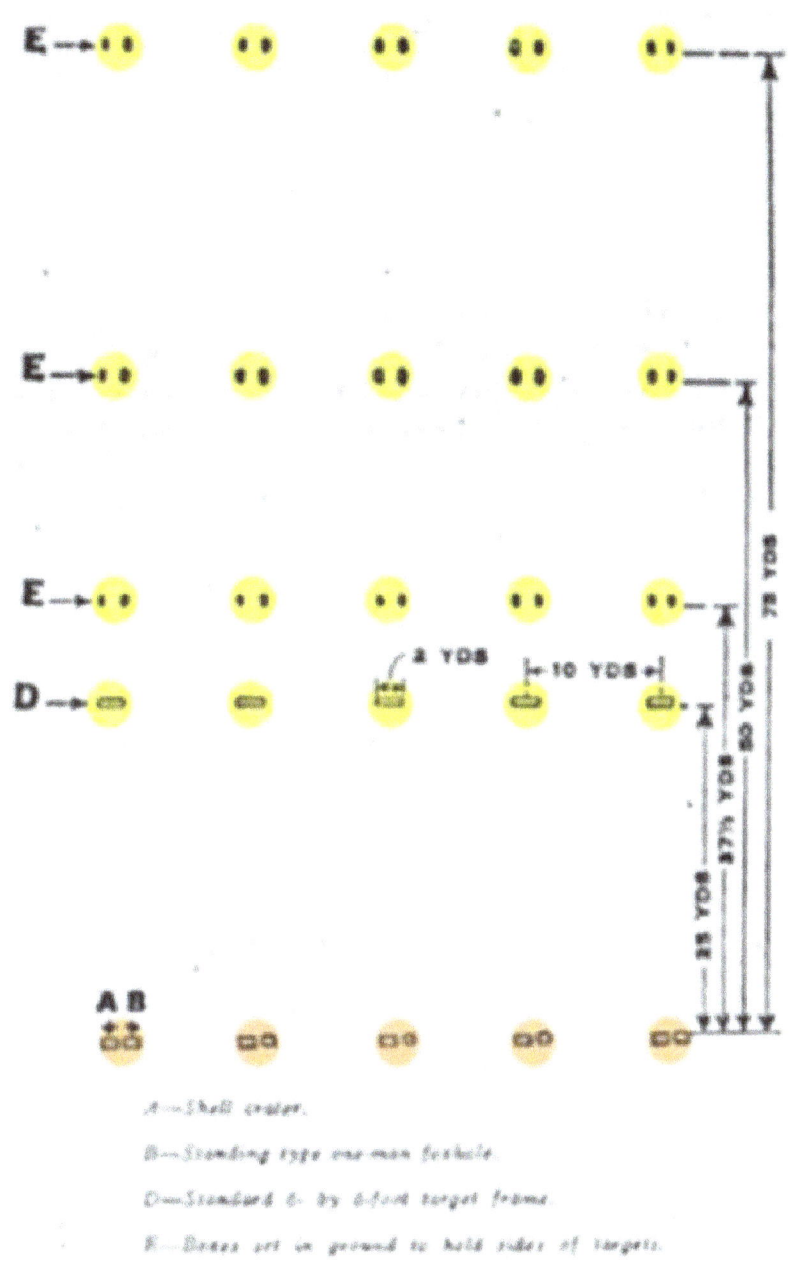

Figure 384. Stationary target rifle grenade course, circa 1942-1949 (FM 23-30, Hand and rifle grenades, rocket, AT, HE, 2.36-in., 14 February 1944, pp 85, 90, FM 23-30, Hand and rifle grenades, 14 April 1949, pp 86, 90).

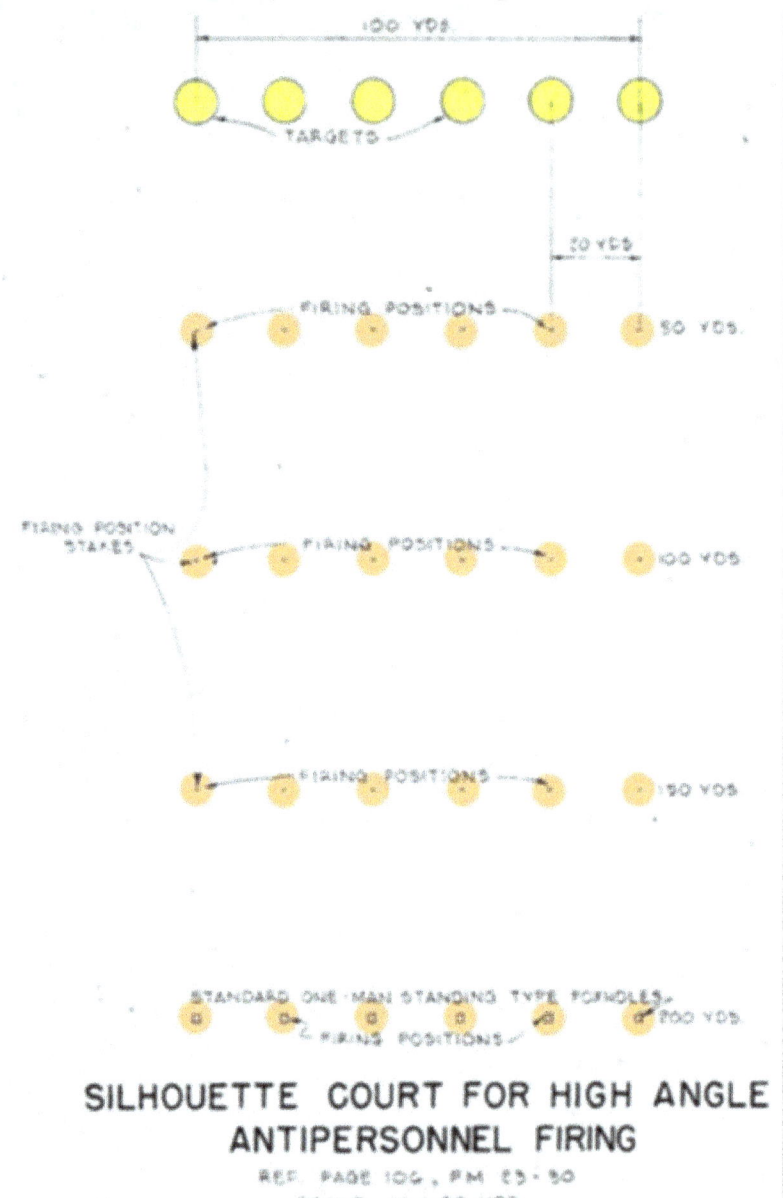

Figure 385. Practice rifle grenade course, silhouette court for high angle antipersonnel firing, Fort Bragg, NC, 1952 (Standard drawing 28-13-45 sheet 1 of 2, Practice rifle grenade course plans, 30 January 1952).

Vertical target court for flat trajectory Antipersonnel firing

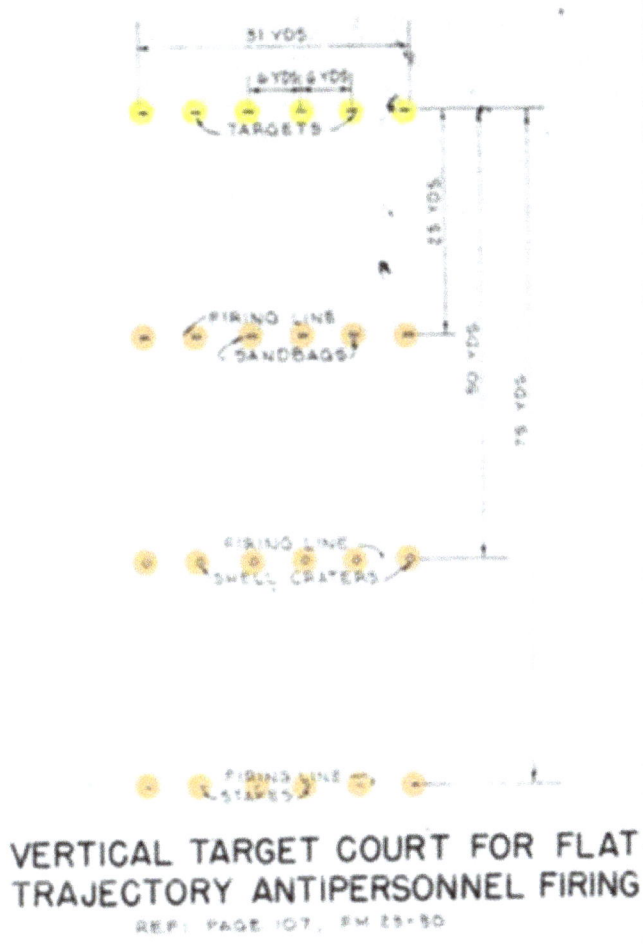

Figure 386. Practice rifle grenade course, vertical target court for flat trajectory antipersonnel firing, Fort Bragg, NC, 1952 (Standard drawing 28-13-45 sheet 1 of 2, Practice rifle grenade course plans, 30 January 1952).

Antitank marksmanship rifle grenade range

"The court for known-distance firing at stationary targets consisted of a single firing line using shell craters and/or 1-man foxholes for firing positions. Targets were located at 25 yds, 37½ yds, and 75 yds" ("RO-14" 31). "The court for firing at moving targets was constructed on the same stationary target court. Moving targets were either towed parallel or perpendicular to the firing line. Each firing position was allowed to use one target when towed parallel to the firing line. However, only two firing positions were allowed to use one target when towed perpendicular to the firing line" ("RO-14" 31).

Moving target rifle grenade range

"Moving target rifle grenade courses were designed for use on or adjacent to a stationary target range. The two ranges were normally co-located. Training requirements may have mandated that separate ranges be maintained in order to achieve more efficiency in training personnel. The layout of a moving target rifle grenade range is illustrated in Figure 387 below. During the World War II period, only practice antitank grenades were available for training. However, the regulations do not address other rifle grenades or hand grenades fired as rifle grenades with the use of rifle grenade adapters except to say that the general provisions for hand grenades apply" ("RO-14" 35).

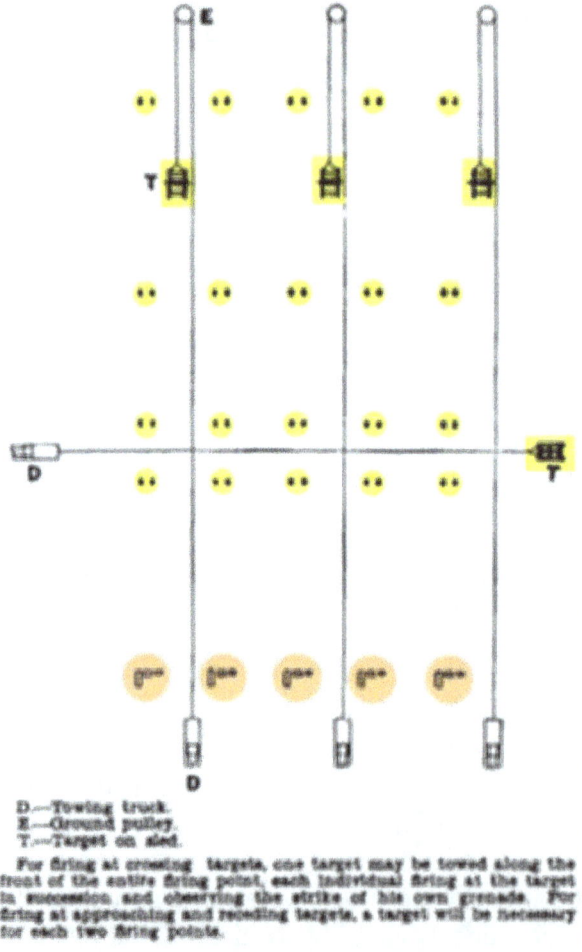

Figure 387. Moving target rifle grenade course, circa 1942-1964 (FM 23-30, Grenades, 15 June 1942, pp 54, 56, FM 23-30, Hand and rifle grenades, rocket, AT, HE, 2.36-in., 14 February 1944, pp 85, 89, 91, FM 23-30, Hand and rifle grenades, 14 April 1949, pp 88 – 91, AR 210-21, Training areas and facilities for ground troops, 18 December 1964, p 6).

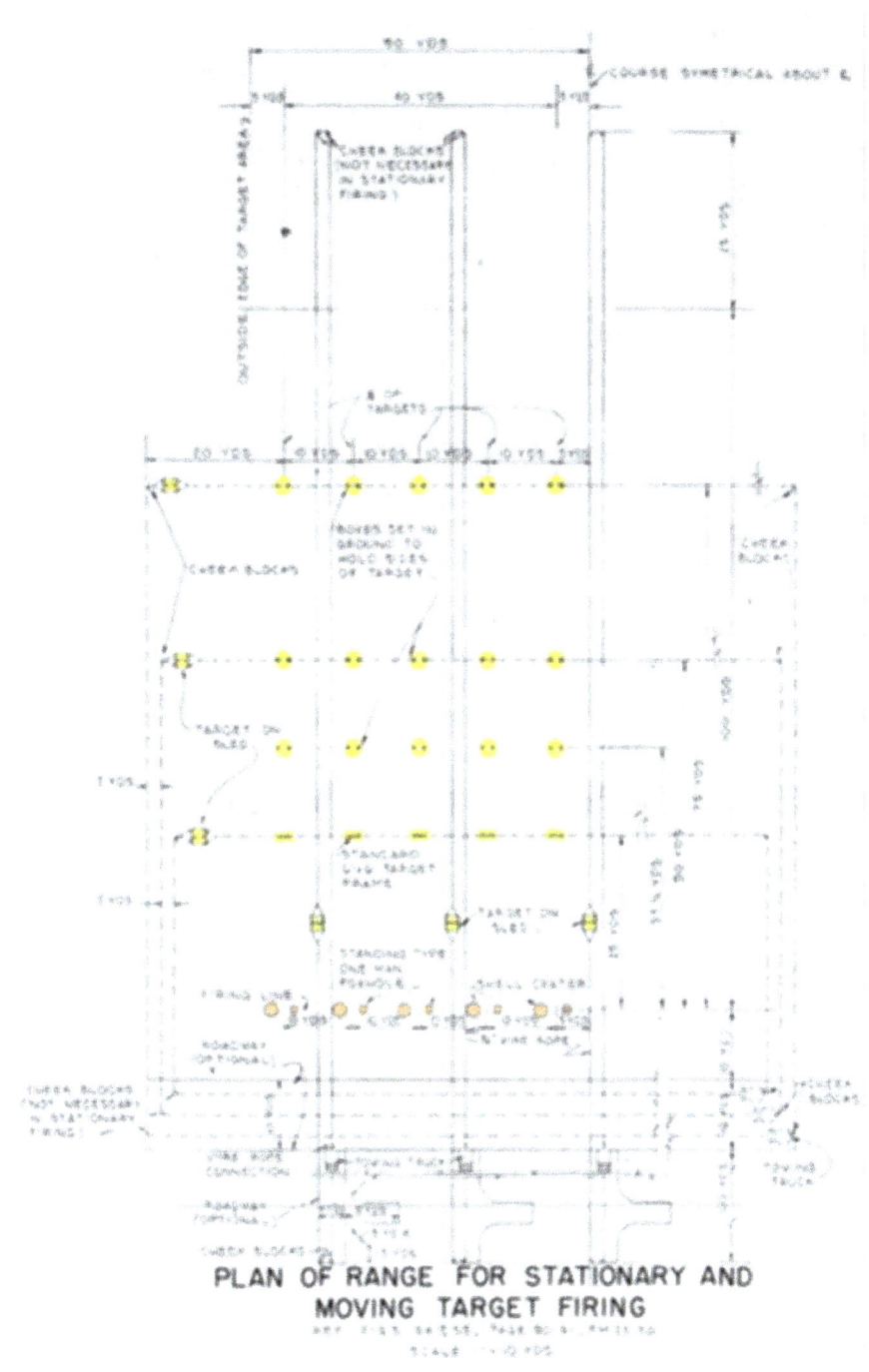

Figure 388. Practice rifle grenade course, plan of range for stationary and moving target firing, Fort Bragg, NC, 1952 (Standard drawing 28-13-45 sheet 1 of 2, Practice rifle grenade course plans, 30 January 1952).

Antitank field firing rifle grenade range

"This ground towed target court was utilized for field firing exercises with practice antitank grenades. Several standing-type, 1-man foxholes or shell craters were constructed at a location near the center of the range" ("RO-14" 31). "Since practice antitank rifle grenades did not contain an explosive charge, training in their use was authorized in any area where troops were not within a distance of 200 yds to the rear of the target. In order to minimize damage to fin assemblies, an area free from trees, stumps, rocks, and other hard objects was preferred" ("RO-14" 31). "Firing rifle grenades required a grenade launcher affixed to a standard service rifle. Most launchers also required a special rifle grenade cartridge. When firing grenades with impact fuses, personnel within 125 yds of the target wore a helmet and assumed a prone position, face down, or used cover during the time a grenade was in flight. When firing other explosive grenades for instructional purpose, the grenades were fired only from a trench or from behind adequate protection equivalent to a screen of sandbags 20 in. thick" ("RO-14" 31).

37-MM antitank range

"A 37-mm antitank range could also be utilized for field firing of AT practice rifle grenades by constructing several individual prone shelters or standing type one-man foxholes. These shelters or foxholes would be located as indicated by the X in Figure 389 below. Several moving targets would pass within a range of 75 yds from the shelter or foxhole" ("RO-14" 36). "The range for a rifle grenade depended on the rifle being used, the angle of fire, and how the grenade was fitted on the grenade launcher. For example, an M1 rifle using an M7 grenade launcher with the angle of fire at 30 degrees and 6 rings on the grenade launcher exposed would launch a M11 grenade approximately 70 yds. The M9 and M11 grenades fired from M1, M1903, or M1917 rifles at an angle of 45 degrees with no rings exposed on the launcher went approximately 365 yds. Range tables for high angles of fire are provided for the various types of grenades when used with various types of rifles. The longest distance given in these tables is the 365 yds described above. The 1968 edition of AR 385-63 is the first range safety regulation to specify the size and shape of grenade ranges and the surface danger areas" ("RO-14" 36, 37).

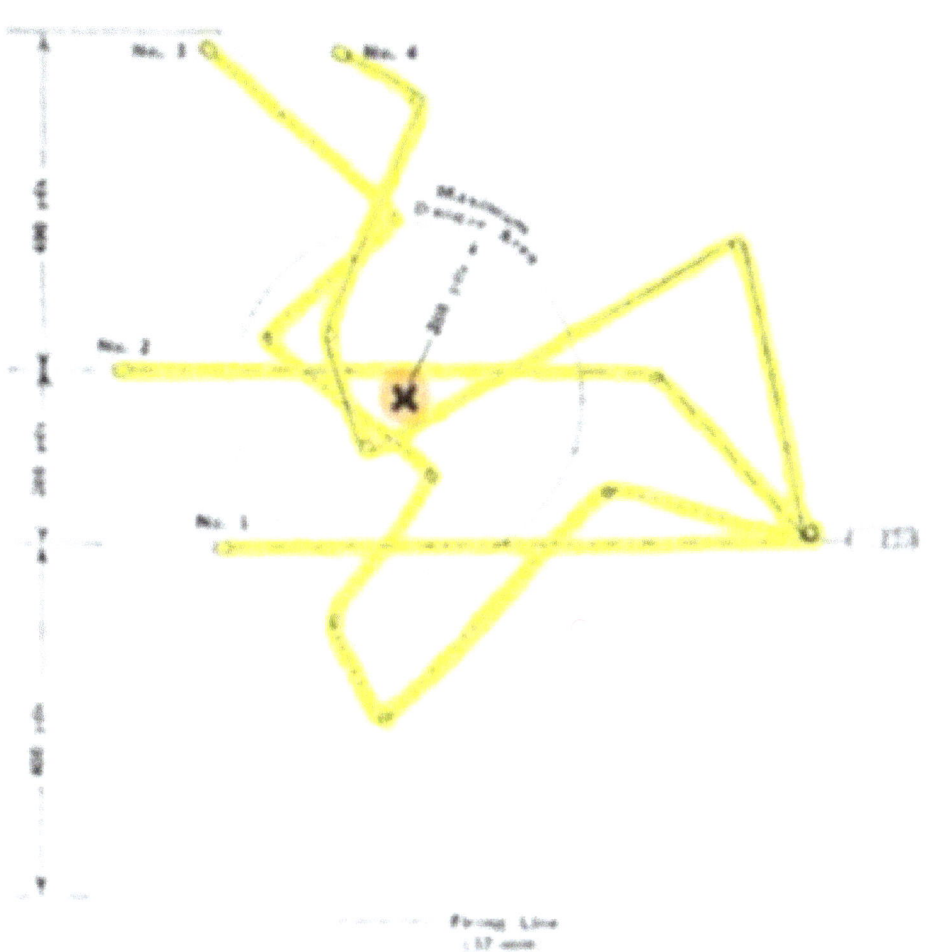

Figure 389. Employment of 37-mm antitank range for antitank practice rifle grenade field firing, circa 1942-1949 (FM 23-30, Grenades, 14 June 1942, p 57, FM 23-30, Hand and rifle grenades, rocket, AT, HE, 2.36-in., 14 February 1944, p 92, FM 23-30, Hand and rifle grenades, 14 April 1949, p 93).

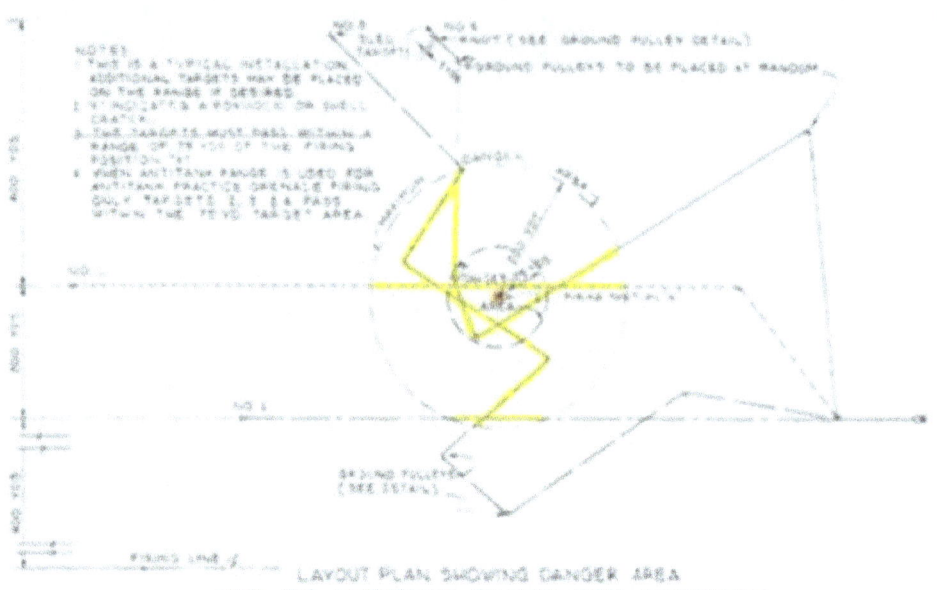

Figure 390. Practice rifle grenade course, use of antitank range for antitank practice grenade field firing, Fort Bragg, NC, 1952 (Standard drawing 28-13-45 sheet 1 of 2, Practice rifle grenade course plans, 30 January 1952).

Firing lines

Firing positions may have included individual prone shelters, foxholes, sandbags, shell craters, and trenches. Some may have been masked with an embankment.

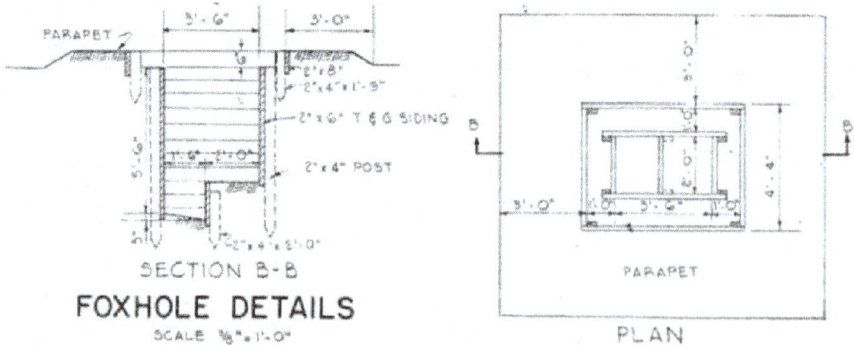

Figure 391. Practice rifle grenade course, foxhole details, Fort Bragg, NC, 1952 (Standard drawing 28-13-45 sheet 2 of 2, Practice rifle grenade course details, 30 January 1952).

Targets

"Prior to concerns about tanks, a rifle grenade court consisted of 10-yd diameter circles located at 150, 200, and 250 yds from the firing line. With the increased emphasis on tanks during World War II, rifle grenade ranges consisted of stationary, moving target, and field firing ranges. For stationary target firing on a rifle grenade court, a standard 6 x 6-ft wooden target frame, covered with target cloth, burlap, or other light material, was used. A 20-in. bull's eye was either painted on or a target repair center (B-C) was pasted on. A sandbag tank silhouette could also serve as a suitable target. For HEAT rifle grenades, objects such as a salvaged vehicle, piece of armor, boiler plate, large stone, or wall were recommended as targets to insure detonation of the grenade" ("RO-14" 57). Some ranges also used silhouette targets. "For a moving target, a wooden sled 6 ft by 6 ft was constructed. The wooden frame was bolted to the sled. A standard 6 x 6-ft wooden target frame, covered with target cloth, burlap, or other light material, and painted with a 20-in. bull's eye [or target repair center (B-C) pasted on] was used. The sled was towed by a motor vehicle. The target and sled are shown in Figure 394 below" ("RO-14" 58).

Stationary targets

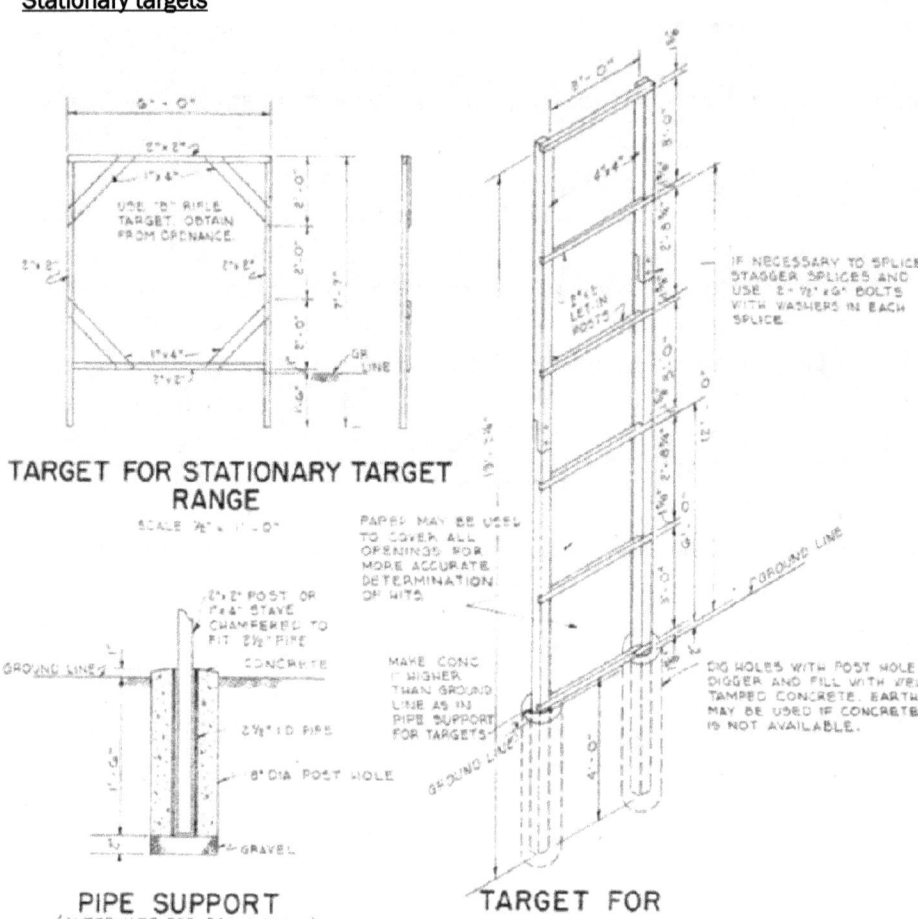

Figure 392. Practice rifle grenade course, targets for stationary target range and vertical target court, Fort Bragg, NC, 1952 (Standard drawing 28-13-45 sheet 2 of 2, Practice rifle grenade course details, 30 January 1952).

Silhouette targets

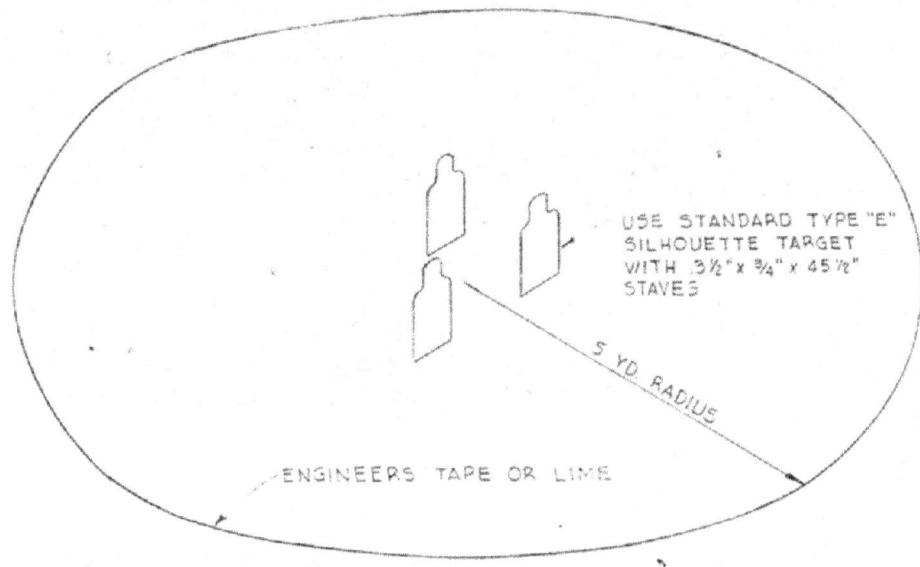

Figure 393. Practice rifle grenade course, high angle antipersonnel target, Fort Bragg, NC, 1952 (Standard drawing 28-13-45 sheet 2 of 2, Practice rifle grenade course details, 30 January 1952).

Moving targets

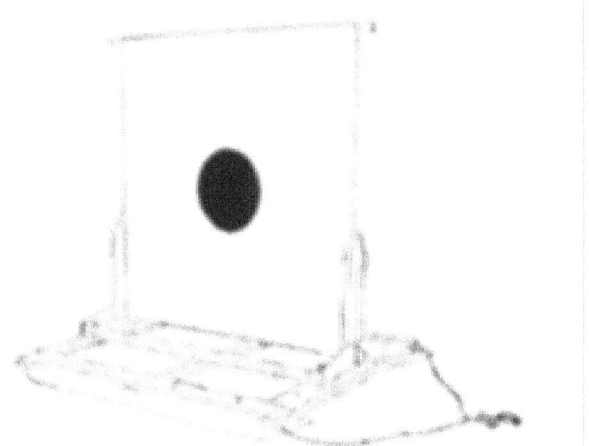

Figure 394. Moving target, circa 1942-1949 (FM 23-30, Hand and rifle grenades, rocket, AT, HE, 2.36-in., 14 February 1944, p 54, FM 23-30, Hand and rifle grenades, rocket, AT, HE, 2.36-in., 14 February 1944, p 89, FM 23-30, Hand and rifle grenades, 14 April 1949, p 92).

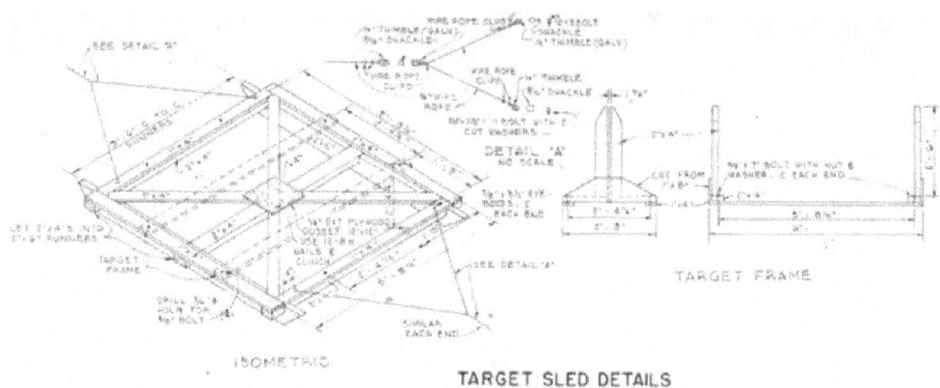

Figure 395. Practice rifle grenade course, target sled details, Fort Bragg, NC, 1952 (Standard drawing 28-13-45 sheet 2 of 2, Practice rifle grenade course details, 30 January 1952).

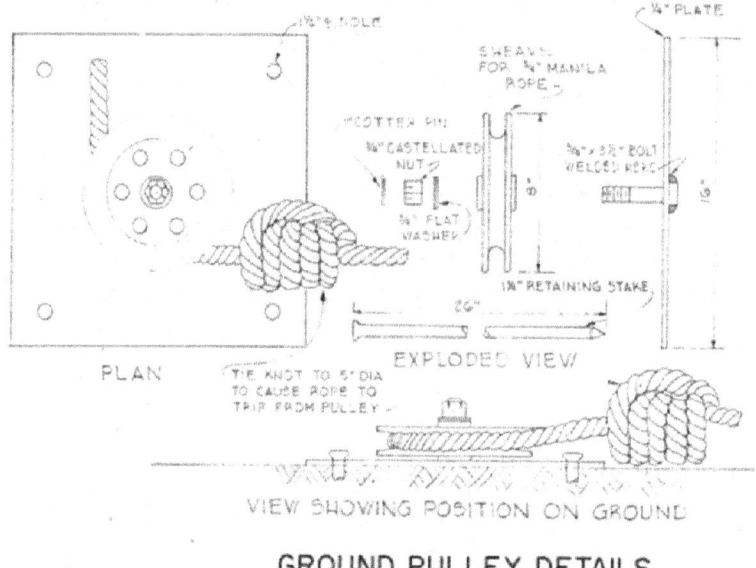

Figure 396. Practice rifle grenade course, ground pulley details, Fort Bragg, NC, 1952 (Standard drawing 28-13-45 sheet 2 of 2, Practice rifle grenade course details, 30 January 1952).

Embankments/trenches/etc.

Embankments of sandbags may have been built around some firing positions for protection, and were a mask for others. Other firing positions included foxholes, trenches, shell craters, and individual prone shelters.

Buildings

A range may have had a control tower, latrine, target storage building, ammunition storage building, other storage sheds, and administrative/maintenance buildings supporting general range functions. The range may also have been part of a larger installation range complex that contained these buildings.

Trainfire ranges (layouts)

Trainfire I Ranges were first developed in 1953 as a result of the poor marksmanship skills shown by soldiers during the Korean War (McFann et al. 9, 54-63). Trainfire called for an entirely different type of range for instruction and used a new concept of teaching during rifle training. Soldiers were trained on these ranges to improve rifle firing in combat situations; including quickly detecting indistinct or fleeting targets, quickly assuming steady firing positions, and hitting silhouette targets (Liwanag). The U.S. Army formally adopted trainfire as its approved method of teaching basic rifle marksmanship in the summer of 1957, and trainfire ranges have been used continuously since that time ("Trainfire" 89).

Multiple range layout plan

Figure 397. Trainfire ranges layout plan, Fort Bragg, NC, 1960 (Standard drawing 28-13-115 sheet 1 of 24, Trainfire ranges, layout plan, 5 October 1960).

Figure 398. Aerial photograph of Clark Trainfire Range, Fort Knox, KY, 1960 (Training 4a, vol. 1, Stock Shot # 263, Fort Knox, KY, Clark Trainfire, Fort Knox, Ky, photo by: SFC John A. Gilstrap, 23 May 1960, USAARMC Photo Branch, Fort Knox, KY).

Figure 399. Aerial photograph of Captain O'Brien Trainfire Range, Fort Knox, KY, 1960 (Training 11, vol. 1, Stock Shot # 281, Fort Knox, KY, O'Brien Trainfire Range, Fort Knox, photo by: SFC John A. Gilstrap, 23 May 1960, USAARMC Photo Branch, Fort Knox, KY).

25m zero range

On the 25m zero range, Trainfire 1, the trainee learns the fundamentals of good marksmanship, positions, steady hold factors, trigger control, sight alignment, sight manipulation, and fires for his 25m zero (see Figures 400 through 402).

Danger area

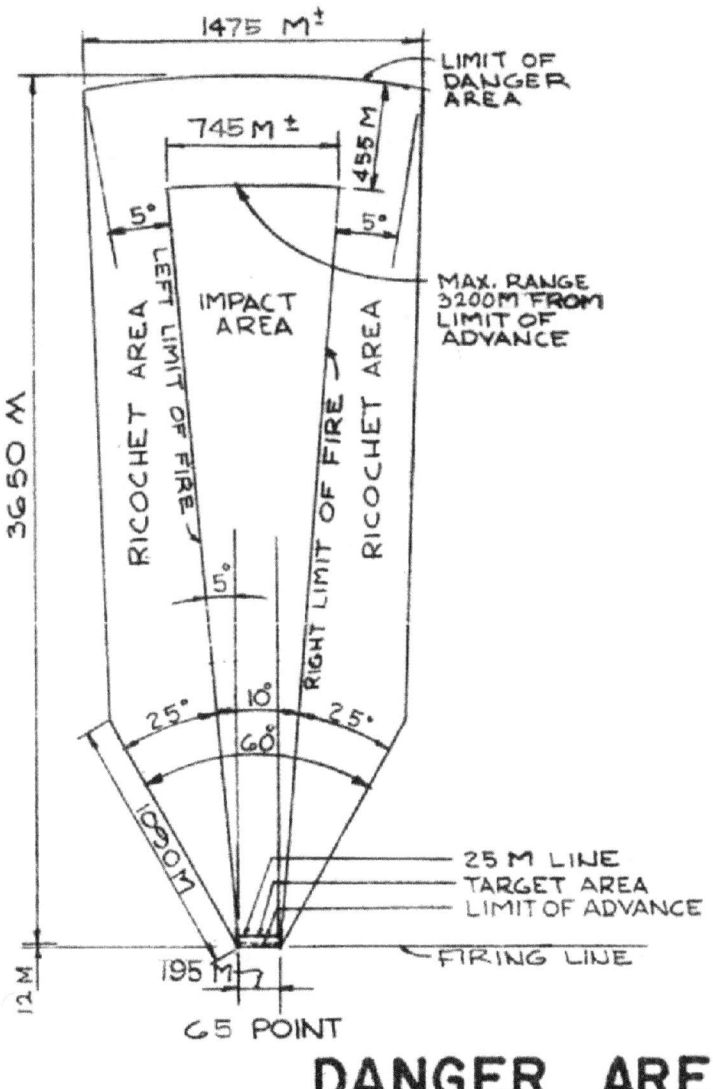

Figure 400. Rifle marksmanship course, trainfire I, 25m range (65 point), danger area, Fort Bragg, NC, 1958 (Standard drawing 28-13-105 sheet 2, Rifle marksmanship course, trainfire i, 25m range (65 point), plan, section and details, 5 August 1958).

Typical layout

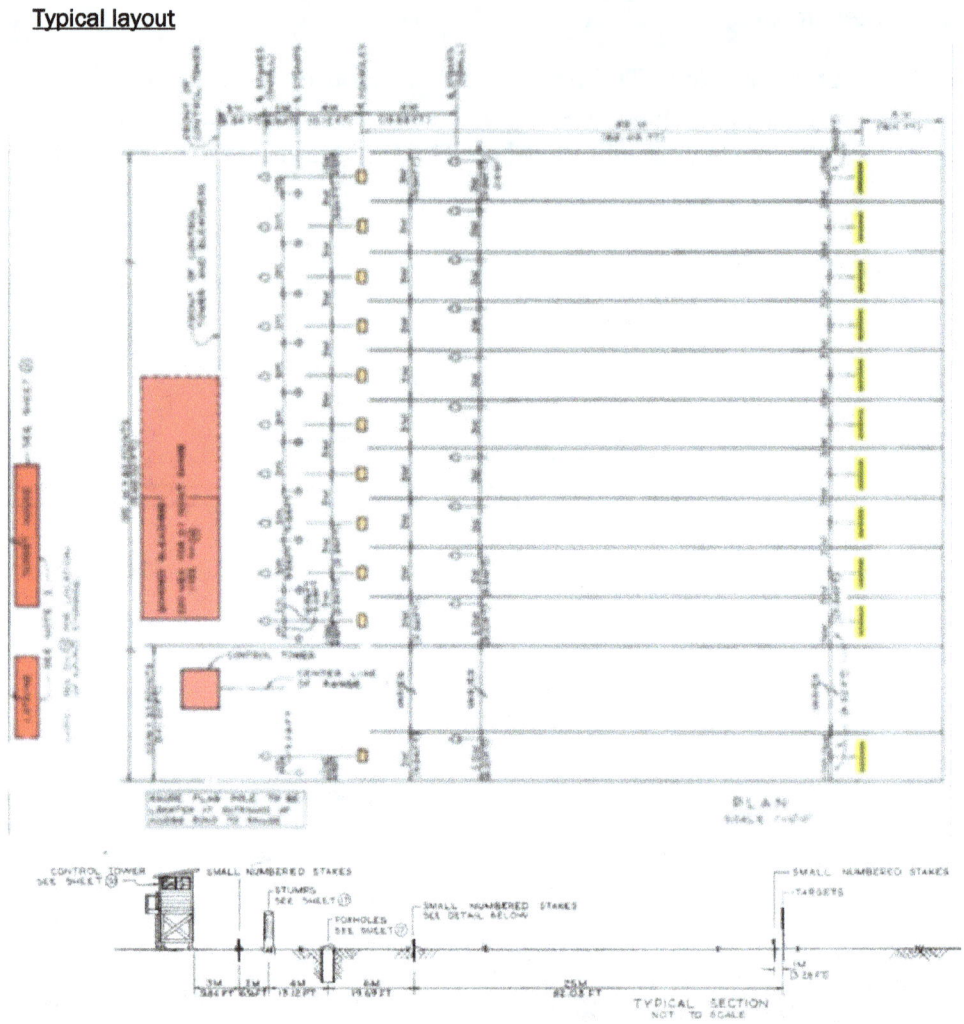

Figure 401. Rifle marksmanship course, trainfire I, 25m range (65 point), plan, Fort Bragg, NC, 1958 (Standard drawing 28-13-105 sheet 2, Rifle marksmanship course, trainfire I, 25m range (65 point), plan, section and details, 5 August 1958).

Figure 402. Twenty-five meter Range at Fort Jackson, SC, 25 October 1965 (NARA College Park, RG 111-SC post-1955, box 396, photo SC624269).

Target detection range

Target Detection Range at Trainfire 1 is used to teach the trainee the various methods of range estimations and detecting enemy targets (see Figures 403 and 404).

Typical layout

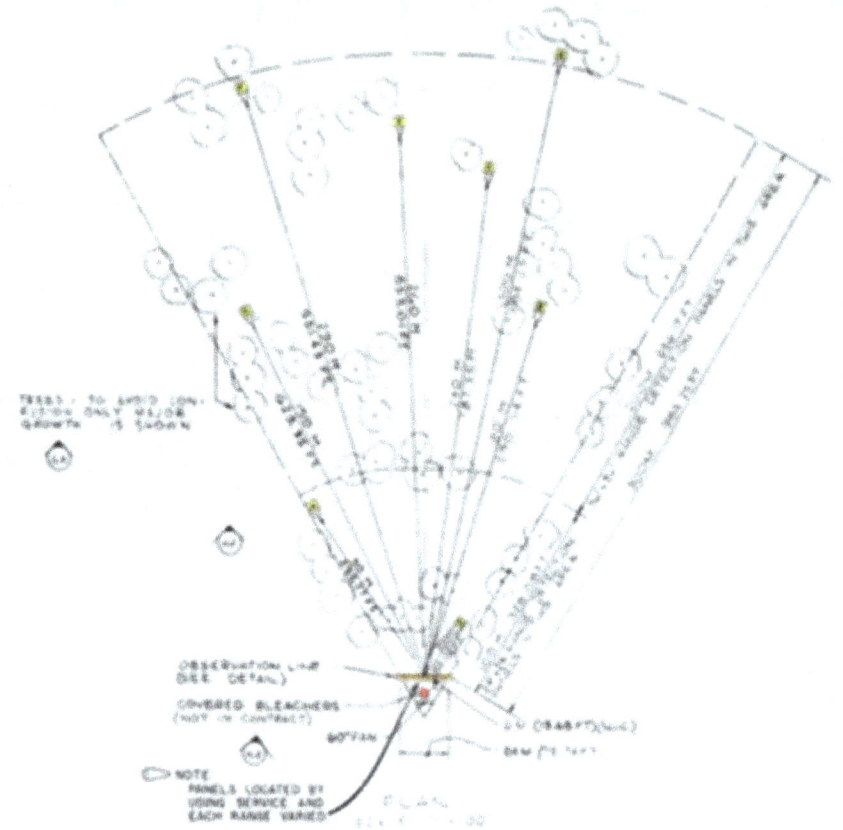

Figure 403. Trainfire ranges, target detection range plan, Fort Bragg, NC, 1960 (Standard drawing 28-13-115 sheet 12 of 24, Rifle marksmanship course, trainfire I, target detection range, plan, section and details, 5 October 1960).

Figure 404. Trainfire ranges, target detection range perspective, Fort Bragg, NC, 1960 (Standard drawing 28-13-115 sheet 12 of 24, Rifle marksmanship course, trainfire I, target detection range, plan, section and details, 5 October 1960).

Field firing range

Field Firing Range, Trainfire 1 where the trainee is introduced to field firing. The trainee fires on simulated advancing targets and surprise targets, and learns to engage lateral and multiple surprise targets. Field firing is more realistic than 25-meter firing and the firer is allowed more freedom in engaging the targets (see Figures 405 through 407).

Danger area

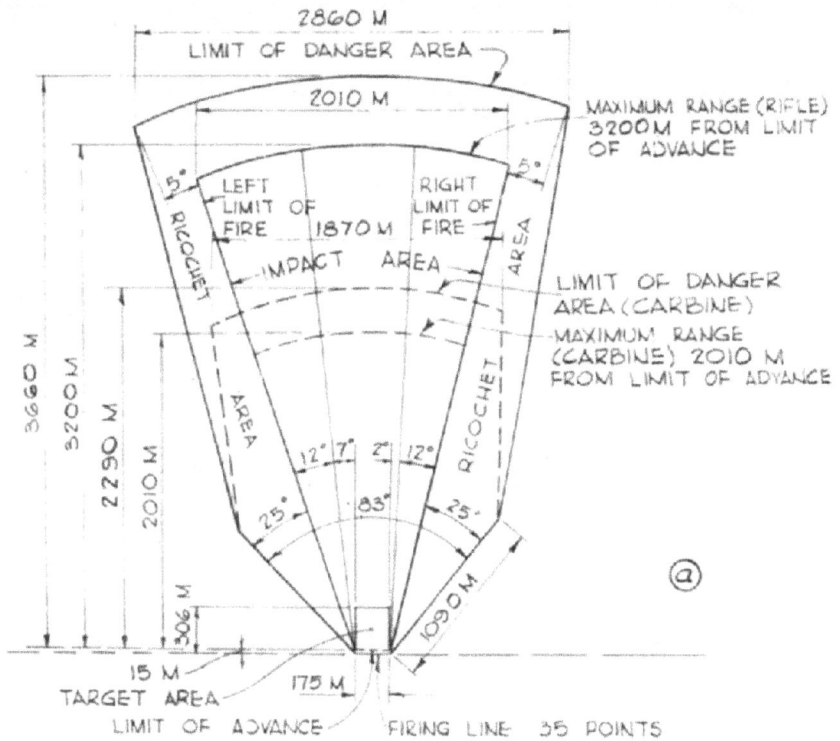

Figure 405. Trainfire field firing range (35 points) danger area, Camp Gordon, GA, 1958 (Standard drawing 28-13-105 sheet 3, Field firing range (35 points), plan, section and details, 5 August 1958).

Typical layout

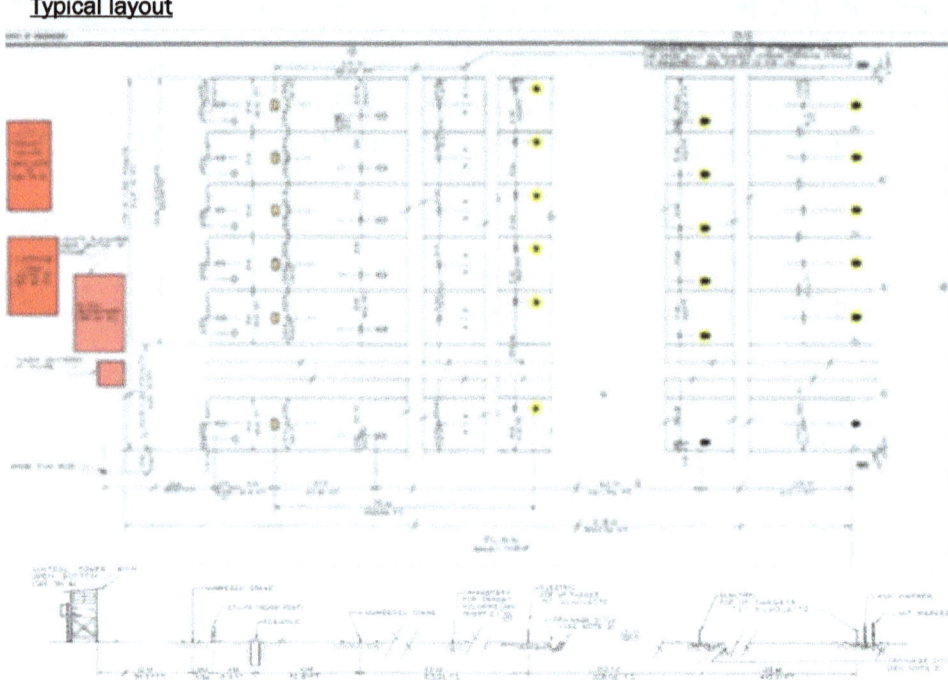

Figure 406. Trainfire field firing range (35 points) typical section, Camp Gordon, GA, 1958 (Standard drawing 28-13-105 sheet 3, Field firing range (35 points), plan, section and details, 5 August 1958).

Figure 407. Trainfire field firing range (35 points) plan legend, Camp Gordon, 1958 (Standard drawing 28-13-105 sheet 3, Field firing range (35 points), plan, section and details, 5 August 1958).

Trainfire 1 record fire range

Trainfire 1 Record Fire Range is laid out to present as near a combat situation as possible using a series of M-31A1 target devices and the "E" and "F" type silhouette well camouflaged, dispersed, and taking advantage of the natural terrain. Each trainee fires this course to establish his degree of qualification (see Figures 408 through 420).

Typical range

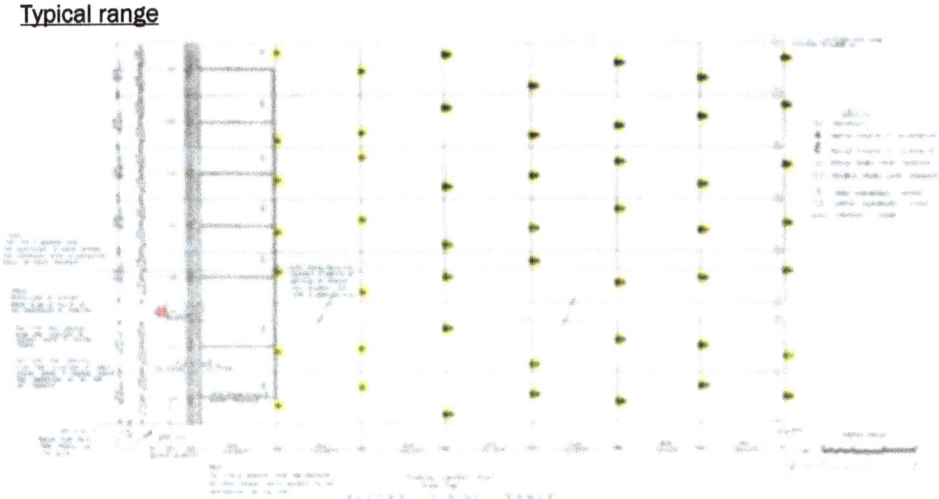

Figure 408. Trainfire I record firing range typical layout plan, Fort Knox, KY, 1960 (Standard drawing 28-13-02 sheet 12, Trainfire I ranges-1960 F.Y., Ditto Hill record firing range, typical layout plan and foxhole profile, 25 May 1960).

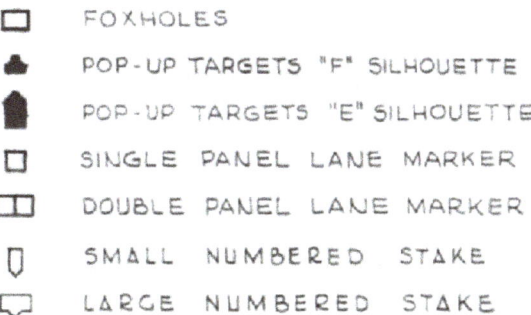

Figure 409. Trainfire ranges record firing range (16 point) plan legend, Fort Gordon, 1966 (Standard drawing 28-13-105 sheet 4, Rifle marksmanship course, trainfire I, record firing range (16 point) plan and section, 7 June 1966).

Twelve point range

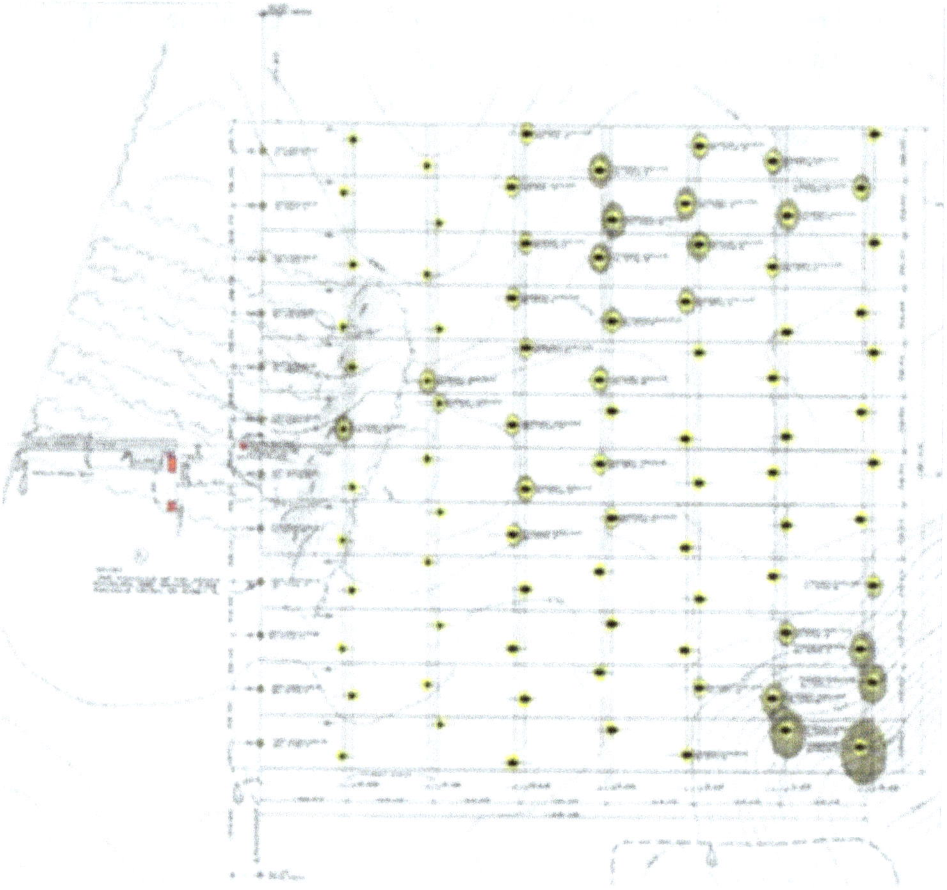

Figure 410. Trainfire ranges record fire No.4 (12 points), layout, Fort Bragg, NC, 1960 (Standard drawing 28-13-115 sheet 5 of 24, Record fire No.4 (12 Points), 5 October 1960).

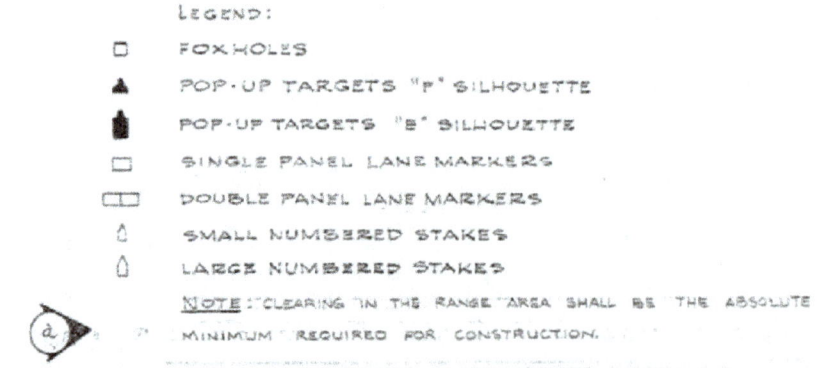

Figure 411. Trainfire ranges record fire (12 points), legend, Fort Bragg, NC, 1960 (Standard drawing 28-13-115 sheet 2 of 24, Record fire No.1 (12 Points), 5 October 1960).

Sixteen point range

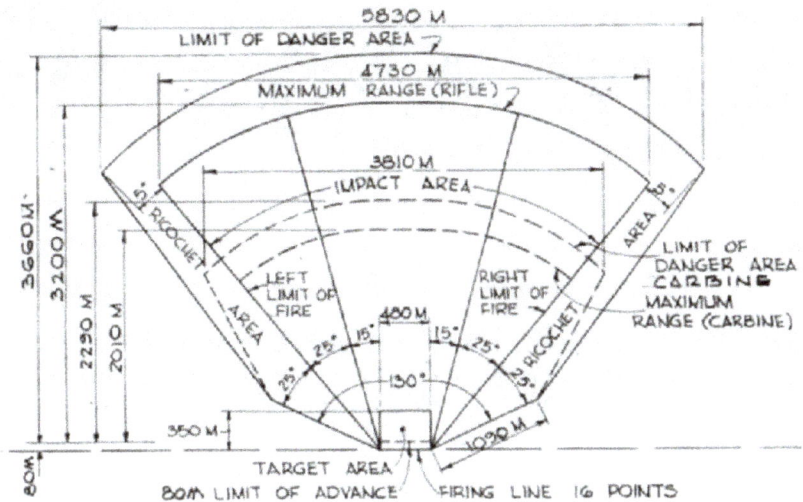

Figure 412. Trainfire ranges record firing range (16 Point) danger area, Fort Gordon, 1966 (Standard drawing 28-13-105 sheet 4, Rifle marksmanship course, trainfire I, Record firing range (16 point) plan and section, 7 June 1966).

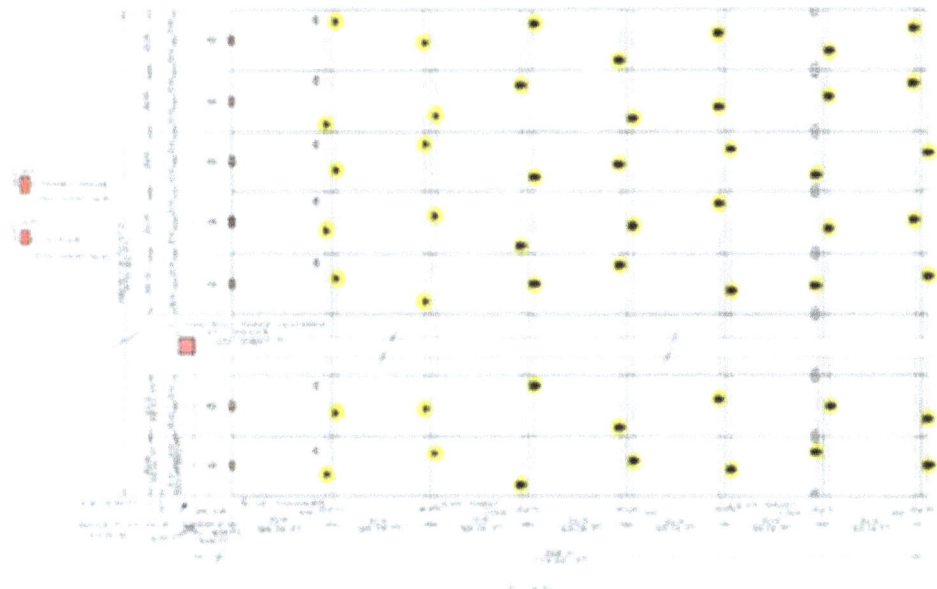

Figure 413. Trainfire ranges record firing range (16 point) plan, Fort Gordon, 1966 (Standard drawing 28-13-105 sheet 4, Rifle marksmanship course, trainfire I, Record firing range (16 point) plan and section, 7 June 1966).

Figure 414. Rifle marksmanship course, trainfire I, record firing range (16 point), perspective view of layout plan, Fort Bragg, NC, 1958 (Standard drawing 28-13-105 sheet 4, Rifle marksmanship course, trainfire I, record firing range (16 point), plan and section, 5 August 1958).

Automated record fire range

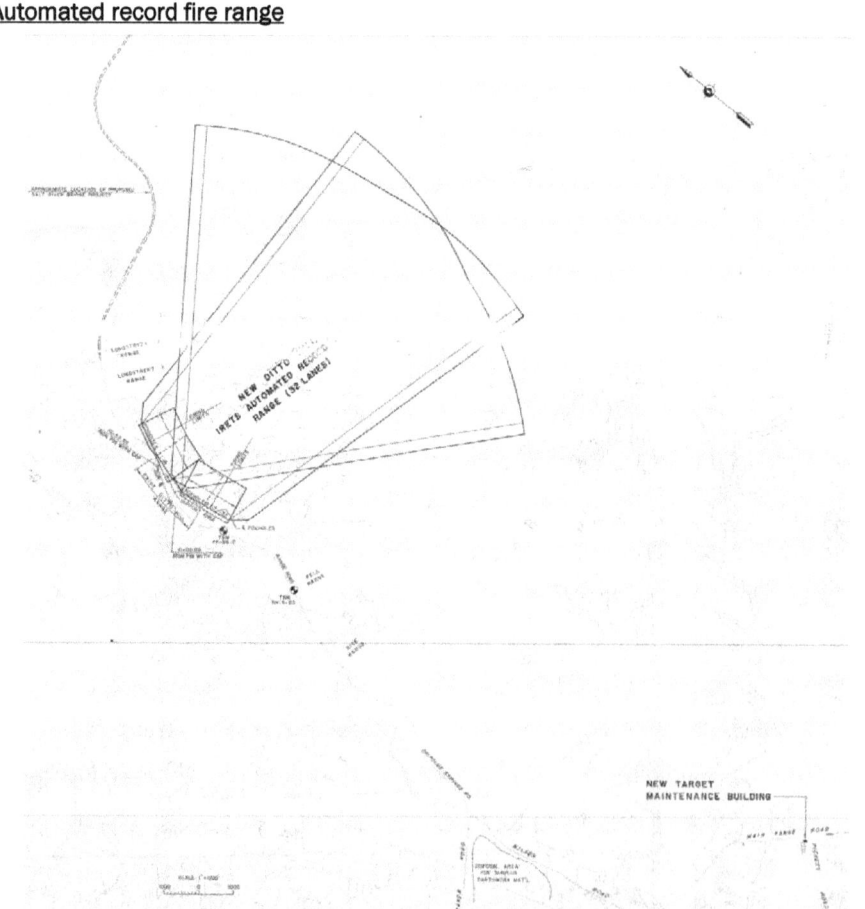

Figure 415. Automated record fire range location map, Fort Knox, KY, 1985 (FK-360-2, Automated record fire range, infantry remote electronic target system, PN 308 FY 85, location map, 1 May 1985).

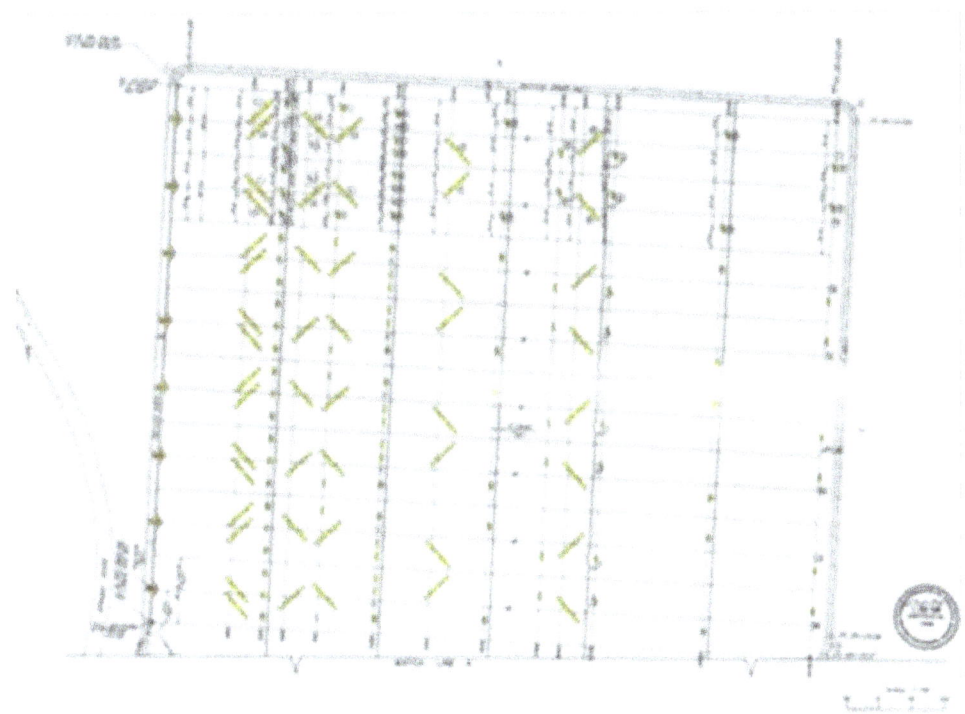

Figure 416. Automated record fire range site layout, Fort Knox, KY, 1985 (FK-360-7, Automated record fire range, infantry remote electronic target system, PN 308 FY 85, site layout, 1 May 1985).

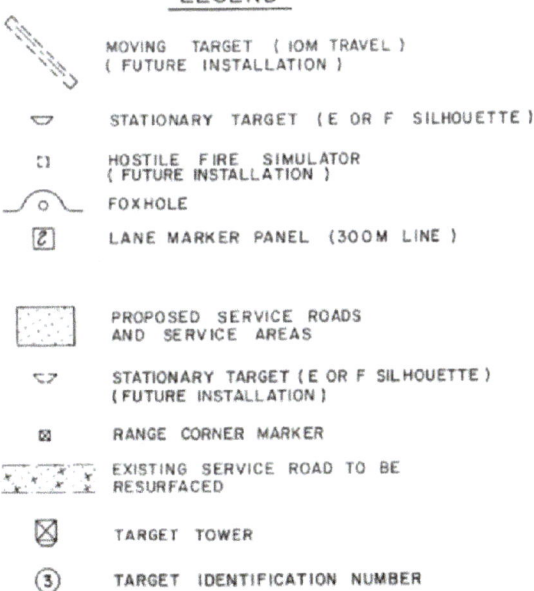

Figure 417. Automated record fire range site layout legend, Fort Knox, KY, 1985 (FK-360-7, Automated record fire range, infantry remote electronic target system, PN 308 FY 85, site layout, 1 May 1985).

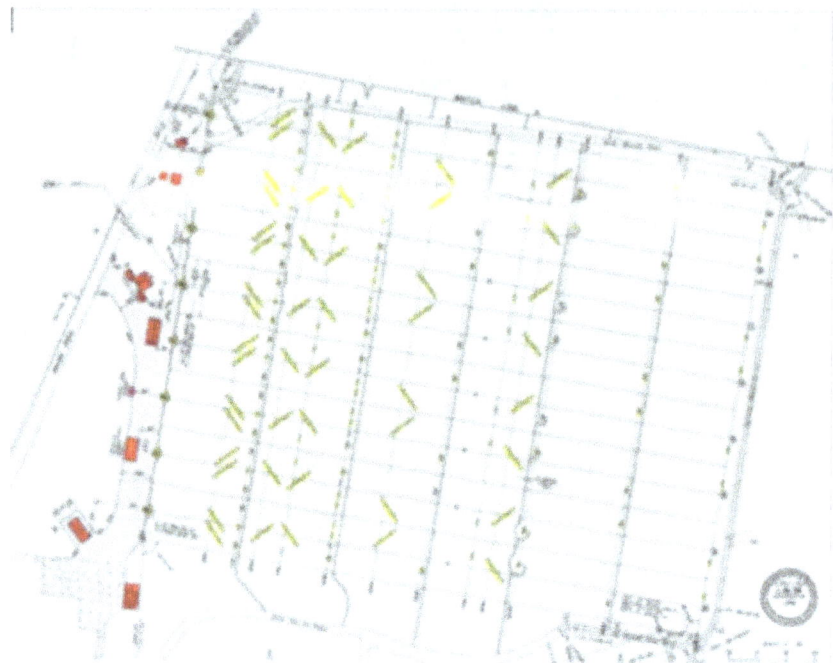

Figure 418. Automated record fire range site layout, Fort Knox, KY, 1985 (FK-360-8, Automated record fire range, infantry remote electronic target system, PN 308 FY 85, site layout, 1 May 1985).

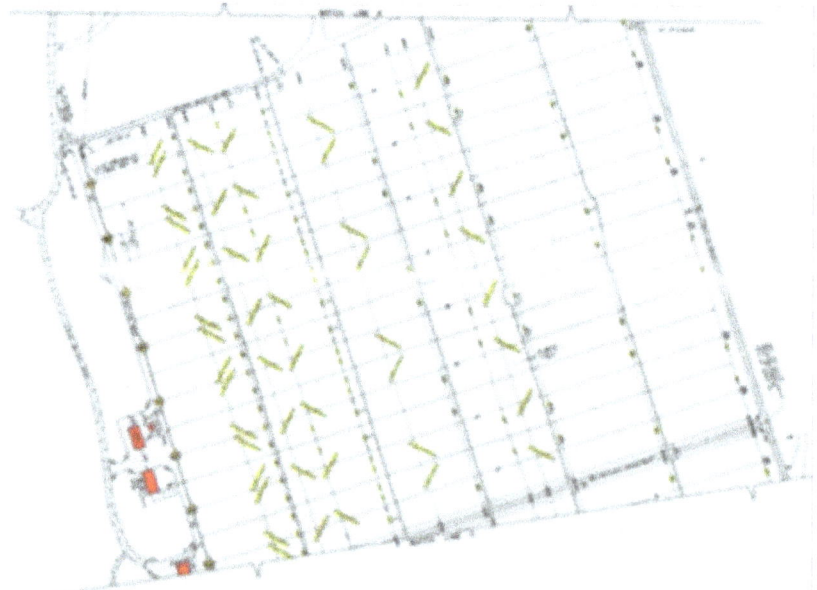

Figure 419. Automated record fire range site layout, Fort Knox, KY, 1985 (FK-360-9, Automated record fire range, infantry remote electronic target system, PN 308 FY 85, site layout, 1 May 1985).

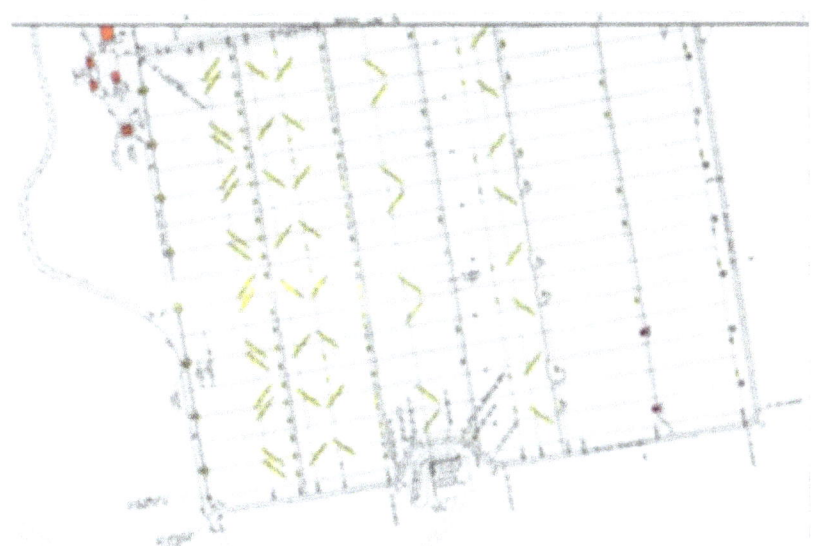

Figure 420. Automated record fire range site layout, Fort Knox, KY, 1985 (FK-360-10, Automated record fire range, infantry remote electronic target system, PN 308 FY 85, Site layout, 1 May 1985).

Firing lines

Soldiers fire from a supported prone position, unsupported prone position, sitting position, or from a foxhole. At the target detection range, firing positions consisted of sighting devices on an observation line. Firing lines were placed on a mound to give a clear view of the range.

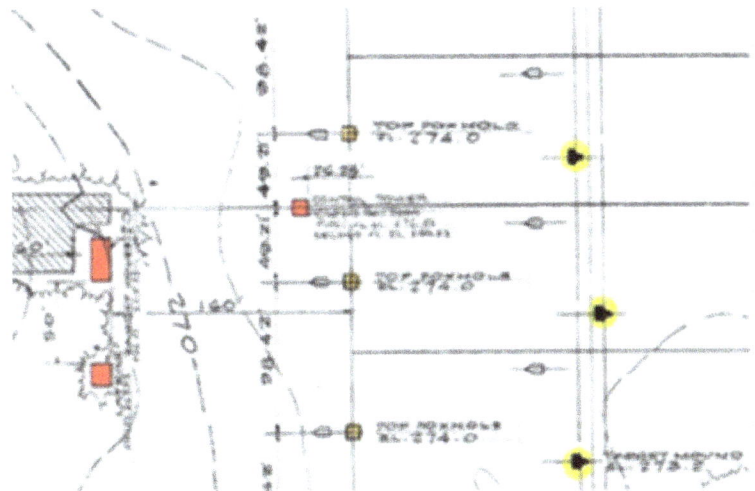

Figure 421. Trainfire ranges record fire No. 1 (12 points), foxholes, Fort Bragg, NC, 1960 (Standard drawing 28-13-115 sheet 2 of 24, Record fire No.1 (12 points), 5 October 1960).

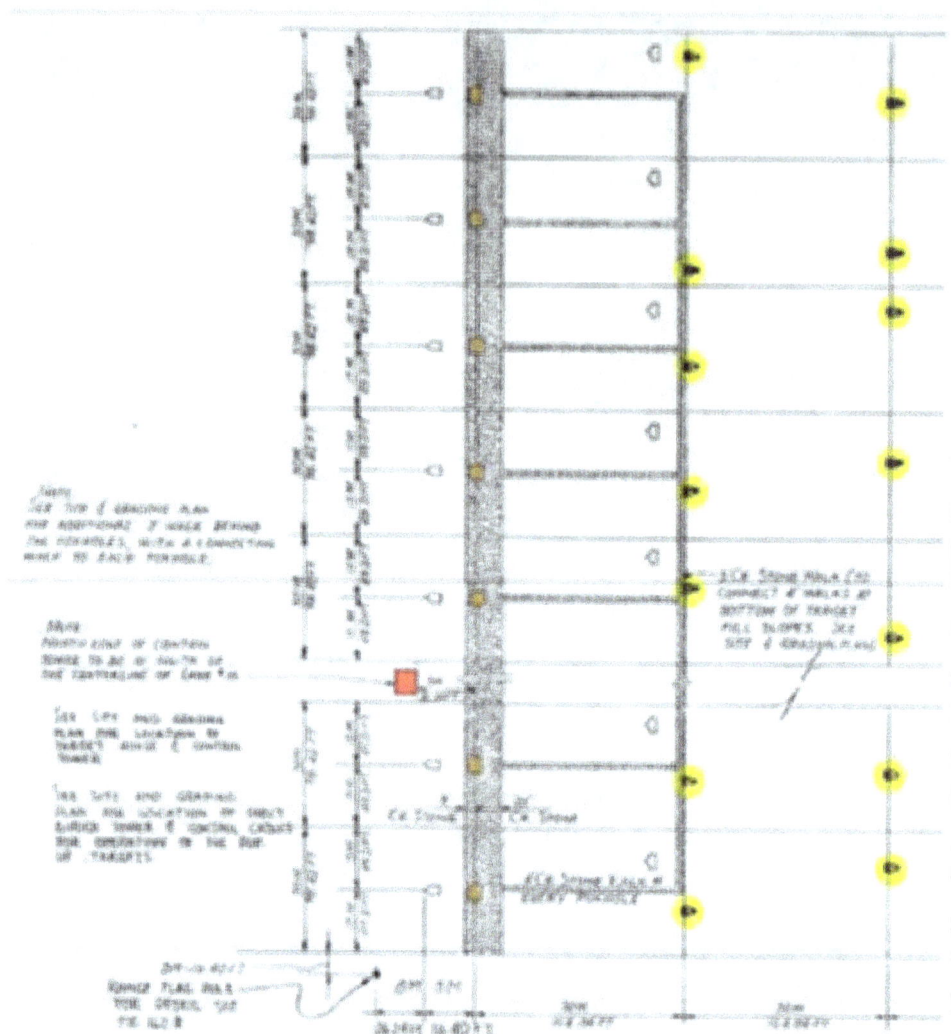

Figure 422. Trainfire I record firing range typical layout plan, Fort Knox, KY, 1960 (Standard drawing 28-13-02 sheet 12, Trainfire I ranges-1960 F.Y., Ditto Hill record firing range, typical layout plan and foxhole profile, 25 May 1960).

Figure 423. Range 24—25m range at Fort Leonard Wood, MO, 27 April 1966 (NARA College Park, RG 111-CRB box 86, photo SC35770).

Foxholes

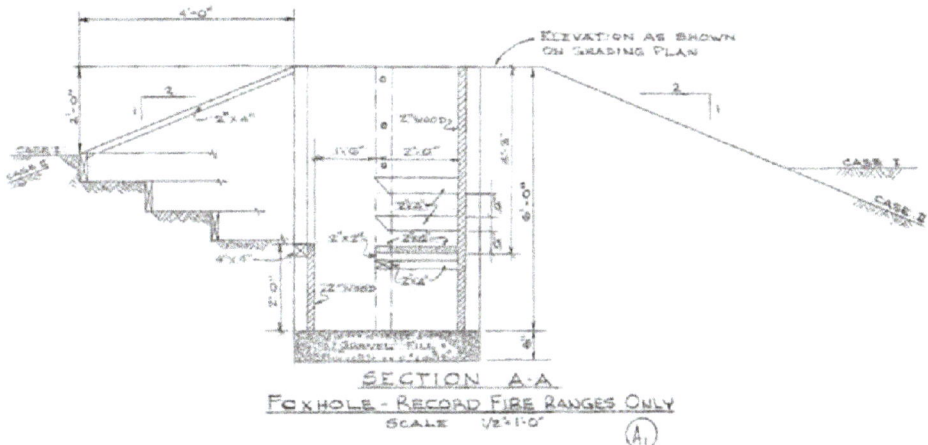

Figure 424. Trainfire ranges, record fire range foxhole section, Fort Bragg, NC, 1960 (Standard drawing 28-13-115 sheet 13 of 24, Trainfire ranges details, 5 October 1960).

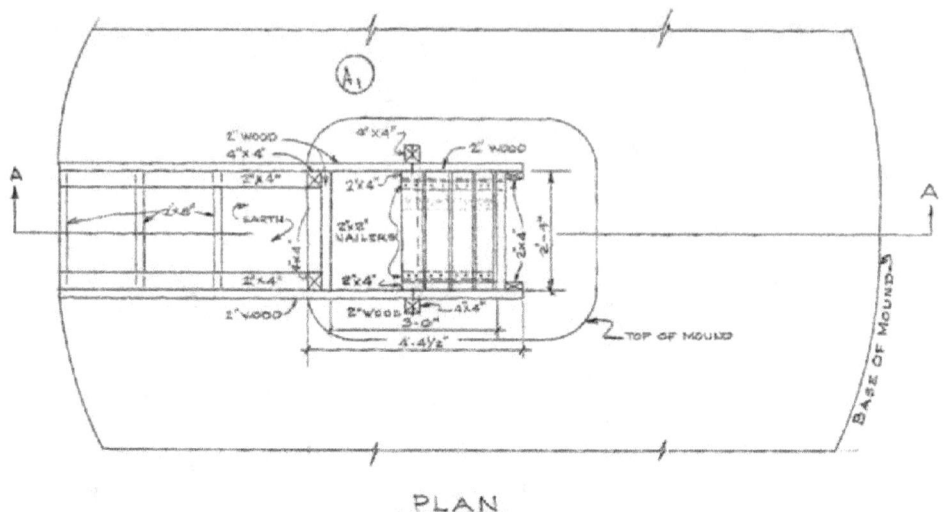

Figure 425. Trainfire ranges, record fire range foxhole plan, Fort Bragg, NC, 1960 (Standard drawing 28-13-115 sheet 13 of 24, Trainfire ranges details, 5 October 1960).

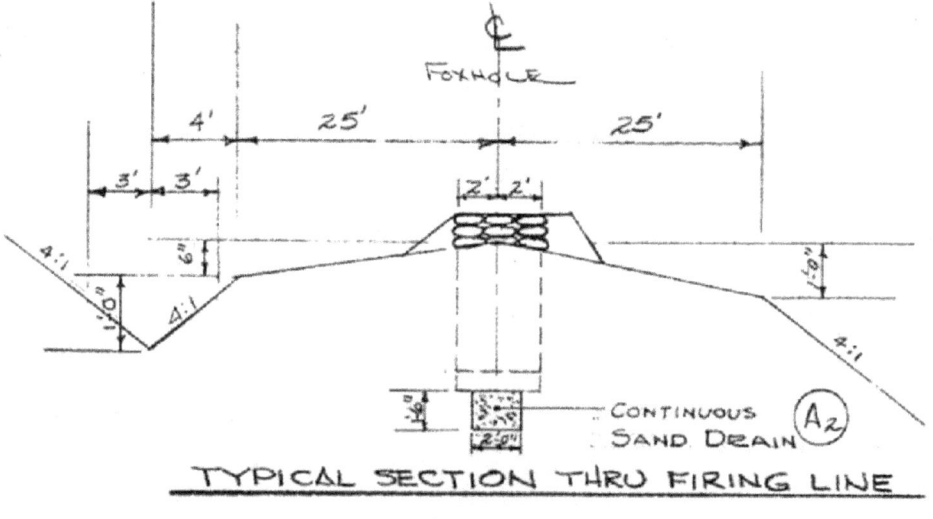

Figure 426. Trainfire ranges field fire No.2 (35 Points), Fort Bragg, NC, 1960 (Standard drawing 28-13-115 sheet 7 of 24, Field fire No.2 (35 Points), 5 October 1960).

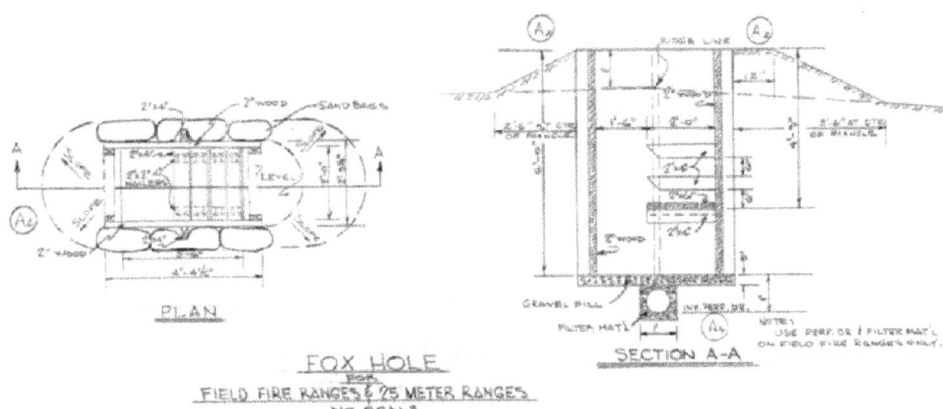

Figure 427. Trainfire ranges field fire No.3 (35 Points), Fort Bragg, NC, 1960 (Standard drawing 28-13-115 sheet 8 of 24, Field fire No.3 (35 Points), 5 October 1960).

Observation line with sighting devices

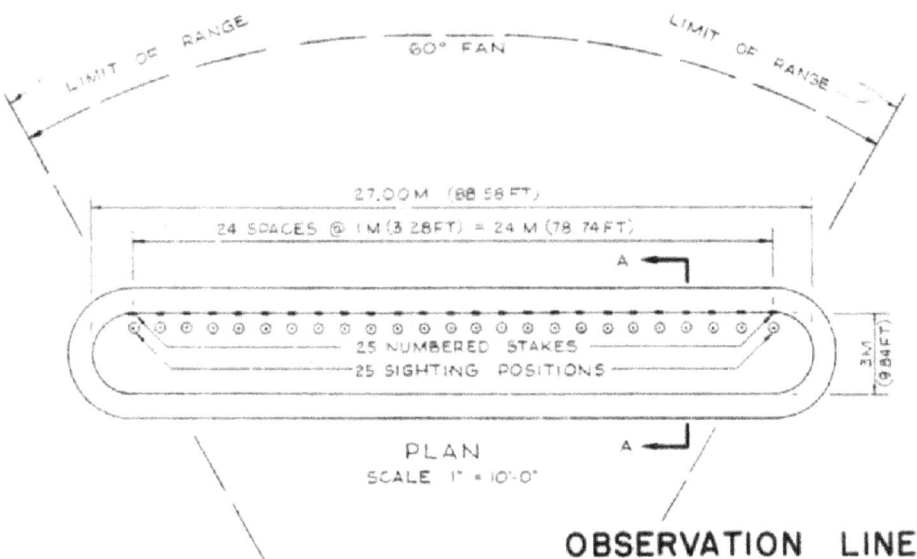

Figure 428. Trainfire ranges, target detection range observation line plan, Fort Bragg, NC, 1960 (Standard drawing 28-13-115 sheet 12 of 24, Rifle marksmanship course, trainfire I, target detection range, plan, section and details, 5 October 1960).

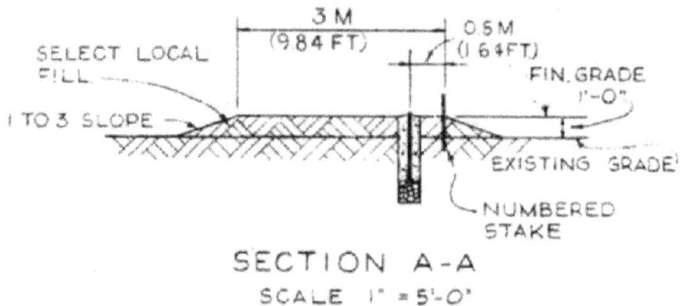

Figure 429. Trainfire ranges, target detection range sighting device, Fort Bragg, NC, 1960 (Standard drawing 28-13-115 sheet 12 of 24, Rifle marksmanship course, trainfire I, target detection range, plan, section and details, 5 October 1960).

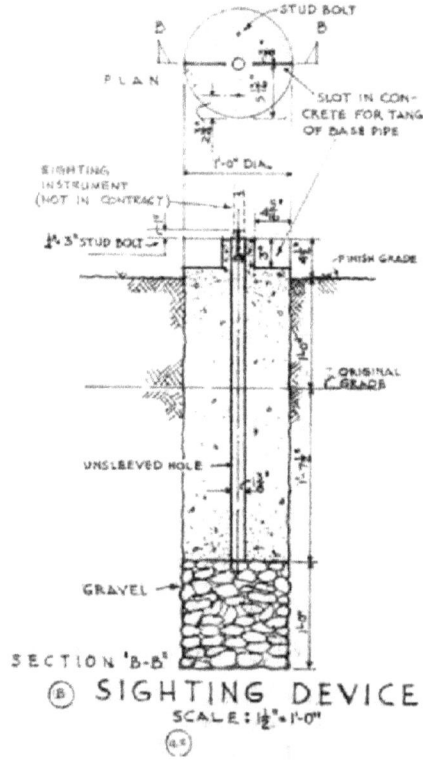

Figure 430. Trainfire ranges, target detection range sighting device, Fort Bragg, NC, 1960 (Standard drawing 28-13-115 sheet 12 of 24, Rifle marksmanship course, trainfire I, target detection range, plan, section and details, 5 October 1960).

Targets

A combination of stationary panel targets and pop-up silhouette targets were used on trainfire ranges. Twenty-five meter ranges used panel targets. On target detection ranges range detection panels were used. On field and record fire ranges hidden targets pop up for a short time in which soldiers must identify them, aim and shoot. Targets were laid out to present as near a combat situation as possible using a series of M-31A1 target devices and the "E" and "F" type silhouette well camouflaged, dispersed, and taking advantage of the natural terrain. Targets are set at 50, 100, 150, 200, 250,300 and 350 yds. One hit will nock the target down.

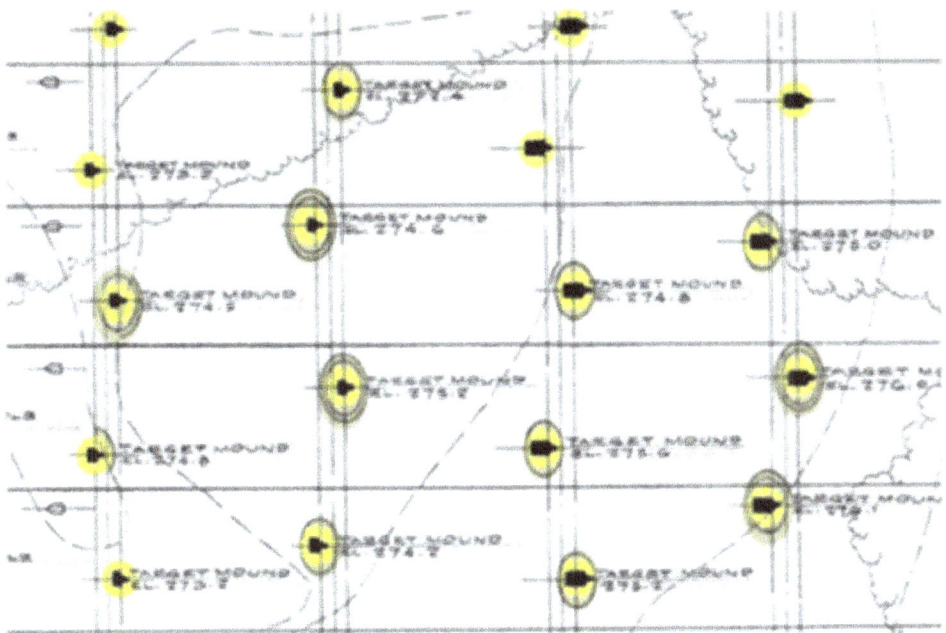

Figure 431. Trainfire ranges record fire No.1 (12 points), targets, Fort Bragg, NC, 1960 (Standard drawing 28-13-115 sheet 2 of 24, Record fire No.1 (12 Points), 5 October 1960).

Panel targets

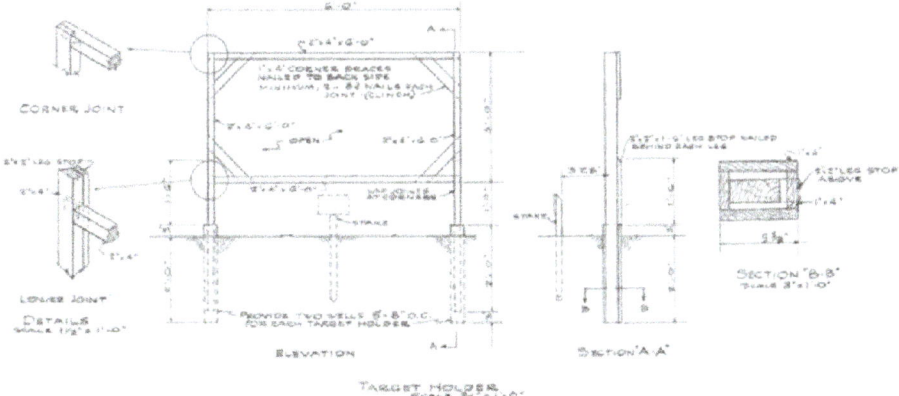

Figure 432. Trainfire ranges, target holder details, Fort Bragg, NC, 1960 (Standard drawing 28-13-115 sheet 13 of 24, Trainfire ranges details, 5 October 1960).

Range detection panels

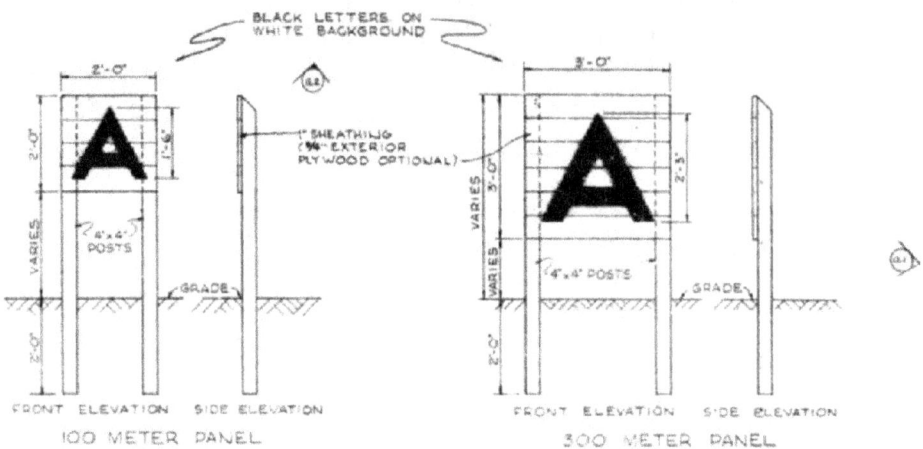

Figure 433. Trainfire ranges, target detection range detection panels, Fort Bragg, NC, 1960 (Standard drawing 28-13-115 sheet 12 of 24, Rifle marksmanship course, trainfire I, target detection range, plan, section and details, 5 October 1960).

E and F type silhouette targets

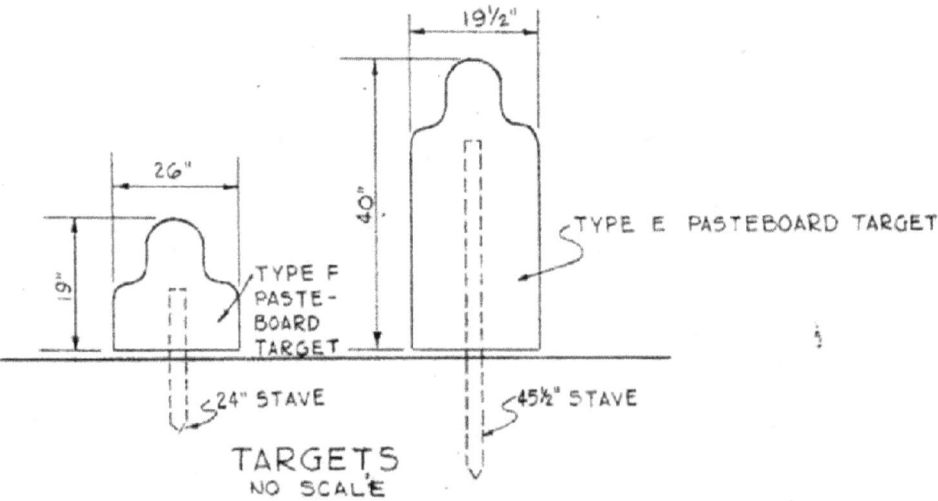

Figure 434. Type F and E silhouette targets, Fort Bragg, NC, 1951 (Standard drawing 28-13-21 sheet 1 of 1, Range, moving vehicle, submachine gun, plan and details, 21 November 1951).

Pop-up targets

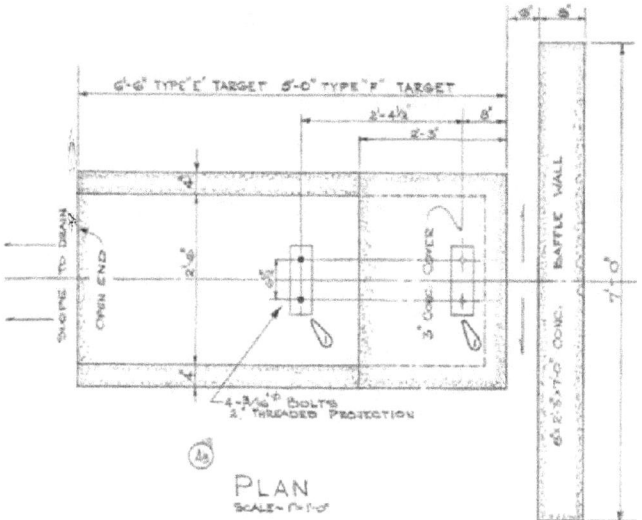

Figure 435. Trainfire ranges, concrete pop-up target emplacement plan, Fort Bragg, NC, 1960 (Standard drawing 28-13-115 sheet 13 of 24, Trainfire ranges details, 5 October 1960).

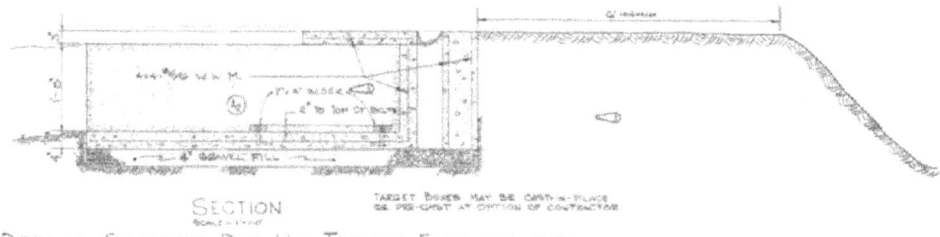

Figure 436. Trainfire ranges, concrete pop-up target emplacement section, Fort Bragg, NC, 1960 (Standard drawing 28-13-115 sheet 13 of 24, Trainfire ranges details, 5 October 1960).

Figure 437. Electrical targets at Camp Lejeune, NC, 2 May 1958 (NARA College Park, RG 127-GC, box 22, photo 340848).

M31A1 pop-up target device

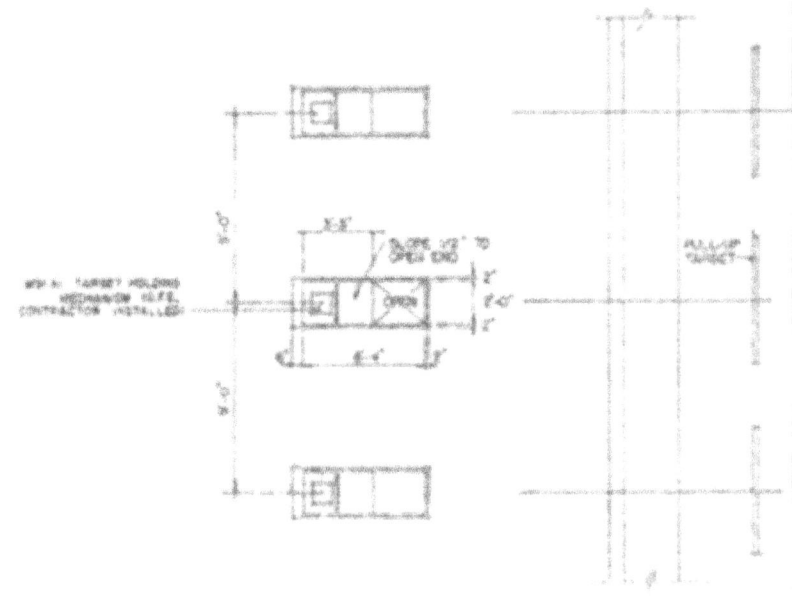

Figure 438. Typical rifle range, m31a1 target device, plan of pop-up target enclosure, Fort Bragg, NC, 1963 (Standard drawing No. 28-13-09 drawing 5 of 7, Range, rifle, known distance, layout and details for installation of M31A1 target device, 6 November 1963).

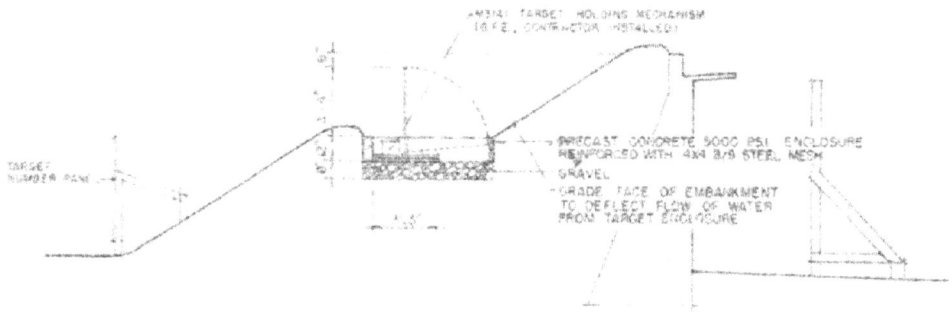

Figure 439. Typical rifle range, M31A1 target device, section through target butt and pop-up target enclosure, Fort Bragg, NC, 1963 (Standard drawing No. 28-13-09 drawing 5 of 7, Range, rifle, known distance, layout and details for installation of M31A1 target device, 6 November 1963).

Embankments/trenches/etc.

Embankments were placed at firing positions and often at target positions, although the natural terrain was used as often as possible to camouflage targets and to create more realistic combat conditions. Foxholes were often used for firing positions.

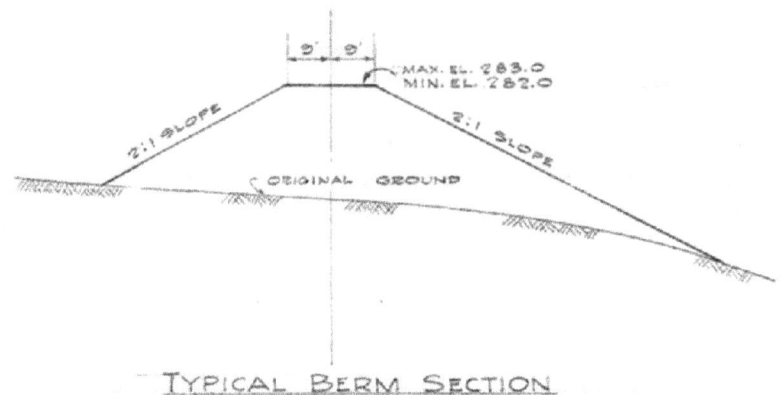

Figure 440. Trainfire ranges record fire No.4 (12 points), typical berm section, Fort Bragg, NC, 1960 (Standard drawing 28-13-115 sheet 5 of 24, Record fire No.4 (12 Points), 5 October 1960).

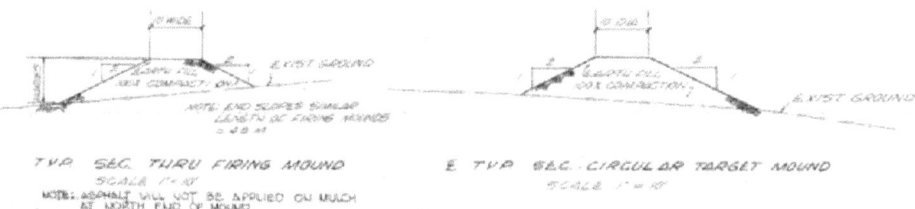

Figure 441. ATC facilities, record firing range (16 point), range "K," Fort Bragg, NC, 1966 (Standard drawing 28-13-117 sheet 5, Construction of ranges, phase 1, U.S. ATC facilities, Fort Bragg, NC, range "K," record firing range (16 Point), 28 June 1966).

Figure 442. Rifle marksmanship course, trainfire I, record firing range (16 point), layout plan, Fort Bragg, NC, 1958 (Standard drawing 28-13-105 sheet 4, Rifle marksmanship course, trainfire I, record firing range (16 point), plan and section, 5 August 1958).

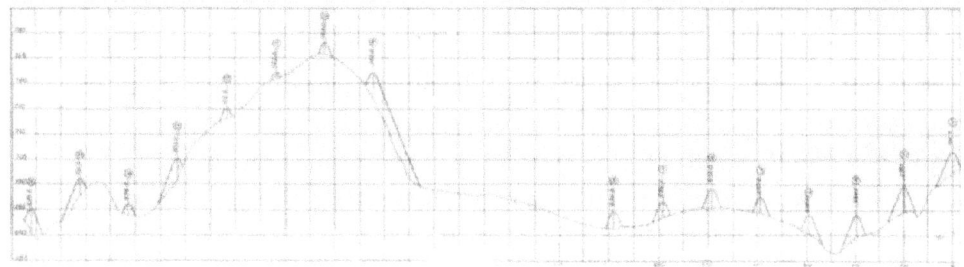

Figure 443. Trainfire I record firing range foxhole profile, Fort Knox, KY, 1960 (Standard drawing 28-13-02 sheet 12, Trainfire I Ranges-1960 F.Y., Ditto Hill Record Firing Range, Typical layout plan and foxhole profile, 25 May 1960).

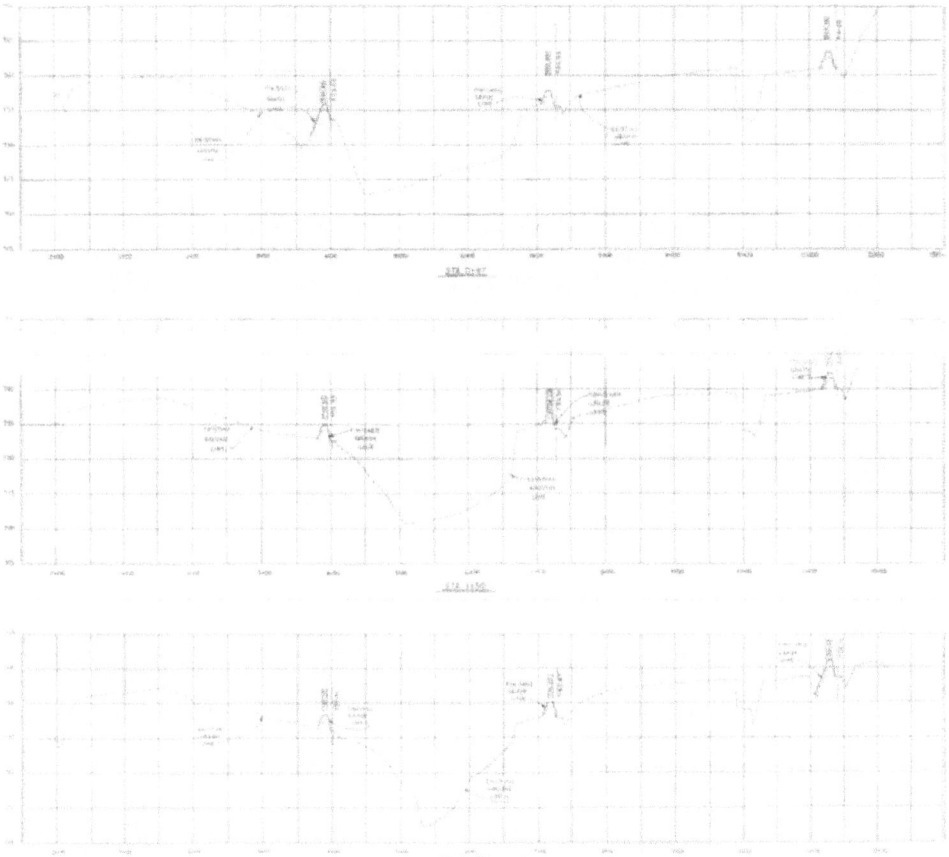

Figure 444. Trainfire I field firing range cross sections, Fort Knox, KY, 1960 (Standard drawing 28-13-02 sheet 4, Trainfire I ranges-1960 F.Y., addition to Dripping Springs-Field Firing Range, Cross Sections, 25 May 1960).

Buildings

A range may have had a control tower, latrine, target storage building, ammo storage building, bleachers, other storage sheds, administrative, and maintenance buildings in support of general range functions. The range may also have been part of a larger installation range complex that contained these buildings.

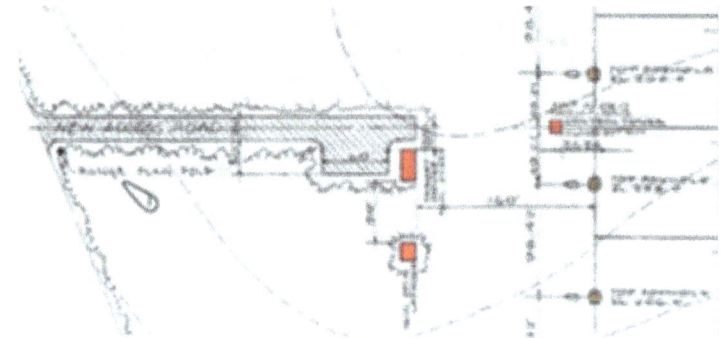

Figure 445. Trainfire ranges record fire No.3 (12 Points), Buildings and foxholes, Fort Bragg, NC, 1960 (Standard drawing 28-13-115 sheet 4 of 24, Record fire No.3 (12 points), 5 October 1960).

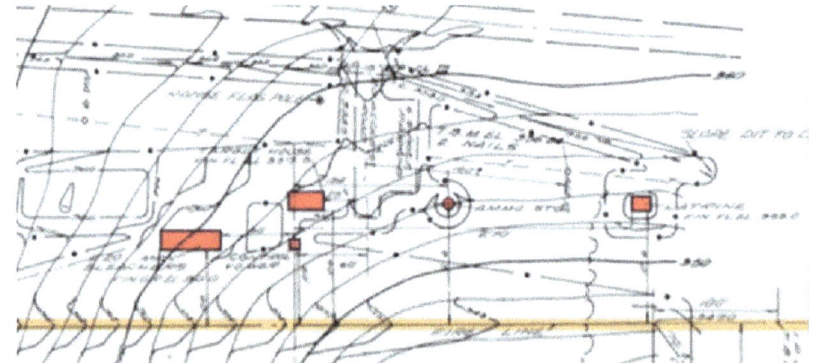

Figure 446. Rifle marksmanship course, trainfire I, record firing range (16 point), layout plan, Fort Bragg, NC, 1958 (Standard drawing 28-13-105 sheet 4, Rifle marksmanship course, trainfire I, record firing range (16 point), plan and section, 5 August 1958).

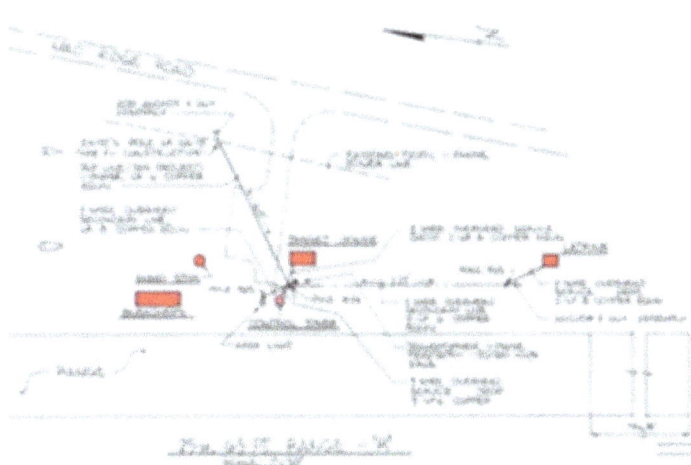

Figure 447. 25m range "R" electrical layout, Fort Bragg, NC, 1966 (Standard drawing 28-13-117 sheet 28, Electrical distribution, 28 June 1966).

Control tower

Closed bottom

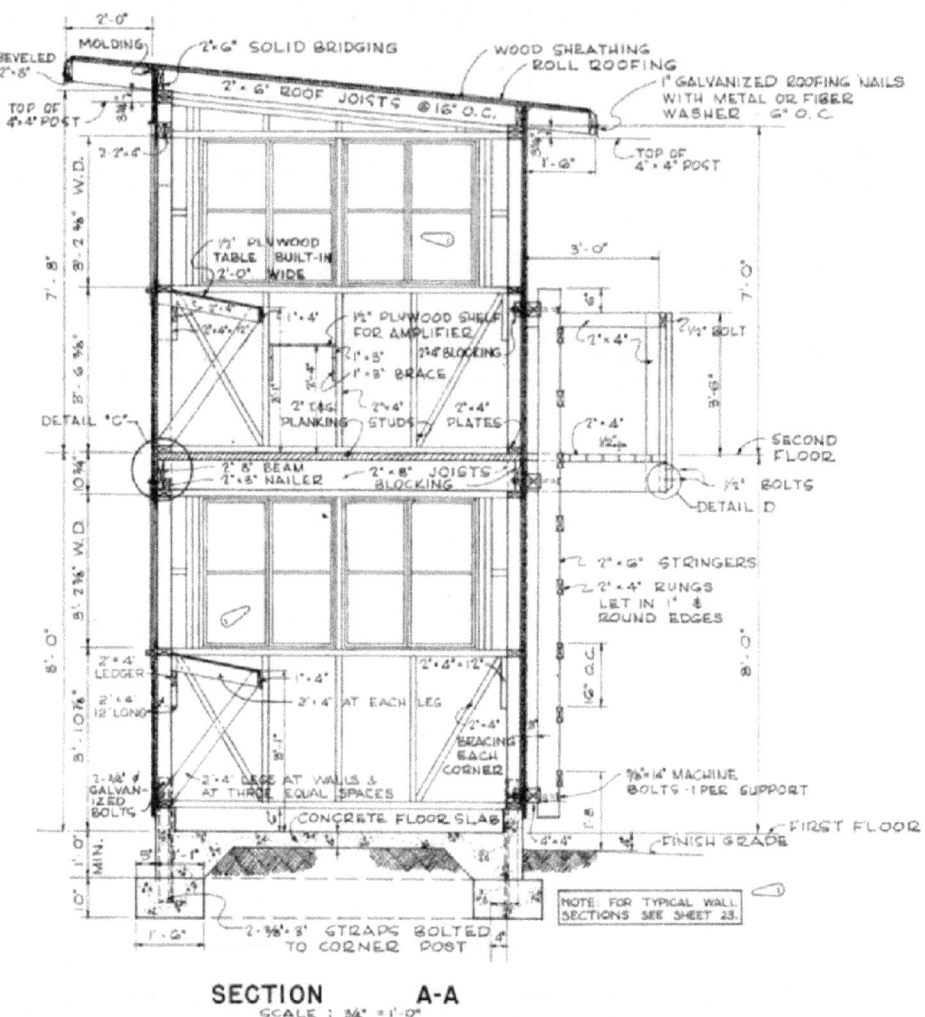

Figure 448. Rifle marksmanship course, trainfire I, control tower (closed bottom), section, Fort Bragg, NC, 1958 (Standard drawing 28-13-105 sheet 7, Rifle marksmanship course, trainfire I, control tower (closed bottom), plans, elevations, section and details, 5 August 1958).

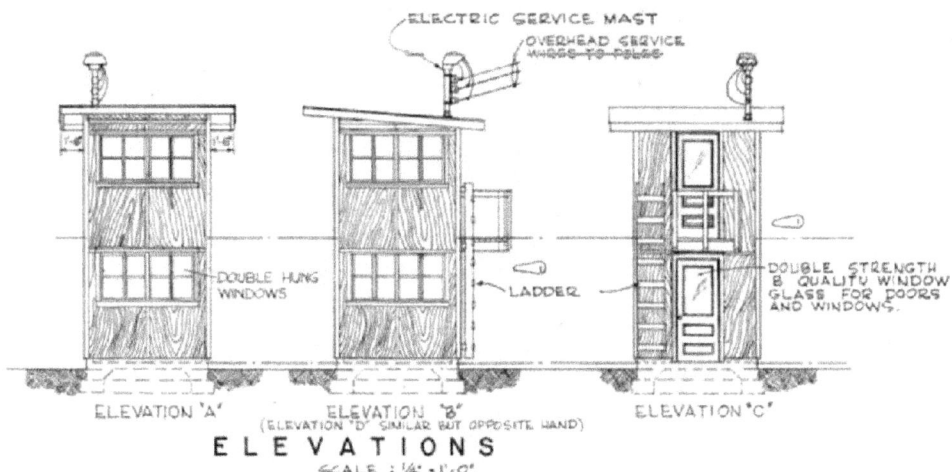

Figure 449. Rifle marksmanship course, trainfire I, control tower (closed bottom), elevations, Fort Bragg, NC, 1958 (Standard drawing 28-13-105 sheet 7, Rifle marksmanship course, trainfire I, control tower (closed bottom), plans, elevations, section and details, 5 August 1958).

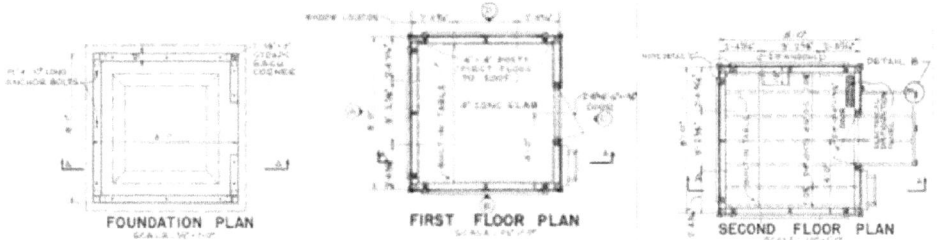

Figure 450. Rifle marksmanship course, trainfire I, control tower (closed bottom), floor plans, Fort Bragg, NC, 1958 (Standard drawing 28-13-105 sheet 7, Rifle marksmanship course, trainfire I, control tower (closed bottom), plans, elevations, section and details, 5 August 1958).

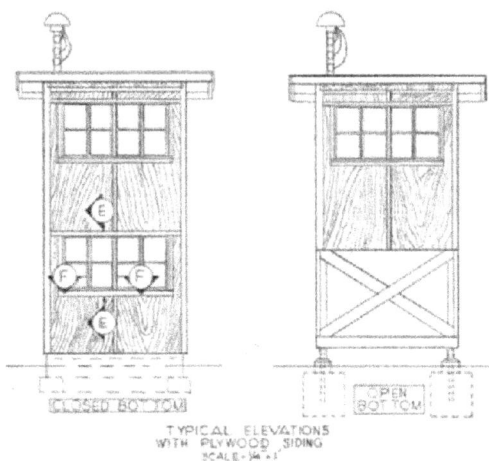

Figure 451. ATC facilities, control towers, alternate details, Fort Bragg, NC, 1966 (Standard drawing 28-13-117 sheet 23, Construction of ranges, phase 1, U.S. ATC facilities, Fort Bragg, NC, option details, range buildings, 28 June 1966).

Open bottom

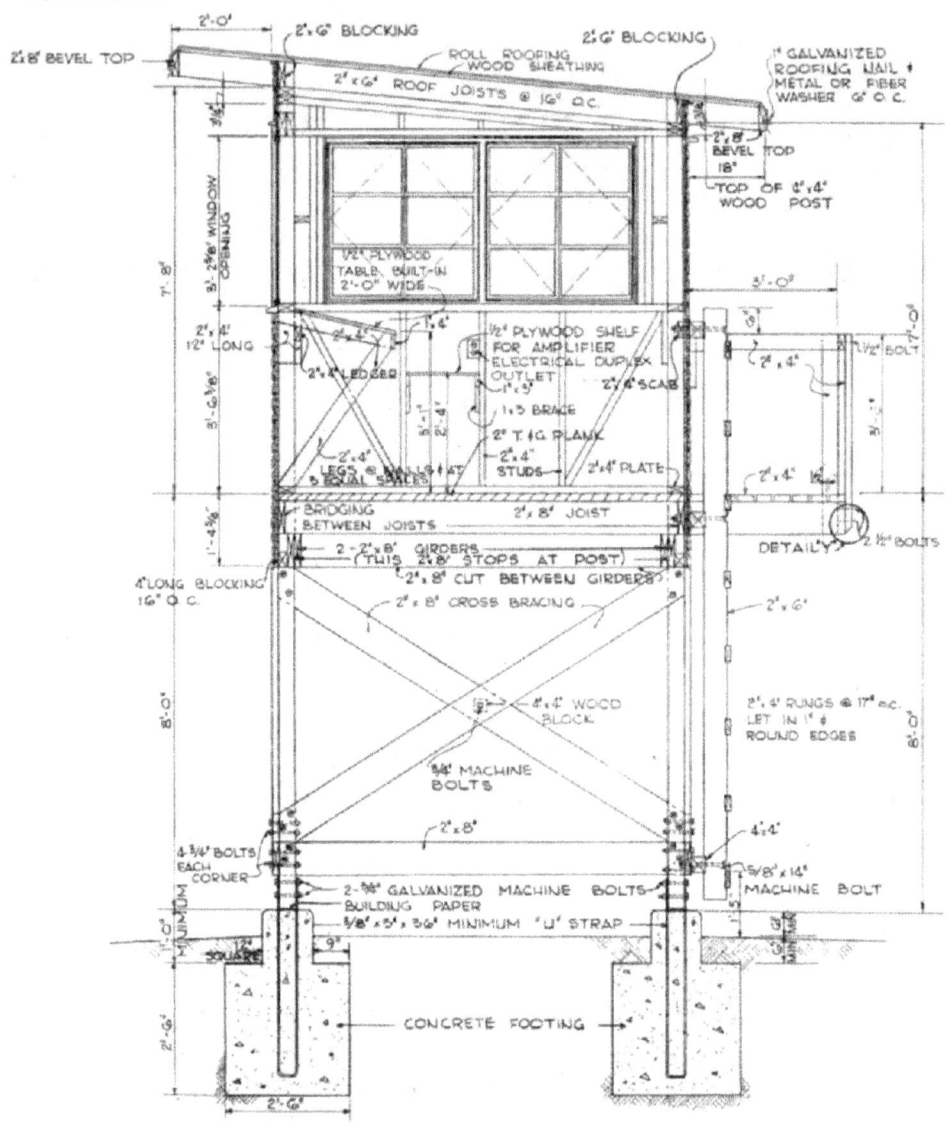

Figure 452. Rifle marksmanship course, trainfire i, control tower, section A, Fort Bragg, NC, 1958 (Standard drawing 28-13-105 sheet 6, Rifle marksmanship course, trainfire 1, control tower (open bottom), plans, elevations, sections and details, 5 August 1958).

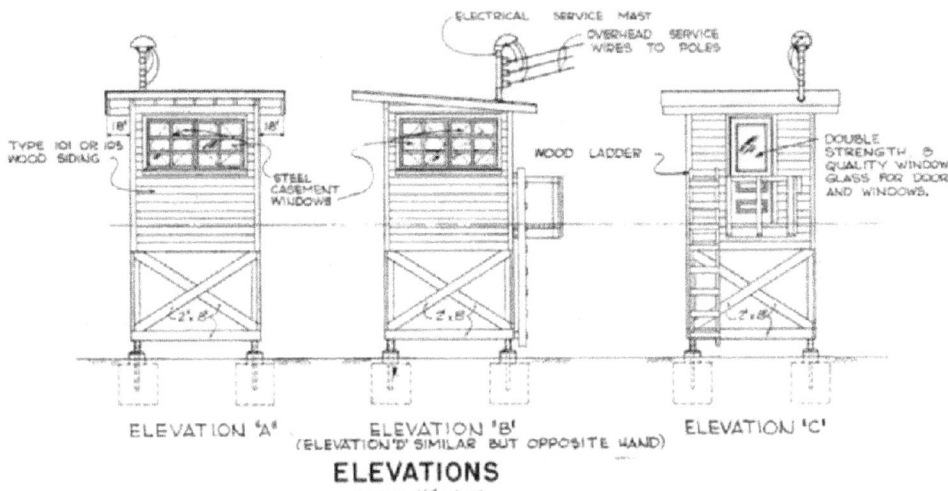

Figure 453. Rifle marksmanship course, trainfire I, control tower, elevations, Fort Bragg, NC, 1958 (Standard drawing 28-13-105 sheet 6, Rifle marksmanship course, trainfire 1, control tower (open bottom), plans, elevations, sections and details, 5 August 1958).

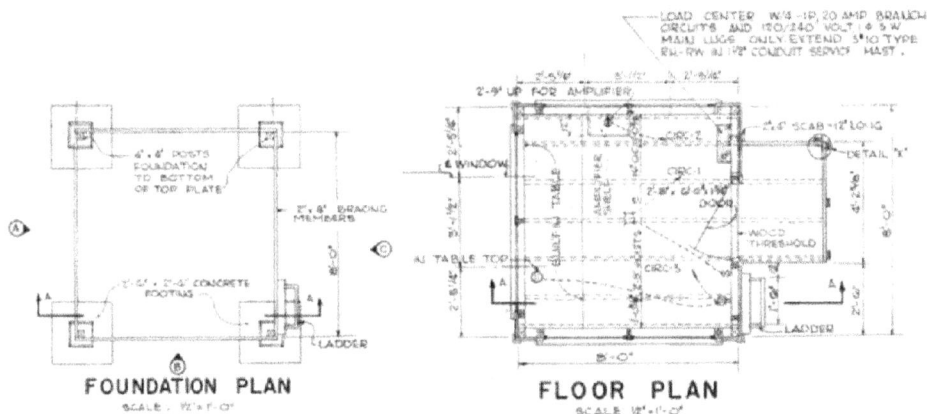

Figure 454. Rifle marksmanship course, trainfire I, control tower, foundation and floor plans, Fort Bragg, NC, 1958 (Standard drawing 28-13-105 sheet 6, Rifle marksmanship course, trainfire 1, control tower (open bottom), plans, elevations, sections and details, 5 August 1958).

Target house

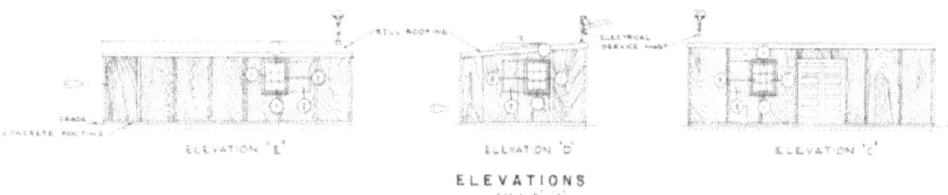

Figure 455. Rifle marksmanship course, trainfire I, target house, elevations, Fort Bragg, NC, 1958 (Standard drawing 28-13-105 sheet 8, Rifle marksmanship course, trainfire I, target house, elevations, plans and details, 5 August 1958).

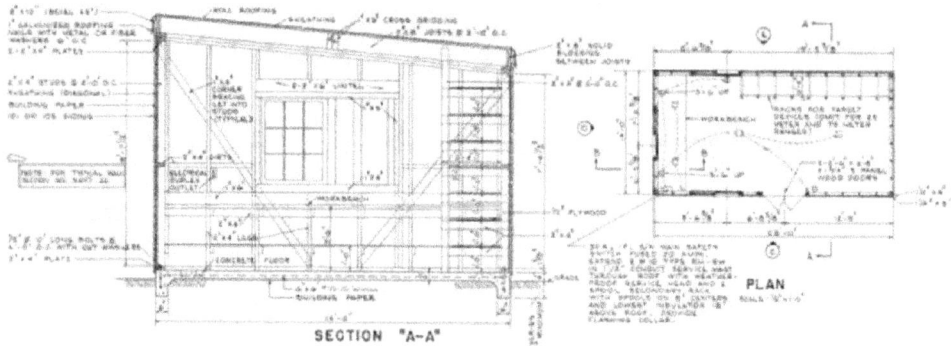

Figure 456. Rifle marksmanship course, trainfire I, target house, section and plan, Fort Bragg, NC, 1958 (Standard drawing 28-13-105 sheet 8, Rifle marksmanship course, trainfire i, target house, elevations, plans and details, 5 August 1958).

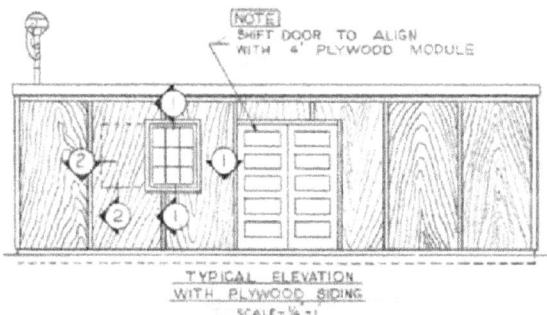

Figure 457. ATC facilities, target house, alternate details, Fort Bragg, NC, 1966 (Standard drawing 28-13-117 sheet 23, Construction of ranges, phase 1, U.S. ATC facilities, Fort Bragg, NC, option details, range buildings, 28 June 1966).

Latrine

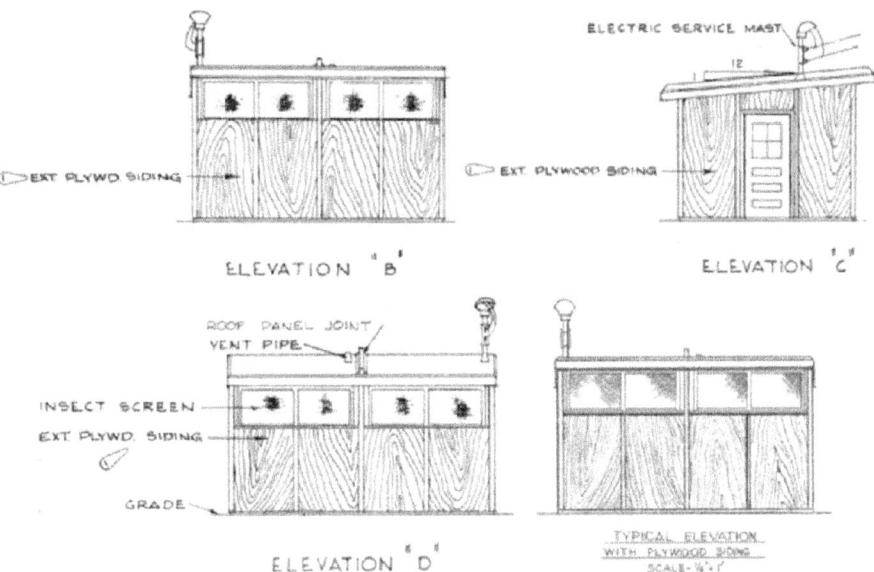

Figure 458. Rifle marksmanship course, trainfire i, latrine, elevations, Fort Bragg, NC, 1958 (Standard drawing 28-13-105 sheet 9, Rifle marksmanship course, trainfire I, latrine, plans, elevations, section and details, 5 August 1958, "Typical elevation with plywood siding" Standard drawing 28-13-117 sheet 23, Construction of ranges, phase 1, U.S. ATC facilities, Fort Bragg, NC, option details, range buildings, 28 June 1966).

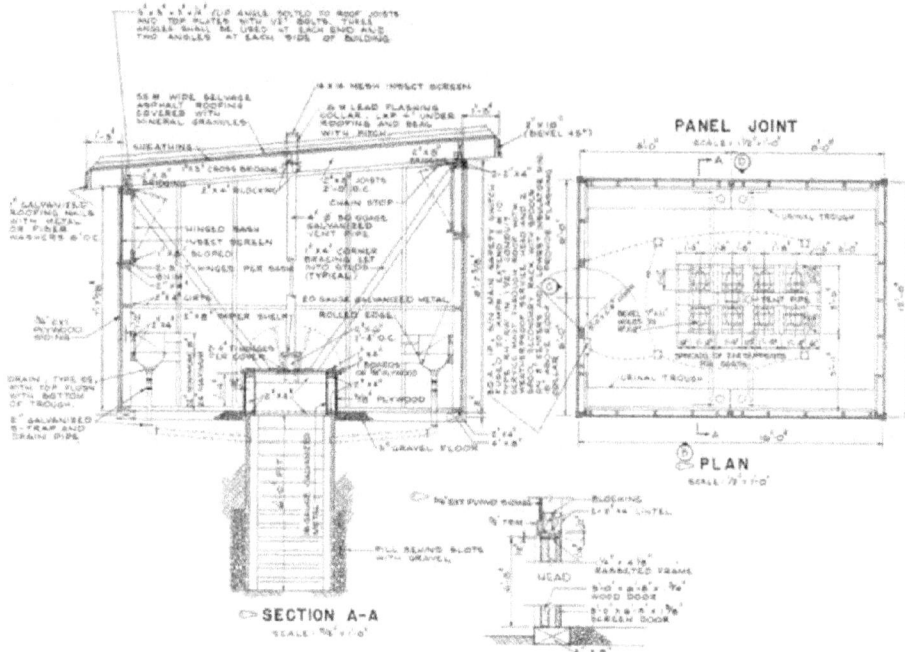

Figure 459. Rifle marksmanship course, trainfire I, latrine, section and plan, Fort Bragg, NC, 1958 (Standard drawing 28-13-105 sheet 9, Rifle marksmanship course, trainfire I, latrine, plans, elevations, section and details, 5 August 1958).

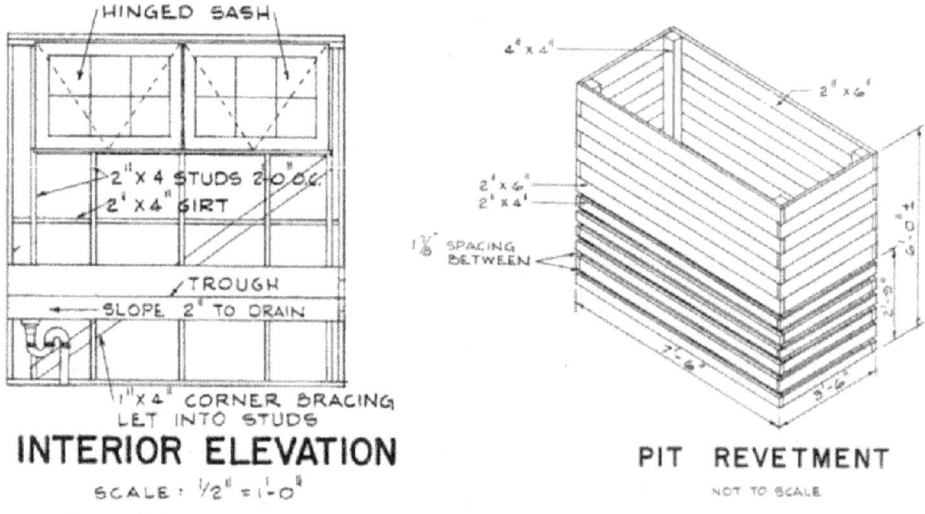

Figure 460. Rifle marksmanship course, trainfire I, latrine, interior elevation and pit revetment, Fort Bragg, NC, 1958 (Standard drawing 28-13-105 sheet 9, Rifle marksmanship course, trainfire i, latrine, plans, elevations, section and details, 5 August 1958).

3 Evaluating Properties Under the Military Training Lands Historic Context

Cultural resources are identified and managed within the Department of Defense (DoD) in accordance with Federal laws and military regulations. The identification of historically significant properties and resources can be achieved only through evaluation of their position within the larger historic context. According to the NRHP, historic contexts are defined as "… the patterns, themes, or trends in history by which a specific occurrence, property, or site is understood and its meaning (and ultimately its significance) within prehistory or history is made clear" (NRB #15, 7). A historic property is determined to be significant or not significant based on the application of standardized National Register Criteria within the property's historical context.

Criteria for evaluation

The NRHP Criteria for Evaluation (36 CFR Part 60.4) describe how properties and districts are significant for their association with important events or persons (Criterion A and Criterion B), for their importance in design or construction (Criterion C), or for their information potential (Criterion D). The following is a brief description of each of the four NRHP Criteria for Evaluation (excerpted from National Register Bulletin #15: How to Apply the National Register Criteria for Evaluation):

A. **Event**—associated with events that have made a significant contribution to the broad patterns of our history; or

B. **Person**—associated with the lives of persons significant in our past; or

C. **Design/Construction**—embody the distinctive characteristics of a type, period, or method of construction, or that represent the work of a master, or that possess high artistic values, or that represents a significant and distinguishable entity whose components may lack individual distinction; or

D. **Information Potential**—yielded, or is likely to yield, information important in prehistory or history.

Criterion consideration G

Generally, buildings, structures, landscapes, etc. constructed within the last 50 years are not eligible for the National Register unless they can be classified as exceptionally important under Criterion Consideration G in the National Register Bulletin #15. "The National Register Criteria for Evaluation excludes properties that achieved significance within the past 50 years unless they are of exceptional importance. Fifty years is a general estimate of the time needed to develop historical perspective and to evaluate significance. This consideration guards against the listing of prosperities of passing contemporary interest and ensures that the National Register is a list of truly historic places."

Although the National Register Criteria do not explicitly define the term exceptional importance, National Register Consideration G and the National Register Bulletin #22: *Guidelines for Evaluating and Nominating Properties that have Achieved Significance within the Past Fifty Years* offers guidance for identifying and evaluating properties that have achieved significance in the past 50 years. Both of these sources stress that, for such properties, sufficient historical perspective must exist to make justifiable determinations of exceptional importance. Proof that sufficient historical perspective exists usually comes in the form of scholarly research and other sources of historical evidence associated with a particular historic context. The significance of Cold War era properties may lie at the national level in association with military themes directly tied to the Cold War, or at the state or local level under other themes.

The Army and Air Force have all issued interim guidelines for managing Cold War resources. The Navy is still working on draft version of guidance. These guidelines are not meant to replace the NHPA and its implementing regulations (Sections 106 and 110). The intent of the guidance is to set up an initial framework for the inventory and evaluation of the Cold War historic properties.

Army cold war guidelines and contexts

The Army developed its "interim Policy for Cold War Era Properties" in 1995. Applying to Army, Army National Guard, and Army Reserve installations, this policy stated that in applying the criteria of exceptional importance, the Army would "focus on the production and combat subsystems of the Army and their associated Real Property and technology that is of unmistakable and extraordinary importance by virtue of a direct and influential relationship to Cold War tactics, strategy, and events" (Department of the Army Cultural Resources Interim Policy Statements, 1995).

The Interim Policy was set into guidance with *The Thematic Study and Guidelines: Identification and Evaluation of U.S. Army Cold War Era Military-Industrial Historic Properties* in 1997. This guidance is a thematic study on historic properties associated with the military-industrial theme of the Cold War and provides guidelines for the identification and evaluation of Cold War era military-industrial historic properties in the Army. The context focuses in on what the Army did in direct response to the Cold War and directly associated with a major Army mission.

The Cold War context states that only "properties that are directly related to the Cold War military-industrial context" are exceptionally important. They must meet "any or all" of the following conditions:

1. They were specifically constructed or used prior to 1989 to:
   a. Meet the perceived Soviet/communist military threat;
   b. Project a force designed to influence Soviet policy; and
   c. Affect global opinion of the relationship between the superpowers.
2. Through the architectural or engineering design, they clearly reflect one of the Cold War themes:
   a. Basic Scientific Research (Laboratories)
   b. Materiel Development (Research, Development, Engineering Centers, and Proving Grounds)
   c. Wholesale Logistical Operations (Ammunition Production Facilities)
   d. Air Defense, Ballistic Missile Defense, and Army Missiles
   e. Command and Control, Communications, Computer, and Intelligence
   f. Army School System
   g. Operational Forces
   h. Army Medical Activities
   i. Miscellaneous (Nuclear and Aviation).
3. They are directly related to the United States/Soviet relationship through association with a milestone event of the period.
4. They are directly related to the United States/Soviet relationship through association with the life of a person during the Cold War period.

Air Force cold war guidelines and context

The U.S. Air Force recognizes five property type groups in the Interim Guidance that may convey important aspects of the Cold War. These five properties include:

1. Operational and Support Installations
   a. Air Force bases, including Command Centers

b. Missile Stations
  c. Launch Complexes
2. Combat Weapons Systems and Combat Support Systems
   a. Missiles
   b. Aircraft (Fixed Wing and Rotary)
   c. Ground Vehicles and Equipment
3. Training Facilities
   a. Warfighting, Combat Support, and Intelligence Schools
   b. Launch Complexes
   c. Combat Training Ranges
   d. Impact Areas; Targets
   e. POW (Prisoner of War) Training Camps
4. Materiel Development Facilities
   a. Research Laboratories
   b. Manufacturing Sites
   c. Test Sites
   d. Proving Grounds
5. Intelligence Facilities
   a. Radar Sites
   b. Listening Posts.

Significance

Military training ranges need to be researched and evaluated as a whole landscape, including all the buildings/structures, firing lines, target mechanisms, etc. and not evaluated as individual elements that sit on the range. Military training ranges were originally designed and intended to be utilized as a whole complex. Each structure/element provides a vital role in the functioning of the range and the overall effectiveness of the training procedures for the soldiers.

The overall importance of particular ranges depends on the mission of whichever installation the research is focusing on. The mission critical ranges are what is important and need to be evaluated as a historic district. For example, a large arms range like a tank range needs to be examined and evaluated from the parking lot all the way out to the target butt, regardless of individual building or range element construction date. Thus just looking at an individual observation tower, latrine, firing targets, etc. should not be done. Look at the entire range. But go one step further and look at all of the ranges and training lands on the installation as one large group to see if there is even information for a large district. No individual building/structure/element will ever be individually significant.

Once the training range is inventoried and evaluated as a complex, the next step is to determine if a particular range/buildings are significant to the individual installation being researched. For example, all ranges at Fort Jackson, SC could possibly be evaluated as one large district because Fort Jackson is the home of basic training; whereas the tank ranges located at Fort Knox, KY would be important to the mission because Fort Knox was the home of the Armor division. Ultimately, the researcher needs to look at the overall mission of the installation before deciding what is important on the ranges.

For instance, a large arms range, like the field artillery range, may have been constructed in 1944 but may contain buildings and structures from the entire stretch of the Cold War. As individual building elements and training mechanisms wore out they typically were replaced with new materials and technologies. The ranges will always be ranges and used for training, therefore, continue use of the landscape and structures are important. It is important to evaluate the location of replacement elements. Is the newer observation tower in the same location as the original? Are the replacement latrines, bleachers, and storage buildings located in the same spot on the range landscape?

Properties considered under the Large Arms Range Context are training ranges that the War Department, Navy Department, and Department of Defense constructed for their personnel and are associated with one of the following military training periods:

- Pre-Civil War (up to 1861)
- Civil War (1861-1865)
- National Expansion (1865-1916)
- World War I (1917-1920)
- Interwar (1921-1940)
- World War II (1941-1945)
- Early Cold War (1946-1955)
- Late Cold War (1956-1989).

The researcher still has to be able to identify that firing range to what period it is significant for no matter if there are replacement structures or elements located on the range.

Aspects of Integrity

In addition to possessing historical significance, training ranges must also retain sufficient physical integrity of the features that convey their significance to be eligible to the NRHP (NRB #15, 44).

Training lands/ranges will either retain integrity (that is, convey their significance) or they will not. Within the concept of integrity, the National Register criteria recognize seven aspects or qualities that, in various combinations, define integrity.

To retain historic integrity a property will always possess several, and usually most, of the aspects. The retention of specific aspects of integrity is paramount for training lands/ranges to convey their significance. Determining which of these aspects are most important to a particular training land/range requires knowing which association is significant.

Although some training lands/ranges may not meet integrity standards for individual eligibility to the National Register, they may meet a standard as a contributing resource to a larger training district. Training lands/ranges are considered to be significant if they possess a majority of the following Seven Aspects of Integrity (NRB #15, 44-45):

1. **Location.** Location is the place where the historic property was constructed or the place where the historic event occurred.
2. **Design.** Design is the combination of elements that create the form, plan, space, structure, and style of a property. It results from conscious decisions made during the original conception and planning of a property (or its significant alteration) and applies to activities as diverse as community planning, engineering, architecture, and landscape architecture. Design includes such elements as organization of space, proportion, scale, technology, ornamentation, and materials.
3. **Setting.** Setting is the physical environment of a historic property. Setting refers to the character of the place in which the property played its historical role. It involves how, not just where, the property is situated and its relationship to surrounding features and open space.
4. **Materials.** Materials are the physical elements that were combined or deposited during a particular period of time and in a particular pattern or configuration to form a historic property.
5. **Workmanship.** Workmanship is the physical evidence of the crafts of a particular culture or people during any given period in history or prehistory.

6. **Feeling.** Feeling is a property's expression of the aesthetic or historic sense of a particular time period.
7. **Association.** Association is the direct link between an important historic event or person and a historic property.

Character defining features

The character defining features of a range depend on the associated NRHP Criteria and the associated property type. A large arms range typically was designed and constructed with the following:

- a set of cleared and leveled firing points laid out on a firing line and associated features (foxholes, trenches, sandbags, embankments, etc)
- stationary or moving targets (cables, pulleys, tracks, pop-up targets, miniature airplanes, etc)
- embankments or walls (built up behind targets to catch ammunition, in front of targets for concealment and protection, at firing lines for firing support, between ranges to protect from adjacent fire)
- buildings (control or observation tower, bleachers, latrines, target storage houses, ammunition storage buildings)
- typical features include multiple range layouts, firing lines, targets, embankments/trenches, and buildings.

Context example photographs

Two members of the research team conducted a site visit to Fort Bragg, NC. Fort Bragg was chosen for the site visit because it had one of the largest groupings of different training lands in the Department of Defense; the complexity of its training lands; and the level of historical background that Fort Bragg had on its training lands. There are few examples gathered from other installations. In addition to the photographs taken at Fort Bragg, the researchers searched the previous ERDC/CERL pertaining to training lands and used some of these for examples in the evaluation chapter.

When the researcher is tasked to research and inventory items on a military training range, the researcher is going to find things that are on the real property list, items that are not listed on the real property list, abandoned structures, and foundations. It is the task of the researcher to inventory and document all elements of the range, the role of the elements and the condition of the elements.

Below, are photographic representations of a variety of examples of small arms range elements. The examples should be used as a guide to help identify key character defining features which will ultimately help determine the integrity of each range.

Trainfire ranges

Figure 461. A 25m Range, Fort Bragg, NC, 17 May 2006.

Figure 462. A 25m Range, Fort Bragg, NC, 17 May 2006.

Figure 463. Trainfire Range, Fort Bragg, NC, 17 May 2006.

Figure 464. Trainfire Range, Fort Bragg, NC, 17 May 2006.

Transition range

Figure 465. Transition Range, Fort Bragg, NC, 17 May 2006.

Figure 466. Transition Range, Fort Bragg, NC, 17 May 2006.

Figure 467. Transition Range, Fort Bragg,, NC, 17 May 2006.

Firing lines

Cleared firing positions

Figure 468. Cleared firing positions on 25m range, Fort Bragg, NC, 17 May 2006.

Sandbag supports

Figure 469. Sandbag support on firing line of trainfire range, Fort Bragg, NC, 17 May 2006.

Foxholes

Figure 470. Foxholes on firing line of 25m range, Fort Bragg, NC, 17 May 2006.

Figure 471. Abandoned known distance range (overgrown foxholes on firing lines), Fort Bragg, NC, 17 May 2006.

Covered firing lines

Figure 472. Covered firing line on Range 2, Fort Gordon, January 2004.

Firing pits

Figure 473. Grenade firing pit, Fort Bragg, NC, 17 May 2006.

Firing trenches

Figure 474. Grenade firing trench, Fort Bragg, NC, 17 May 2006.

Machine gun firing platforms

Figure 475. Machine gun firing platform, Fort Bragg, NC, 17 May 2006.

Simulated window, door, and rooftop firing positions

Figure 476. Transition range firing line with window, foxhole, door, rooftop and other simulated firing positions, Fort Bragg, NC, 17 May 2006.

Targets

Stationary panel targets

Figure 477. Stationary target frames on 25m range, Fort Bragg, NC, 17 May 2006.

Stationary silhouette targets

Figure 478. Stationary silhouette targets on grenade range, Fort Bragg, NC, 17 May 2006.

Raised panel targets

Figure 479. Raised panel targets with pulleys, Fort Bragg, NC, 17 May 2006.

Figure 480. Raised panel targets with weights, Fort Bragg, NC, 17 May 2006.

Figure 481. Firing line view of raised panel targets (in lowered positions), Fort Bragg, NC, 17 May 2006.

Pop-up silhouette targets

Figure 482. Remnant of pop-up targets, Fort Bragg, NC, 17 May 2006.

Figure 483. Remnant of pop-up targets, Fort Bragg, NC, 17 May 2006.

Figure 484. Transition range target line with pop-up targets, Fort Bragg, NC, 17 May 2006.

Figure 485. Transition range pop-up target, Fort Bragg, NC, 17 May 2006.

Figure 486. Transition range pop-up target, Fort Bragg, NC, 17 May 2006.

Figure 487. Pop-up target, Fort Bragg, NC, 17 May 2006.

Figure 488. Pop-up targets, Fort Bragg, NC, 17 May 2006.

Moving target tracks

Figure 489. Silhouette target mounted to a dolly on a moving target track, Fort Bragg, NC, 17 May 2006.

Figure 490. Abandoned target track, Fort Bragg, NC, 17 May 2006.

Figure 491. Cable pulley system on moving target track, Fort Bragg, NC, 17 May 2006.

Figure 492. Cable pulley system on moving target track, Fort Bragg, NC, 17 May 2006.

Figure 493. Target cars mounted on dollies and pulled by a cable system on a moving target track, Fort Bragg, NC, 17 May 2006.

Figure 494. Moving target track operator and storage building, Fort Bragg, NC, 17 May 2006.

Embankments and Trenches

Embankments

Figure 495. Embankments behind target lines, Fort Bragg, NC, 17 May 2006.

Figure 496. Embankments in front of pop-up targets, Fort Bragg, NC, 17 May 2006.

Figure 497. Walled embankments at grenade range firing lines, Fort Bragg, NC, 17 May 2006.

Trenches

Figure 498. Grenade firing trench, Fort Bragg, NC, 17 May 2006.

Walls

Figure 499. Grenade range protection walls, Fort Bragg, NC, 17 May 2006.

Figure 500. Grenade range protection walls, Fort Bragg, NC, 17 May 2006.

Figure 501. Grenade area on old Range 30, Fort Gordon, July 2004.

Target butts

Figure 502. Remnants of WWII target butt, Fort Gordon, January 2004.

Figure 503. Remnants of WWII known distance range target butt, Fort Jackson (SCARNG), June 2004.

Figure 504. Remnants of WWII known distance range target butt, Fort Jackson (SCARNG), June 2004.

Figure 505. Remnants of WWII known distance range target butt, Fort Jackson (SCARNG), June 2004.

Figure 506. Remnants of WWII known distance range target pulley, Fort Jackson (SCARNG), June 2004.

Buildings

Observation towers

WWII

Figure 507. Remnants of WWII Observation Tower Bldg R122, Fort Gordon, GA, January 2004.

Figure 508. WWII Observation Tower Building 9805, Fort Bliss, TX, November 2005.

Figure 509. Remnants of WWII Observation Tower Building R122, Fort Gordon, GA, January 2004.

Figure 510. WWII Observation Tower Building 9789, Fort Knox, KY, November 2005.

Figure 511. Remnants of WWII Observation Tower Building 9606, Fort Knox, KY, November 2005.

Post WWII

Figure 512. Remnants of 1950's Observation Tower, Fort Jackson, SC (SCARNG), June 2004.

Figure 513. Remnants of 1950's observation tower on Combat II Range, Fort Jackson, SC, June 2004.

Figure 514. Observation tower, Fort Bragg, NC, 17 May 2006.

Figure 515. 1966 Observation tower near Old Range #19, Fort Gordon, GA, January 2004.

Figure 516. Remnants of 1966 Observation tower (near Building 421), Fort Gordon, GA, July 2004.

Figure 517. 1987 Range Bldg 486, Fort Gordon, GA, January 2004.

Figure 518. 1995 Range 10 observation tower and grandstand, Fort Jackson, SC, July 2003.

Figure 519. Control tower and range buildings, Fort Bragg, NC, 17 May 2006.

Figure 520. Observation tower, Fort Bragg, NC, 17 May 2006.

Figure 521. 25m trainfire range observation tower, Fort Bragg, NC, 17 May 2006.

Figure 522. Observation tower, Fort Bragg, NC, 17 May 2006.

Figure 523. Range firing tower, Fort Bragg, NC, 17 May 2006.

Figure 524. Abandoned known distance range (foundations of observation tower and range building), Fort Bragg, NC, 17 May 2006.

Storage buildings

WWII

Figure 525. WWII Range Target Storage Bldg 9898, Fort Bliss, TX, November 2005.

Figure 526. WWII Range Target Storage Bldg 9898, Fort Bliss, TX, November 2005.

Figure 527. Remains of WWII known distance range target butt, latrine, and target storage, Fort Jackson, SC (SCARNG), June 2004.

Figure 528. Remains of WWII Range Target Shed Bldg R161, Fort Gordon, GA, January 2004.

Post WWII

Figure 529. 1966 Bldg 421 range target shed, Fort Gordon, GA, January 2004.

Figure 530. Range storage building, Fort Bragg, NC, 17 May 2006.

Figure 531. Range storage building, Fort Bragg, NC, 17 May 2006.

Figure 532. Range storage building, Fort Bragg, NC, 17 May 2006.

Figure 533. Range storage building, Fort Bragg, NC, 17 May 2006.

Figure 534. Range storage building, Fort Bragg, NC, 17 May 2006.

Ammunition storage buildings

Figure 535. 1987 Building 485 range storage, Fort Gordon, January 2004.

Figure 536. Range ammunition storage building, Fort Bragg, NC, 17 May 2006.

Figure 537. Range ammunition storage building, Fort Bragg, NC, 17 May 2006.

Latrines

WWII

Figure 538. Remnants of WWII Latrine Bldg R152, Fort Gordon, GA, January 2004.

Figure 539. Remnants of WWII Latrine Bldg 9347, Fort Knox, KY, November 2005.

Figure 540. WWII Latrine Bldg 9606, Fort Knox, KY, November 2005.

Figure 541. Remnants of WWII Latrine Bldg 9606, Fort Jackson, SC, November 2003.

Figure 542. remnants of wwii known distance range target butt, latrine, and target storage, Fort Jackson, SC (SCARNG), June 2004.

Post WWII

Figure 543. Range latrine, Fort Bragg, NC, 17 May 2006.

Figure 544. Range latrine, Fort Bragg, NC, 17 May 2006.

Figure 545. Range latrine, Fort Bragg, NC, 17 May 2006.

Bleachers

Figure 546. Range bleachers, Fort Bragg, NC, 17 May 2006.

Figure 547. 1995 Range 10 observation tower and grandstand, Fort Jackson, SC, July 2003.

Figure 548. Range bleachers, Fort Bragg, NC, 17 May 2006.

Mess halls

Figure 549. 1966 Bldg 427 Range Mess Hall, Fort Gordon, GA, January 2004.

Figure 550. 1966 Bldg 427 Range Mess Hall Interior, Fort Gordon, GA, January 2004.

Figure 551. Range Mess Facilities O-7906, Fort Bragg, NC, 17 May 2006.

Weapons cleaning point

Figure 552. 1995 Range 10 Weapons cleaning point, Fort Jackson, July 2003.

Figure 553. Weapons cleaning point, Fort Bragg, NC, 17 May 2006.

Firing towers

Figure 554. Range firing tower, Fort Bragg, NC, 17 May 2006.

4 Conclusions

This work developed a historic context for the development of military training lands used by the DOD and its forerunners. This overall project covered five types of military training:

1. Small arms ranges
2. Large arms ranges
3. Training villages and sites
4. Bivouac areas
5. Large-scale operation areas.

This document provides an historic context of small arms ranges on military training lands for the U.S. Army, U.S. Navy, U.S. Army Air Corps/U.S. Air Force, and the U.S. Marines, with a focus on the landscape outside the developed core of military installations. This work concludes that military training lands are significant enough in our nation's history to be surveyed for eligibility to the NRHP. However, training lands must be viewed as a whole; individual buildings on a training range are rarely eligible for the NRHP; buildings in their larger context (and the integrity of that larger context) are important.

References

Air Force Center for Engineering and the Environment. Interim Guidance, Treatment of Cold War Historic Properties for U.S. Air Force Installations. June 1933.

Goodwin, R. Christopher and Associates, Inc. National Historic Context for Department of Defense Installations, 1790-1940, Volume I. Balitimore District, U. S. Army Corps of Engineers, August 1995.

Liwanag, David. "Improving Army marksmanship: regaining the initiative in the infantryman's hale kilometer." Infantry Magazine. July-Aug. 2006.

McFann, Howard H. et al. Trainfire I: A New Course in Basic Rifle Marksmanship. Human Resources Research Office Technical Report 22. Washington: George Washington University, Human Resources Research Office, October 1955: 9, 54-63.

"Trainfire I Adopted," Infantry 47. July 1957: 89.

U.S. Army Corps of Engineers. St. Louis District. Range Operations Report No. 1 (RO-1) Small Arms Range. St. Louis: US Army Corps of Engineers, 2006.

U.S. Army Corps of Engineers. St. Louis District. Range Operations Report No. 14 (RO-14) Flame Thrower Range. St. Louis: US Army Corps of Engineers, 2006.

www.ingramcontent.com/pod-product-compliance
Lightning Source LLC
Chambersburg PA
CBHW082026300426
44117CB00015B/2361